W9-BUX-877

Knocking on Heaven's Door

ALSO BY LISA RANDALL

Warped Passages: Unravelling the Universe's Hidden Dimensions

Knocking on Heaven's Door

HOW PHYSICS AND SCIENTIFIC THINKING ILLUMINATE THE UNIVERSE AND THE MODERN WORLD

Lisa Randall

THE BODLEY HEAD
LONDON

Published by The Bodley Head 2011

2 4 6 8 10 9 7 5 3

Copyright © Lisa Randall 2011

Lisa Randall has asserted her right under the Copyright, Designs
and Patents Act 1988 to be identified as the author of this work

First published in Great Britain in 2011 by
The Bodley Head
Random House, 20 Vauxhall Bridge Road,
London SW1V 2SA

www.bodleyhead.co.uk
www.vintage-books.co.uk

Addresses for companies within The Random House Group Limited
can be found at: www.randomhouse.co.uk/offices.htm

The Random House Group Limited Reg. No. 954009

A CIP catalogue record for this book
is available from the British Library

ISBN 9781847920690

The Random House Group Limited supports The Forest Stewardship Council
(FSC®), the leading international forest certification organisation. Our books
carrying the FSC label are printed on FSC® certified paper. FSC is the only forest
certification scheme endorsed by the leading environmental organisations,
including Greenpeace. Our paper procurement policy can be found at:
www.randomhouse.co.uk/environment

Grateful acknowledgment is made to reprint the following:
"Come Together 1969" from Sony/ATV Music Publishing LLC, 8 Music Square West,
Nashville, TN 37203. All rights reserved. Used by permission.
Jet Song by Leonard Bernstein and Stephen Sondheim © 1956, 1958, 1959 by Amberson
Holdings LLC and Stephen and Stephen Sondheim. Copyright renewed. Leonard
Bernstein Music Publishing Company LLC, publisher. Boosey & Hawkes, agent for
rental. International copyright secured. Reprinted by Permission.

Printed and bound in Great Britain by
Clays Ltd. St Ives PLC

CONTENTS

PART V
SCALING THE UNIVERSE

Part VI
ROUNDUP

LIST OF ILLUSTRATIONS

INTRODUCTION

We are poised on the edge of discovery. The biggest and most exciting experiments in particle physics and cosmology are under way and many of the world's most talented physicists and astronomers are focused on their implications. What scientists find within the next decade could provide clues that will ultimately change our view of the fundamental makeup of matter or even of space itself—and just might provide a more comprehensive picture of the nature of reality. Those of us who are focused on these developments don't anticipate that they will be mere postmodern additions. We look forward to discoveries that might introduce a dramatically different twenty-first-century paradigm for the universe's underlying construction—altering our picture of its basic architecture based on the insights that lie in store.

September 10, 2008, marked the historic first trial run of the Large Hadron Collider (LHC). Although the name—Large Hadron Collider—is literal but uninspired, the same is not true for the science we expect it to achieve, which should prove spectacular. The "large" refers to the collider—not to hadrons. The LHC contains an enormous 26.6 kilometer[1] circular tunnel deep underground that stretches between the Jura Mountains and Lake Geneva and crosses the French-Swiss border. Electric fields inside this tunnel accelerate two beams, each consisting of billions of protons (which belong to a class of particles called hadrons—hence the collider's name), as they go around—about 11,000 times each second.

The collider houses what are in many respects the biggest and most impressive experiments ever built. The goal is to perform detailed studies of the structure of matter at distances never before measured and at energies higher than have ever been explored before. These energies should generate an array of exotic fundamental particles and reveal interactions that occurred early in the universe's evolution—roughly a trillionth of a second after the time of the Big Bang.

The design of the LHC stretched ingenuity and technology to their limits and its construction introduced even further hurdles. To the great frustration of physicists and everyone else interested in a better understanding of nature, a bad solder connection triggered an explosion a mere nine days after the LHC's auspicious initial run. But when the LHC came back on line in the fall of 2009—working better than anyone had dared anticipate—a quarter-century promise emerged as a reality.

In the spring of that same year, the Planck and Herschel satellites were launched in French Guiana. I learned about the timing from an excited group of Caltech astronomers who met May 13 at 5:30 A.M. in Pasadena, where I was visiting, to witness remotely this landmark event. The Herschel satellite will give insights into star formation, and the Planck satellite will provide details about the residual radiation from the Big Bang—yielding fresh information about the early history of our universe. Launches such as this are usually thrilling but very tense—since two to five percent fail, destroying years of work on customized scientific instruments in those satellites that fall back to Earth. Happily this particular launch went very well and sent information back throughout the day, attesting to just how successful it had been. Even so, we will have to wait several years before these satellites give us their most valuable data about stars and the universe.

*   *

Physics now provides a solid core of knowledge about how the universe works over an extremely large range of distances and energies. Theoretical and experimental studies have provided scientists with a deep understanding of elements and structures, ranging from the extremely tiny to the very large. Over time, we have deduced a detailed and comprehensive story about how the pieces fit together. Theories successfully describe how the cosmos evolved from tiny constituents that formed atoms, which in turn coalesced into stars that sit in galaxies and in larger structures spread throughout our universe, and how some stars then exploded and created heavy elements that entered our galaxy and solar system and which are ultimately essential to the formation of life. Using the results from the LHC and from such satellite explorations as those mentioned above, today's physicists hope to build on this solid and extensive base to expand our understanding to smaller distances and higher energies, and to achieve greater precision than has ever been reached before. It's an adventure. We have ambitious goals.

You have probably heard very clear, apparently precise definitions of science, particularly when it is being contrasted with belief systems such as religion. However, the real story of the evolution of science is complex. Although we like to think of it—at least I did when first starting out—as a reliable reflection of external reality and the rules by which the physical world works, active research almost inevitably takes place in a state of indeterminacy where we hope we are making progress, but where we really can't yet be sure. The challenge scientists face is to persevere with promising ideas while all the time questioning them to ascertain their veracity and their implications. Scientific research inevitably involves balancing delicately on the edge of difficult and sometimes conflicting and competing—but often exciting—ideas. The goal is to expand the boundaries of knowledge. But when first juggling data, concepts, and equations, the correct interpretation can be uncertain to everyone—including those most actively involved.

My investigations focus on the theory of elementary particles (the study of the smallest objects we know of), with forays into string theory as well as cosmology—the study of the largest. My colleagues and I try to understand what's at the core of matter, what's out there in the universe, and how all the fundamental quantities and properties that experimenters discover are ultimately connected. Theoretical physicists like myself don't do the actual experiments that determine which theories apply in the real world. We try instead to predict possible outcomes for what experiments might find and help devise innovative means for testing ideas. In the foreseeable future, the questions we try to answer will likely not change what people eat for dinner each day. But these studies could ultimately tell us about who we are and where we came from.

Knocking on Heaven's Door is about our research and the most important scientific questions we face. New developments in particle physics and cosmology have the potential to revise radically our understanding of the world: its makeup, its evolution, and the fundamental forces that drive its operation. This book describes experimental research at the Large Hadron Collider and theoretical studies that try to anticipate what they will find. It also describes research in cosmology—how we go about trying to deduce the nature of the universe, and in particular that of the dark matter hidden throughout the universe.

But *Knocking on Heaven's Door* also has a wider scope. This book explores more general questions that pertain to all scientific investigations. Along with describing the frontiers of today's research, clarifying the nature of science is at the core of what this book is about. It describes how we go about deciding which are the right questions to pose, why scientists don't always agree even on that, and how correct scientific ideas ultimately prevail. This book explores the real ways in which science advances and the respects in which it contrasts with other ways of seeking truth, giving some of the philosophical underpinnings of science and describing the intermediate stages at which it is uncertain where we will end up or who is right. Also, and as importantly, it shows how scientific ideas and methods might apply outside science, thus encouraging more rational decision-making in other spheres as well.

Knocking on Heaven's Door is intended for an interested lay reader who would like to have a greater understanding of current theoretical and experimental physics and who wants a better appreciation of the nature of modern science—as well as the principles of sound scientific thought. Often people don't really understand what science is and what we can expect it to tell us. This book is my attempt to correct some of the misconceptions—and perhaps vent a little of my frustration with the way science is currently understood and applied.

The last few years have provided me with some unique experiences and with conversations that have taught me a great deal, and I want to share these as launching points to explore some important ideas. Although I'm not a specialist in all the areas I cover and there is not enough space to do them all full justice, my hope is that this book will lead readers in more productive directions, while elucidating some exciting new developments along the way. It should also help readers identify the most reliable sources of scientific information—or misinformation—when they look for further answers in the future. Some of the ideas this book presents might appear very basic, but a more thorough understanding of the reasoning that underlies modern science will help pave a better approach both to research and to important issues the modern world currently faces.

In this era of movie prequels, you can think of *Knocking on Heaven's Door* as the origin story to my previous book, *Warped Passages,* combined with an update of where we are now and what we are anticipating. It fills in the gaps—going over the basics about science that underlie new ideas and new discoveries—and explains why we're on the edge of our seats waiting for new data to emerge.

The book alternates between details of science being done today and reflections on the underlying themes and concepts that are integral to science but that are useful for understanding the broader world as well. The first part of the book, Chapters 11 and 12 in the second part, Chapters 15 and 18 in the third part, and the final (Roundup) part are more about scientific thinking, whereas the remaining chapters focus more on physics—where we are today and how we got there. In some respects, it is two books in one—but books that are best read together. Modern physics might appear to some to be too far removed from our daily lives to be relevant or even readily comprehensible, but an appreciation of the philosophical and methodological underpinnings that guide our thinking should clarify both the science and the relevance of scientific thinking—as we'll see in many examples. Conversely, one will only fully grasp the basic elements of scientific thinking with some actual science to ground the ideas. Readers with a greater taste for one or the other might choose to skim or skip one of the courses, but the two together make for a well-balanced meal.

A key refrain throughout the book will be the notion of scale. The laws of physics provide a consistent framework for how established theoretical and physical descriptions fit together into a coherent whole, from the infinitesimal lengths currently explored at the LHC to the enormous size of the entire cosmos.[2] The rubric of scale is critical to our thinking, as well as to the specific facts and ideas we will encounter. Established scientific theories apply to accessible scales. But those theories become absorbed in increasingly precise and more fundamental ones as we add newly gained knowledge from previously unexplored distances—small or large. The first chapter focuses on the defining element of scale, explaining how categorizing by length is essential to physics and to the way in which new scientific developments build upon prior ones.

The first part also presents and contrasts different ways of approaching knowledge. Ask people what they think about when they think about science, and the answers are likely to be as varied as the individuals you ask. Some will insist on rigid, immutable statements about the physical world. Others will define it as a set of principles that are constantly being replaced, and still others will respond that science is nothing more than another belief system, not qualitatively different from philosophy or religion. And they would all be wrong.

The evolving nature of science is at the heart of why there can be so much debate—even within the scientific community itself. This part presents a little of the history that informs how today's research is rooted in seventeenth-century intellectual advances and then continues with a couple of less-featured aspects of the science-religion debate—a confrontation that in some respects originated at that time. It also looks into the materialist view of matter and its thorny implications for the science-religion question, as well as the issue of who gets to answer fundamental questions and how they go about it.

Part II turns to the physical makeup of the material world. It charts the terrain for the book's scientific journey, touring matter from familiar scales down to the smallest ones, all the while partitioning according to scale. This path will take us from recognizable territory down to submicroscopic sizes whose internal structure can be probed only by giant particle accelerators. The section closes with an introduction to some of the major experiments being performed today—the Large Hadron Collider (LHC) and astronomical probes into the early universe—which should broaden the extreme edges of our understanding.

As with any exciting development, these bold and ambitious enterprises have the potential to alter radically our scientific worldview. In Part III, we'll start to dig down into the LHC's operations and explore how this machine creates and collides proton beams to produce new particles that should tell us about the smallest accessible scales. This section also explains how experimenters will interpret what is found.

CERN (as well as the hilariously misleading Hollywood blockbuster *Angels and Demons*) has gone a long way toward publicizing the experimental side of particle physics. Many have now heard of the giant particle accelerator that will smash together very energetic protons that will be focused in a tiny region of space to create forms of matter never seen before. The LHC is now running and is poised to change our view of the fundamental nature of matter and even of space itself. But we don't yet know what it will find.

In the course of our scientific journey, we'll reflect on scientific uncertainty and what measurements can truly tell us. Research is by its nature at the edge of what we know. Experiment and calculation are designed to reduce or eliminate as many uncertainties as possible and precisely determine those that remain. Nonetheless, though it might sound paradoxical, in practice, on a day-to-day basis, science is fraught with uncertainty. Part III examines how scientists address the challenges intrinsic to their difficult explorations and how everyone can benefit from scientific thinking when interpreting and understanding statements that are made in an increasingly complex world.

Part III also considers black holes at the LHC, and how the fears that were raised about them contrast with some real dangers we currently face. We'll consider the important issues of cost-benefit analysis and risk, and how people might better approach thinking about them—both in and out of the lab.

Part IV describes the Higgs boson search as well as specific models, which are educated guesses for what exists and are search targets for the LHC. If LHC experiments confirm some of the ideas theorists have proposed—or even if they uncover something unforeseen—the results will change the way we think about the world. This section explains the Higgs mechanism responsible for elementary particle masses as well as the hierarchy problem that tells us we should find more. It also investigates models that address this problem and the exotic new particles they predict, such as those associated with supersymmetry or extra dimensions of space.

Along with presenting specific hypotheses, this part explains how physicists go about constructing models and the efficacy of guiding principles such as "truth through beauty" and "top-down" versus "bottom-up." It explains what the LHC is searching for, but also how physicists anticipate what it might find. This part describes how scientists will try to connect the seemingly abstract data the LHC will produce to some of the deep and fundamental ideas that we currently investigate.

Following our tour of research into the interior of matter, we'll look outward in Part V. At the same time as the LHC probes the tiniest scales of matter, satellites and telescopes explore the largest scales in the cosmos—studying the rate at which its expansion accelerates—and also study details of the relic radiation from the time of the Big Bang. This era could witness astounding new developments in *cosmology*, the science of how the universe evolved. In this section, we'll explore the universe out to larger scales and discuss the particle physics–cosmology connection, as well as the elusive dark matter and experimental searches for it.

The final roundup in Part VI reflects on creativity, and the rich and varied elements of thought that enter into creative thinking. It examines how we attempt to answer the big questions through the somewhat smaller seeming activities we engage in on a day-to-day basis. We'll conclude with some final thoughts on why science and scientific thinking are so important today, as well as the symbiotic relationship between technology and scientific thinking that has produced so much progress in the modern world.

I am frequently reminded how tricky it can be for non-scientists to appreciate the sometimes remote ideas that modern science addresses. This challenge became apparent when I met with a class of college students following a public lecture I gave about extra dimensions and physics. When I was told they all had the same pressing question, I expected some confusion about dimensions, but instead learned that they were eager to know my age. But lack of interest isn't the only challenge—and the students actually did go on to engage with the scientific ideas. Still, there is no denying that fundamental science is often abstract, and justifying it can be difficult—a hurdle I had to face at a congressional hearing about the importance of basic science that I attended in the fall of 2009 along with Dennis Kovar, director of High Energy Physics at the U.S. Department of Energy; Pier Oddone, director of the Fermi National Accelerator Laboratory; and Hugh Montgomery, director of Jefferson Lab, a nuclear physics facility. This was my first time in the halls of government since my congressman, Benjamin Rosenthal, took me around when I was a high school finalist in the Westinghouse Science Competition many years before. He generously provided me with more than the mere photo op that the other finalists had received.

During my more recent visit, I again enjoyed observing the offices where policy is made. The room dedicated to the House Committee on Science and Technology is in the Rayburn House Office Building. The representatives sat in the back and we "witnesses" sat facing them. Inspirational plaques hung above the representatives' heads, the first of which read "WHEN THERE IS NO VISION THE PEOPLE PERISH. PROVERBS 29:18."

It seems American government must refer to scripture even in the congressional room explicitly dedicated to science and technology. The line nonetheless expresses a noble and accurate sentiment, which we all would like to apply.

The second plaque contained a more secular quote from Tennyson: "FOR I DIPPED INTO THE FUTURE, FAR AS MY EYES COULD SEE / SAW THE VISION OF THE WORLD AND ALL THE WONDER THAT WOULD BE."

That was also a nice thought to bear in mind while describing our research goals.

The irony was that the room was arranged so that we "witnesses" from the science world—who already were sympathetic to these statements—faced the plaques, which hung directly in our line of view. The representatives, on the other hand, sat underneath the words so they couldn't see them. Congressman Lipinski, who in opening statements said that discoveries inspire more questions—and large metaphysical inquiries—acknowledged that he used to notice the plaques but they were now all too easy to forget. "Few of us ever look up there." He expressed his gratitude for being reminded.

Moving on from the decor, we scientists turned to the task at hand—explaining what it is that makes this such an exciting and unprecedented era for particle physics and cosmology. Although the representatives' questions were occasionally pointed and skeptical, I could appreciate the resistance they constantly face in explaining to their constituents why it would be a mistake to stop funding scientific work—even in the face of economic uncertainties. Their questions ranged from details about the purposes of specific experiments to broader issues concerning the role of science and where it is heading.

In between the absences of the representatives, who periodically had to leave to vote, we gave some examples of the side benefits accrued by advancing fundamental science. Even science intended as basic research often proves fruitful in other ways. We talked about Tim Berners-Lee's development of the World Wide Web as a means of letting physicists in different countries collaborate more readily on their joint experiments at CERN. We discussed medical applications, such as PET scans—positron emission tomography—a way of probing internal body structure with the electron's antiparticle. We explained the role of the industrial-scale production of superconducting magnets that were developed for colliders but now are used for magnetic resonance imaging as well, and finally the remarkable application of general relativity to precision predictions, including the global positioning systems we use daily in our cars.

Of course significant science doesn't necessarily have any immediate benefit in practical terms. Even if there is an ultimate pay-off, we rarely know about it at the time of the discovery. When Benjamin Franklin realized lightning was electricity, he didn't know electricity soon would change the face of the planet. And when Einstein worked on general relativity, he didn't anticipate it would be used in any practical devices.

So the case we made that day was focused primarily not on specific applications, but rather on the vital importance of pure science. Though the status of science in America might be precarious, many people currently recognize its worth. Society's view of the universe, time, and space changed with Einstein—as the original lyrics of "As Time Goes By" quoted in *Warped Passages* attest to.[3] Our very language and thoughts change as our understanding of the physical world develops and as new ways of thinking progress. What scientists study today and how we go about this will be critical both to our understanding of the world and to a robust and thoughtful society.

We are currently living in an extraordinarily exciting era for physics and cosmology, with some of the edgiest investigations ever proposed. Through a wide-ranging set of explorations, *Knocking on Heaven's Door* touches on our different ways of understanding the world—through art, religion, and science—but chiefly with a focus on the goals and methods of modern physics. Ultimately, the very tiny objects we study are integral to discovering who we are and where we came from. The large-scale structures we hope to learn more about could shed light on our cosmic environment as well as on the origin and fate of our universe. This book is about what we hope to find and how it might happen. The journey should be an intriguing adventure—so welcome aboard.

Part I: **SCALING REALITY**

CHAPTER ONE

WHAT'S SO SMALL TO YOU
IS SO LARGE TO ME

Among the many reasons I chose to pursue physics was the desire to do something that would have a permanent impact. If I was going to invest so much time, energy, and commitment, I wanted it to be for something with a claim to longevity and truth. Like most people, I thought of scientific advances as ideas that stand the test of time.

My friend Anna Christina Büchmann studied English in college while I majored in physics. Ironically, she studied literature for the same reason that drew me to math and science. She loved the way an insightful story lasts for centuries. When discussing Henry Fielding's novel *Tom Jones* with her many years later, I learned that the edition I had read and thoroughly enjoyed was the one she helped annotate when she was in graduate school.[4]

Tom Jones was published 250 years ago, yet its themes and wit resonate to this day. During my first visit to Japan, I read the far older *Tale of Genji* and marveled at its characters' immediacy too, despite the thousand years that have elapsed since Murasaki Shikibu wrote about them. Homer created the *Odyssey* roughly 2,000 years earlier. Yet notwithstanding its very different age and context, we continue to relish the tale of Odysseus's journey and its timeless descriptions of human nature.

Scientists rarely read such old—let alone ancient—scientific texts. We usually leave that to historians and literary critics. We nonetheless apply the knowledge that has been acquired over time, whether from

Newton in the seventeenth century or Copernicus more than 100 years earlier still. We might neglect the books themselves, but we are careful to preserve the important ideas they may contain.

Science certainly is not the static statement of universal laws we all hear about in elementary school. Nor is it a set of arbitrary rules. Science is an evolving body of knowledge. Many of the ideas we are currently investigating will prove to be wrong or incomplete. Scientific descriptions certainly change as we cross the boundaries that circumscribe what we know and venture into more remote territory where we can glimpse hints of the deeper truths beyond.

The paradox scientists have to contend with is that while aiming for permanence, we often investigate ideas that experimental data or better understanding will force us to modify or discard. The sound core of knowledge that has been tested and relied on is always surrounded by an amorphous boundary of uncertainties that are the domain of current research. The ideas and suggestions that excite us today will soon be forgotten if they are invalidated by more persuasive or comprehensive experimental work tomorrow.

When the 2008 Republican presidential candidate Mike Huckabee sided with religion over science—in part because scientific "beliefs" change whereas Christians take as their authority an eternal, unchanging God—he was not entirely misguided, at least in his characterization. The universe evolves and so does our scientific knowledge of it. Over time, scientists peel away layers of reality to expose what lies beneath the surface. We broaden and enrich our understanding as we probe increasingly remote scales. Knowledge advances and the unexplored region recedes when we reach these difficult-to-access distances. Scientific "beliefs" then evolve in accordance with our expanded knowledge.

Nonetheless, even when improved technology makes a broader range of observations possible, we don't necessarily just abandon the theories that made successful predictions for the distances and energies, or speeds and densities, that were accessible in the past. Scientific theories grow and expand to absorb increased knowledge, while retaining the reliable parts of ideas that came before. Science thereby incorporates old established knowledge into the more comprehensive picture that emerges from a broader range of experimental and theoretical observations. Such changes don't necessarily mean the old rules are wrong, but they can mean, for example, that those rules no longer apply on smaller scales where new components have been revealed. Knowledge can thereby embrace old ideas yet expand over time, even though very likely more will always remain to be explored. Just as travel can be compelling—even if you will never visit every place on the planet (never mind the cosmos)—increasing our understanding of matter and of the universe enriches our existence. The remaining unknowns serve to inspire further investigations.

My own research field of particle physics investigates increasingly smaller distances in order to study successively tinier components of matter. Current experimental and theoretical research attempt to expose what matter conceals—that which is embedded ever deeper inside. But despite the often-heard analogy, matter is not simply like a Russian matryoshka doll, with similar elements replicated at successively smaller scales. What makes investigating increasingly minuscule distances interesting is that the rules can change as we reach new domains. New forces and interactions might appear at those scales whose impact was too tiny to detect at the larger distances previously investigated.

The notion of scale, which tells physicists the range of sizes or energies that are relevant for any particular investigation, is critical to the understanding of scientific progress—as well as to many other aspects of the world around us. By partitioning the universe into different comprehensible sizes, we learn that the laws of physics that work best aren't necessarily the same for all processes. We have to relate concepts that apply better on one scale to those more useful at another. Categorizing in this way lets us incorporate everything we know into a consistent picture while allowing for radical changes in descriptions at different lengths.

In this chapter, we'll see how partitioning by scale—whichever scale is relevant—helps clarify our thinking—both scientific and otherwise—and why the subtle properties of the building blocks of matter are so hard to notice at the distances we encounter in our everyday lives. In doing so, this chapter also elaborates on the meaning of "right" and "wrong" in science, and why even apparently radical discoveries don't necessarily force dramatic changes on the scales with which we are already familiar.

IT'S IMPOSSIBLE

People too often confuse evolving scientific knowledge with no knowledge at all and mistake a situation in which we are discovering new physical laws with a total absence of reliable rules. A conversation with the screenwriter Scott Derrickson during a recent visit to California helped me to crystallize the origin of some of these misunderstandings. At the time, Scott was working on a couple of movie scripts that proposed potential connections between science and phenomena that he suspected scientists would probably dismiss as supernatural. Eager to avoid major solecisms, Scott wanted to do scientific justice to his imaginative story ideas by having them scrutinized by a physicist—namely me. So we met for lunch at an outdoor café in order to share our thoughts along with the pleasures of a sunny Los Angeles afternoon.

Knowing that screenwriters often misrepresent science, Scott wanted his particular ghost and time-travel stories to be written with a reasonable amount of scientific credibility. The particular challenge that he as a screenwriter faced was his need to present his audience not just with interesting new phenomena, but also with ones that would translate effectively to a movie screen. Although not trained in science, Scott was quick and receptive to new ideas. So I explained to him why, despite the ingenuity and entertainment value of some of his story lines, the constraints of physics made them scientifically untenable.

Scott responded that scientists have often thought certain phenomena impossible that later turned out to be true. "Didn't scientists formerly disbelieve what relativity now tells us?" "Who would have thought randomness played any role in fundamental physical laws?" Despite his great respect for science, Scott still wondered if—given its evolving nature—scientists aren't sometimes wrong about the implications and limitations of their discoveries.

Some critics go even further, asserting that although scientists can

predict a great deal, the reliability of those predictions is invariably suspect. Skeptics insist, notwithstanding scientific evidence, that there
could always be a catch or a loophole. Perhaps people could come back
from the dead or at the very least enter a portal into the Middle Ages or
into Middle-earth. These doubters simply don't trust the claims of science that a thing is definitively impossible.

However, despite the wisdom of keeping an open mind and recognizing that new discoveries await, a deep fallacy is buried in this logic. The
problem becomes clear when we dissect the meaning of such statements
as those above and, in particular, apply the notion of scale. These questions ignore the fact that although there will always exist unexplored
distance or energy ranges where the laws of physics might change, we
know the laws of physics on human scales extremely well. We have had
ample opportunity to test these laws over the centuries.

When I met the choreographer Elizabeth Streb at the Whitney Mu-
seum, where we both spoke on a panel on the topic of creativity, she too
underestimated the robustness of scientific knowledge on human scales.
Elizabeth posed a similar question to those Scott had asked: "Could the
tiny dimensions proposed by physicists and curled up to an unimaginably
small size nonetheless affect the motion of our bodies?"

Her work is wonderful, and her inquiries into the basic assumptions
about dance and movement are fascinating. But the reason we cannot
determine whether new dimensions exist, or what their role would be
even if they did, is that they are too small or too warped for us to be able
to detect. By that I mean that we haven't yet identified their influence on
any quantity that we have so far observed, even with extremely detailed
measurements. Only if the consequences of extra dimensions for physical phenomena were vastly bigger could they discernibly influence anyone's motion. And if they did have such a significant impact, we would
already have observed their effects. We therefore know that the funda-
mentals of choreography won't change even when our understanding of
quantum gravity improves. Its effects are far too suppressed relative to
anything perceptible on a human scale.

When scientists have turned out to be wrong in the past, it was often because they hadn't yet explored very tiny or very large distances or extremely high energies or speeds. That didn't mean that, like Luddites, they had closed their minds to the possibility of progress. It meant only that they trusted their most up-to-date mathematical descriptions of the world and their successful predictions of then-observable objects and behaviors. Phenomena they thought were impossible could and sometimes did occur at distances or speeds these scientists had never before experienced—or tested. But of course they couldn't yet have known about new ideas and theories that would ultimately prevail in the regimes of those tiny distances or enormous energies with which they were not yet familiar.

When scientists say we know something, we mean only that we have certain ideas and theories whose predictions have been well tested *over a certain range of distances or energies*. These ideas and theories are not necessarily the eternal laws for the ages or the most fundamental of physical laws. They are rules that apply as well as any experiment could possibly test, over the range of parameters available to current technology. This doesn't mean that these laws will never be overtaken by new ones. Newton's laws are instrumental and correct, but they cease to apply at or near the speed of light where Einstein's theory applies. Newton's laws are at the same time both correct and incomplete. They apply over a limited domain.

The more advanced knowledge that we gain through better measurements really is an improvement that illuminates new and different underlying concepts. We now know about many phenomena that the ancients could not have derived or discovered with their more limited observational techniques. So Scott was right that sometimes scientists have been wrong—thinking phenomena impossible that in the end turned out to be perfectly true. But this doesn't mean there are no rules. Ghosts and time-travelers won't appear in our houses, and alien creatures won't suddenly emerge from our walls. Extra dimensions of space might exist, but they would have to be tiny or warped or otherwise currently hidden from view in order for us to explain why they have not yet yielded any noticeable evidence of their existence.

Exotic phenomena might indeed occur. But such phenomena will happen only at difficult-to-observe scales that are increasingly far from our intuitive understanding and our usual perceptions. If they will always remain inaccessible, they are not so interesting to scientists. And they are less interesting to fiction writers too if they won't have any observable impact on our daily lives.

Weird things are possible, but the ones non-physicists are understandably most interested in are the ones we can observe. As Steven Spielberg pointed out in a discussion about a science fiction movie he was considering, a strange world that can't be presented on a movie screen—and which the characters in a film would never experience—is not so interesting to a viewer. (Figure 1 shows amusing evidence.) Only a new world that we can access and be aware of could be. Even though both require imagination, abstract ideas and fiction are different and have different goals. Scientific ideas might apply to regimes that are too remote to be of interest to a film, or to our daily observations, but they are nonetheless essential to our description of the physical world.

[**FIGURE 1**] An XKCD comic that captures the hidden nature of tiny rolled-up dimensions.

WRONG TURNS

Despite this neat separation by distances, people too often take shortcuts when trying to understand difficult science and the world. And that can easily lead to an overzealous application of theories. Such misapplication of science is not a new phenomenon. In the eighteenth century, when scientists were busy studying magnetism in laboratories, others conjured up the notion of "animal magnetism"— a hypothesized magnetic "vital fluid" in animate beings. It took a French royal commission set up by Louis XVI in 1784, which included Benjamin Franklin among others, to formally debunk the hypothesis.

Today such misguided extrapolations are more likely to be made about quantum mechanics—as people try to apply it on macroscopic scales where its consequences usually average away and leave no measurable signatures.[5] It's disturbing how many people trust ideas such as those in Rhonda Byrne's bestselling book *The Secret,* about how positive thoughts attract wealth, health, and happiness. Equally disquieting is Byrne's claim that "I never studied science or physics at school, and yet when I read complex books on quantum physics I understood them perfectly because I wanted to understand them. The study of quantum physics helped me to have a deeper understanding of *The Secret,* on an energetic level."

As even the Nobel Prize–winning pioneer of quantum mechanics Niels Bohr noted, "If you are not completely confused by quantum mechanics, you do not understand it." Here's another secret (at least as well protected as those in a bestselling book): quantum mechanics is notoriously misunderstood. Our language and intuition derive from *classical reasoning*, which doesn't take quantum mechanics into account. But this doesn't mean that any bizarre phenomenon is possible with quantum logic. Even without a more fundamental, deeper understanding, we know how to use quantum mechanics to make predictions. Quantum mechanics will certainly never account for Byrne's "secret" about the so-called principle of attraction between people and distant things or phenomena. At those large distances, quantum mechanics doesn't play this kind of role. Quantum mechanics has nothing to do with many of the tantalizing ideas people often attribute to it. I cannot affect an experiment by staring at it, quantum mechanics does not mean there are no reliable predictions, and most measurements are constrained by practical limitations and not by the uncertainty principle.

Such fallacies were the chief topic in a surprising conversation I had with Mark Vicente, the director of the movie *What the Bleep Do We Know!?*—a film that is the bane of scientists—in which people claim that human influence matters for experiments. I wasn't sure where this conversation would lead, but I had time to spare since I was sitting on the tarmac at the Dallas/Fort Worth airport for several hours waiting for mechanics to repair a dent in the wing (which first was described as too small to matter—but then was "measured by technology" before the plane could depart, as one crewmember helpfully informed us).

Even with the delay, I realized if I was going to talk to Mark at any length, I had to know where he stood on his film—which I was familiar with from the numerous people at my lectures who asked me off-the-wall questions based on what they had seen in it. Mark's answer caught me by surprise. He had made a rather striking about-face. He confided that he had initially approached science with preconceived notions that he didn't sufficiently question, but that he now viewed his previous thinking as more religious in nature. Mark ultimately concluded that what he had presented in his film was not science. Placing quantum mechanical phenomena at a human level was perhaps superficially satisfying to many of his film's viewers, but that didn't make it right.

Even if new theories require radically different assumptions—as was certainly the case with quantum mechanics—valid scientific arguments and experiments ultimately determined that they were true. It wasn't magic. The scientific method, along with data and searches for economy and consistency, had told scientists how to extend their knowledge beyond what is intuitive at immediately accessible scales to very different ideas that apply to phenomena that are not.

The next section tells more about how the notion of scale systematically bridges different theoretical concepts and allows us to incorporate them into a coherent whole.

EFFECTIVE THEORIES

Our size happens to fall pretty much randomly close to the middle in terms of powers of ten when placed on a scale between the smallest imaginable length and the enormity of the universe.[6] We are very big compared to the internal structure of matter and its minuscule components, while we are extremely tiny compared to stars, galaxies, and the universe's expanse. The sizes that we most readily comprehend are simply the ones that are most accessible to us—through our five senses and through the most rudimentary measuring tools. We understand more distant scales through observations combined with logical deductions. The range of sizes might seem to involve increasingly abstract and hard-to-keep-track-of quantities as we move further from directly visible and accessible scales. But technology combined with theory allows us to establish the nature of matter over a vast stretch of lengths.

Known scientific theories apply over this huge range—spanning distances as small as the tiny objects explored by the Large Hadron Collider (LHC) up to the enormous length scales of galaxies and the cosmos. And for each possible size of objects or distance between them, different aspects of the laws of physics can become relevant. Physicists have to cope with the abundance of information that applies over this enormous span. Although the most basic laws of physics that apply to tiny lengths are ultimately responsible for those that are relevant to larger scales, they are not necessarily the most efficient means of making a calculation. When the extra substructure or underpinnings are irrelevant to a sufficiently precise answer, we want a more practical way to calculate and efficiently apply simpler rules.

One of the most important features of physics is that it tells us how to identify the range of scales relevant to any measurement or prediction—according to the precision we have at our disposal—and then calculate accordingly. The beauty of this way of looking at the world is that we can focus on the scales that are relevant to whatever we are interested in, identify the elements that operate at those scales, and discover and apply the rules that govern how these components relate. Scientists average over or even ignore (sometimes unwittingly) physical processes that occur on immeasurably small scales when formulating theories or setting up calculations. We select relevant facts and suppress details when we can get away with it and focus on the most useful scales. Doing so is the only way to cope with an impossibly dense set of information.

When appropriate, it makes sense to ignore minutiae in order to focus on the topic of interest and not to obscure it with inessential details. A recent lecture by the Harvard psychology professor Stephen Kosslyn reminded me how scientists—and everyone else—prefer to keep track of information. In a cognitive science experiment that he performed on the audience, he asked us all to keep track of line segments he presented on a screen one after the other. Each of the segments could go "north" or "southeast," and so on, and together they formed a zigzagging line. (See Figure 2.) We were asked to close our eyes and say what we had seen. We noticed that even though our brains allow us to keep track of only a few individual segments at a time, we could remember longer sequences by grouping them into repeatable shapes. By thinking on the scale of the shape rather than the individual line segment, we could keep the figure in our heads.

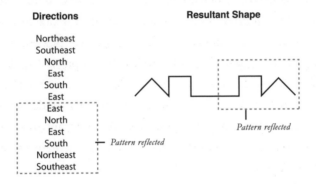

[**FIGURE 2**] You might choose as your component the individual line segment or a larger unit, such as the group of six segments that appears twice.

For almost anything you see, hear, taste, smell, or touch, you have a choice between examining details by looking very closely or examining the "big picture" with its other priorities. Whether staring at a painting, tasting wine, reading philosophy, or planning your next trip, you automatically parcel your thoughts into the categories of interest—be they sizes or flavor categories, ideas, or distances—and the categories that you don't find relevant at the time.

The utility of focusing on the pertinent questions and ignoring structures too small to be relevant applies in many contexts. Think about what you do when you use MapQuest or Google maps or look at the small screen on your iPhone. If you were traveling from far away, you would first get some rough idea where your destination is. Subsequently, when you have the big picture, you would zoom into a map with more resolution. You don't need the additional detailed information in your first pass. You just want to have some sense of location. But as you begin to map out the details of your journey—as your resolution becomes finer in seeking out the exact street you will need—you will care about the details on the finer scale that were inessential to your first exploration.

Of course, the degree of precision you want or need determines the scale you choose. I have friends who don't pay much attention to hotel location when visiting New York City. For them, the gradations in character of the city's blocks is irrelevant. But for anyone who knows New York, those details matter. It's not enough to know you are staying downtown. New Yorkers care if they are above or below Houston Street, or east or west of Washington Square Park, or even whether they are two or five blocks away.

Although the precise choice of scale might differ among individuals, no one would display a map of the United States in order to find a restaurant. The necessary details won't be resolvable on a computer screen displaying such an overly large scale. On the other hand, you don't need the details of a floor plan just to know that the restaurant is there in the first place. For any question you ask, you choose the relevant scale. (See Figure 3 for another example.)

The Eiffel Tower

Scale too small

Relevant scale

Scale too big

[FIGURE 3] Different information becomes more obvious when viewed at different scales.

In a similar manner, we categorize by size in physics so we can focus on the questions of interest. Our tabletop looks solid—and for many purposes we can treat it as such—but in reality it is made up of atoms and molecules that collectively act like the hard impenetrable surface we encounter at the scales we experience in our daily lives. Those atoms aren't indivisible, either. They are composed of nuclei and electrons. And the nuclei are made of protons and neutrons that are in turn bound states of more fundamental objects called quarks. Yet we don't need to know about quarks to understand the electromagnetic and chemical properties of atoms and elements (the field of science known as atomic physics). People studied atomic physics for years before there was even a clue about the substructure beneath. And when biologists study a cell, they don't need to know about quarks inside the proton either.

I remember feeling a tad betrayed when my high school teacher, after devoting months to Newton's Laws, told the class those laws were wrong. But my teacher was not quite right in his statement. Newton's laws of motion work at the distances and speeds that were observable in his time. Newton thought about physical laws that applied, given the accuracy with which he (or anyone else in his era) could make measurements. He didn't need the details of general relativity to make successful predictions about what could be measured then. And neither do we when we make the sorts of predictions relevant to large bodies at relatively low speeds and densities that Newton's Laws apply to. When physicists or engineers today study planetary orbits, they also don't need to know the detailed composition of the Sun. The laws that govern the behavior of quarks don't noticeably affect the predictions relevant to celestial bodies either.

Understanding the most basic components is rarely the most efficient way to understand the interactions at larger scales, where tiny substructure generally plays very little role. We would be hard pressed to make progress in atomic physics by studying the even tinier quarks. It is only when we need to know more detailed properties of nuclei that the quark substructure becomes relevant. In the absence of unfathomable precision, we can safely do chemistry and molecular biology while ignoring any internal substructure in a nucleus. Elizabeth Streb's dance movements won't change no matter what happens at the quantum gravity scale. Choreography relies only on classical physical laws.

Everyone, including physicists, is happy to use a simpler description when the details are beyond our resolution. Physicists formalize this intuition and organize categories in terms of the distance or energy that is relevant. For any given problem, we use what we call an *effective theory*. The effective theory concentrates on the particles and forces that have "effects" at the distances in question. Rather than delineating particles and interactions in terms of unmeasurable parameters that describe more fundamental behavior, we formulate our theories, equations, and observations in terms of the things that are actually relevant to the scales we might detect.

The effective theory we apply at larger distances doesn't go into the details of an underlying physical theory that applies to shorter distance scales. It asks only about things you could hope to measure or see. If something is beyond the resolution of the scales at which you are working, you don't need its detailed structure. This practice is not scientific fraud. It is a way of disregarding the clutter of superfluous information. It is an "effective" way to obtain accurate answers efficiently and keep track of what is in your system.

The reason effective theories work is that it is safe to ignore the unknown, as long as it won't make any measurable differences. If the only unknown phenomena occur at scales, distances, or resolutions where the influence is still indiscernible, we don't need to know about them to make successful predictions. Phenomena beyond our current technical reach, by definition, won't have any measurable consequences aside from those that are already taken into account.

This is why, even without knowing about phenomena as substantial as the existence of relativistic laws of motion or a quantum mechanical description of atomic and subatomic systems, people could still make accurate predictions. This is fortunate, since we simply can't think about everything at once. We'd never get anywhere if we couldn't suppress irrelevant details. When we concentrate on questions we can experimentally test, our finite resolution makes this jumble of information on all scales inessential.

"Impossible" things can happen—but only in environments that we have not yet observed. Their consequences are irrelevant at scales we know—or at least those scales we have so far explored. What is happening at these small distances remains hidden until higher-resolution tools are developed to look directly or until sufficiently precise measurements differentiate and identify the underlying theory through the minuscule distinguishing features it provides at larger distances.

Scientists can legitimately ignore anything too small to be observed when we make predictions. Not only is it impossible to distinguish among the consequences of overly tiny objects and processes, but the physical effects of processes at these scales are interesting only insofar as they determine the physically measurable parameters. Physicists therefore characterize the objects and properties on measurable scales in an effective theory and use these to do science relevant to the scales at hand. When you do know the short-distance details, or the microstructure of a theory, you can derive the quantities in the effective description from more fundamental detailed structure. Otherwise these quantities are just unknowns to be experimentally determined. The observable larger-scale quantities in the effective theory are not giving the fundamental description, but they are a convenient way of organizing observations and predictions.

An effective description can summarize the consequences of any shorter-distance theory that reproduces larger-scale observations but whose direct effects are too tiny to see. This has the advantage of letting us study and evaluate processes using fewer parameters than we would need if we took every detail into account. This smaller set is completely sufficient to characterize the processes that interest us. Furthermore, the set of parameters we use are *universal*—they are the same independently of the more detailed underlying physical processes. To know their values we just have to measure them in any of the many processes in which they apply.

Over a large range of lengths and energies, a single effective theory applies. After its few parameters have been determined by measurements, everything appropriate to this range of scales can be calculated.

It gives a set of elements and rules that can explain a large number of observations. At any given time, the theory we think of as fundamental is likely to turn out to be an effective theory—since we never have infinitely precise resolution. Yet we trust the effective theory because it successfully predicts many phenomena that apply over a range of length and energy scales.

Effective theories in physics not only keep track of short distance information—they can also summarize large distance effects whose consequences might also be too minute to observe. For example, the universe we live in is very slightly curved—in a way that Einstein taught us was possible when he developed his theory of gravity. This curvature applies to larger scales involving the large-scale structure of space. Yet we can systematically understand why such curvature effects are too small to matter for most of the observations and experiments that we perform locally, on much smaller scales. Only when we include gravity in our particle physics description do we need to consider such effects—which are too tiny to matter for much of what I will describe. In that case too, the appropriate effective theory tells us how to summarize gravity's effects in a few unknown parameters to be experimentally determined.

One of the most important aspects of an effective theory is that while it describes what we can see, it also categorizes what is missing—be it small scale or large. With any effective theory, we can determine how big an effect the unknown (or known) underlying dynamics could possibly have on any particular measurement. Even in advance of new discoveries at different scales, we can mathematically determine the maximal size of the influence any new structure can have on the effective theory at the scale at which we are working. As we will explore further in Chapter 12, it is only when the underlying physics is discovered that anyone fully understands the effective theory's true limitations.

One familiar example of an effective theory might be thermodynamics, which tells us how refrigerators or engines work and was developed long before atomic or quantum theory. The thermodynamic state of a system is well characterized by its pressure, temperature, and volume. Though we know that fundamentally the system consists of a gas of atoms and molecules—with much more detailed structure than the preceding three quantities can possibly describe—for many purposes we can concentrate on these three quantities to characterize the system's readily observable behavior.

Temperature, pressure, and volume are real quantities that can be measured. The theory behind their relationships is fully developed and can be used to make successful predictions. The effective theory of a gas makes no mention of the underlying molecular structure. (See Figure 4.) The behavior of those underlying elements determines temperature and pressure, but scientists happily used these quantities to do calculations even before atoms or molecules were discovered.

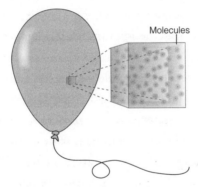

[**FIGURE 4**] Pressure and temperature can be understood at a more fundamental level in terms of the physical properties of individual molecules.

Once the fundamental theory is understood, we can relate temperature and pressure to properties of the underlying atoms and also understand when the thermodynamic description should break down. But we can still use thermodynamics for a wide variety of predictions. In fact, many phenomena are only understood from a thermodynamic point of view, since without huge computing power and memory, well beyond what exists, we can't track the paths of all the individual atoms. The effective theory is the only way at this point to understand some important physical phenomena that are pertinent to solid and liquid *condensed matter*.

This example teaches us another critical aspect of effective theories. We sometimes treat "fundamental" as a relative term. From the perspective of thermodynamics, the atomic and molecular description is fundamental. But from a particle physics description that details the quarks and electrons inside the atoms, the atom is *composite*—made up of smaller elements. Its use from a particle physics perspective is as an effective theory.

This description of the clean developmental progression in science from the well understood to regimes at the frontier of knowledge applies best to fields such as physics and cosmology, where we have a clear understanding of the functional units and their relationships. Effective theories won't necessarily work for newer fields such as systems biology, where the relationships between activities at the molecular and more macroscopic levels, as well as the relevant feedback mechanisms, are yet to be fully understood.

Nonetheless, the effective theory idea applies in a broad range of scientific contexts. The mathematical equations that govern the evolution of species won't change in response to new physics results, as I discussed with the mathematical biologist Martin Nowak in response to a question he had asked. He and his colleagues can characterize the parameters independently of any more fundamental description. They might ultimately relate to more basic quantities—physical or otherwise—but that doesn't change the equations that mathematical biologists use to evolve the behavior of populations over time.

For particle physicists, effective theories are essential. We isolate simple systems at different scales and relate them to each other. In fact, the very invisibility of underlying structure that allows us to focus on observable scales and ignore more fundamental effects keep underlying interactions so well hidden that only with tremendous resources and effort can we ferret them out. The tininess of effects of more fundamental theories on observable scales is the reason that physics today is so challenging. We need to directly explore smaller scales or make increasingly precise measurements if we are to perceive the effects of the more fundamental nature of matter and its interactions. Only with advanced technology can we access very tiny or extremely vast length scales. That is why we need to conduct elaborate experiments—such as those at the Large Hadron Collider—to make advances today.

PHOTONS AND LIGHT

The story of theories of light nicely exemplifies the ways in which effective theories are used as science evolves, with some ideas being discarded while others are retained as approximations appropriate to their specified domains. From the time of the ancient Greeks, people studied light with geometrical optics. It is one of the topics any aspiring physics graduate student is tested on when taking the physics GRE (the exam that is a prerequisite for graduate school). This theory assumes that light travels in rays or lines and tells you how those rays behave as they travel through different media, as well as how instruments use and detect them.

The strange thing is that virtually no one—at least no one at Harvard where I now teach and was once a student—actually studies classical and geometrical optics. Maybe geometrical optics is taught a little bit in high school, but it is certainly no big part of the curriculum.

Geometrical optics is an old-fashioned subject. It hit its heyday several centuries ago with Newton's famous *Opticks*, continuing into the 1800s when William Rowan Hamilton made what is perhaps the first real mathematical prediction of a new phenomenon.

The classical theory of optics still applies to areas such as photography, medicine, engineering, and astronomy, and is used to develop new mirrors, telescopes, and microscopes. Classical optical scientists and engineers work out different examples of various physical phenomena. However, they are simply applying optics—not discovering new laws.

In 2009, I was honored to be asked to give the Hamilton lecture at the University of Dublin—a lecture several of my most respected colleagues had given before me. It is named after Sir William Rowan Hamilton, the remarkable nineteenth-century Irish mathematician and physicist. I confess that the name Hamilton is so universally present in physics that I foolishly didn't initially make the connection with an actual person who was in fact Irish. But I was fascinated by the many areas of math and physics that Hamilton had revolutionized, including, among them, geometrical optics.

The celebration of Hamilton Day is really quite something. The day's activities include a procession down the Royal Canal in Dublin where everyone stops at the Broom Bridge to watch the youngest member of the party write down the same equations on the bridge that Hamilton, in the excitement of discovery, had many years past carved into the bridge's side. I visited the College Observatory of Dunsink where Hamilton lived and got to see the pulleys and wooden structure of a telescope from two centuries ago. Hamilton arrived there after his graduation from Trinity College in 1827, when he was made the chair of astronomy and Astronomer Royal of Ireland. Locals joke that despite Hamilton's prodigious mathematical talent, he had no real knowledge of or interest in astronomy, and that despite his many theoretical advances, he might have set back observational astronomy in Ireland fifty years.

Hamilton Day nonetheless pays homage to this great theorist's many accomplishments. These included advances in optics and dynamics, the invention of the mathematical theory of *quaternions* (a generalization of complex numbers), as well as definitive demonstrations of the predictive power of math and science. The development of quaternions was no small advance. Quaternions are important for vector calculus, which underlies the way we mathematically study all three-dimensional phenomena. They are also now used in computer graphics and hence in the entertainment industry and video games. Anyone with a PlayStation or Xbox can thank Hamilton for some of the fun.

Among his numerous and substantial contributions, Hamilton significantly advanced the field of optics. In 1832, he showed that light falling at a certain angle on a crystal that has two independent axes would be refracted to form a hollow cone of emergent rays. He thereby made predictions about *internal* and *external* conical refraction of light through a crystal. In a tremendous—and perhaps the first—triumph of mathematical science, this prediction was verified by Hamilton's friend and colleague Humphrey Lloyd. It was a very big deal to see verified a mathematical prediction of a never-before-seen phenomenon and Hamilton was knighted for his achievement.

When I visited Dublin, the locals proudly described this mathematical breakthrough—worked out purely on the basis of geometrical optics. Galileo helped pioneer observational science and experiments, and Francis Bacon was an initial advocate of *inductive science*—where one predicts what will happen based on what came before. But in terms of using math to describe a never-before-seen phenomenon, Hamilton's prediction of conical refraction was probably the first. For this reason, at the very least, Hamilton's contribution to the history of science is not to be ignored.

Nonetheless, despite the significance of Hamilton's discovery, classical geometrical optics is no longer a research subject. All the important phenomena were worked out long ago. Soon after Hamilton's time, in the 1860s, the Scottish scientist James Clerk Maxwell, among others, developed the electromagnetic description of light. Geometrical optics, though clearly an approximation, is nonetheless a good description for a wave with wavelength small enough for interference effects to be irrelevant, and for the light to be treated as a linear ray. In other words, geometrical optics is an effective theory, valid in a limited regime.

That doesn't mean we keep every idea that has ever been developed. Sometimes ideas are just proved wrong. Euclid's initial description of light, resurrected in the Islamic world in the ninth century by Al-Kindi, which claimed that light was emitted by our eyes, was one such example. Although others, such as the Persian mathematician Ibn Sahl, correctly described phenomena such as refraction based on this false premise, Euclid and Al-Kindi's theory—which predates science and modern scientific methods—was simply incorrect. It wasn't absorbed into future theories. It was simply abandoned.

Newton didn't anticipate a different aspect of the theory of light. He had developed a "corpuscular" theory that was inconsistent with the wave theory of light developed by his rival Robert Hooke in 1664 and Christian Huygens in 1690. The debate between them lasted a long time. In the nineteenth century, Thomas Young and Augustin-Jean Fresnel measured light interference, providing a clear verification that light had the properties of a wave.

Later developments in quantum theory demonstrated that Newton was correct in some sense too. Quantum mechanics now tells us that light is indeed composed of individual particles called *photons* that are responsible for communicating the electromagnetic force. But the modern theory of photons is based on light quanta, the individual particles of which light is made, that have a remarkable property. Even an individual particle of light, a photon, acts like a wave. That wave gives the probability of a single photon being found in any region of space. (See Figure 5.)

Geometrical Optics	Wave Optics	Photons
Light travels in straight lines.	*Light travels in waves.*	*Light is transmitted by photons, particles that can act like waves.*

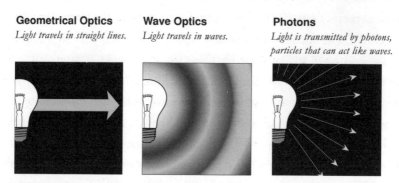

[**FIGURE 5**] Geometrical optics and waves were precursors to our modern understanding of light, and still apply under appropriate conditions.

Newton's corpuscular theory reproduces results from optics. Nonetheless, Newton's corpuscles, which don't have any wavelike nature, are not the same as photons. So far as we now know, the theory of photons is the most basic and correct description of light, which consists of particles that can also accommodate a wave description. Quantum mechanics gives our currently most fundamental description of what light is and how it behaves. It is fundamentally correct and survives.

Quantum mechanics is now much more of a frontier research area than optics. If people continue to think about new science with optics, they are primarily thinking about new effects possible only with quantum mechanics. Modern science, though no longer advancing the science of classical optics, does therefore include a field of quantum optics, which studies the quantum mechanical properties of light. Lasers rely on quantum mechanics, as do light detectors such as photomultipliers, and photovoltaic cells that convert sunlight into electricity.

Modern particle physics also encompasses the theory of quantum electrodynamics (QED), which Richard Feynman and others developed and which includes not only quantum mechanics but also special relativity. With QED, we study individual particles including photons—particles of light—as well as electrons and other particles that carry electric charge. We can understand the rates at which such particles interact and at which they can be created and destroyed. QED is one of the theories that is heavily used in particle physics. It also has made the most accurately verified predictions in all of science. QED is a far cry from geometrical optics, yet both are true in their appropriate domain of validity.

Every area of physics reveals this effective theory idea at work. Science evolves as old ideas get incorporated into more fundamental theories. The old ideas still apply and can have practical applications. But they aren't the domain of frontier research. Though the end of this chapter has focused on the particular example of the physical interpretation of light through the ages, all of physics has developed in this manner. Science proceeds with uncertainty at the edges, but it is advancing methodically overall. Effective theories at a given scale legitimately ignore effects that we can prove won't make a difference for any particular measurement. The wisdom and methods we acquired in the past survive. But theories evolve as we better understand a larger range of distances and energies. Advances give us new insights into what fundamentally accounts for the phenomena we see.

Understanding this progression helps us better interpret the nature of science and appreciate some of the major questions that physicists (and others) are asking today. In the following chapter, we'll see that in many respects, today's methodology began in the seventeenth century.

CHAPTER TWO

UNLOCKING SECRETS

The methods scientists use today are the latest incarnation of a long history of measurements and observations that have been developed over time to verify and—as importantly—rule out scientific ideas. This need to go beyond our intuitive apprehension of the world to advance our understanding is reflected in our very language. The root used in Romance languages for the verb "to think"—*pensum*—comes from the Latin verb "to weigh." English speakers, too, "weigh" ideas.

Many of the formative insights that ushered science into its modern expression were developed in Italy in the seventeenth century, and Galileo was a key player. He was among the first to fully appreciate and advance *indirect measurements*—measurements made with an intermediate device—as well as to design and use experiments as a means of establishing scientific truth. Moreover, he conceived abstract thought experiments that helped him create and consistently formulate his ideas.

I learned about Galileo's many insights that fundamentally changed science when I visited Padua in the spring of 2009. One impetus for my visit was a physics conference that the Paduan physics professor Fabio Zwirner had organized. The other was to receive an honorary citizenship of the city. I was delighted to join my fellow physicist attendees as well as the esteemed group of fellow "citizens," including the physicists Steven Weinberg, Stephen Hawking, and Ed Witten. And—as a bonus—I had a chance to learn some science history.

My trip was auspiciously timed, as 2009 was the 400th anniversary of Galileo's first celestial observations. The citizens of Padua were particularly attentive, since Galileo had been lecturing at the university there at the time of his most significant research. To commemorate his famous observations, the town of Padua (as well as Pisa, Florence, and Venice—other towns that figured prominently in the scientific life of Galileo) had arranged exhibits and ceremonies in his honor. The physics talks took place in a hall in the Centro Culturale Altinate (or San Gaetano), the same building that housed a fascinating exhibit that celebrated Galileo's many concrete accomplishments and highlighted his role in changing and defining what science means today.

Most people I met appreciated Galileo's achievements and conveyed their enthusiasm for modern scientific developments. The interest and knowledge of the Paduan mayor, Flavio Zanonato, impressed even the local physicists. The head of the city not only actively engaged in scientific conversation at a dinner following the public lecture I gave, but during the lecture itself he surprised the audience with an astute question about charge flow at the LHC.

As part of the citizenship ceremony, the mayor gave me the key to the city. The key was fantastic—it lived up to my movie images of what such a thing should be. Large and silver and nicely carved, it prompted one of my colleagues to ask if it was out of a Harry Potter story. It was a ceremonial key—it doesn't actually open anything. Yet it was a beautiful symbol of entry—to a city of course but also, in my imagination, to a rich and textured portal of knowledge.

In addition to the key, Massimilla Baldo-Ceolin, a professor at the University of Padua, gave me a Venetian commemorative medal known as an *osella*. It is engraved with a quote from Galileo that is also on display at the physics department of the university: "Io stimo più il trovar un vero, benché di cosa leggiera, che 'l disputar lungamente delle massime questioni senza conseguir verità nissuna." This translates as "I deem it of more value to find out a truth about however light a matter than to engage in long disputes about the greatest questions without achieving any truth."

I shared these words with many colleagues at our conference since this is in fact a guiding principle to this day. Creative advances often originate with tractable problems—a point we will return to later on. Not all the questions we answer have immediately radical implications. Yet advances, even seemingly incremental ones, occasionally lead to major shifts in our understanding.

This chapter describes how the current observations that this book presents are rooted in developments that occurred in the seventeenth century, and how the fundamental advances made at that time helped define the nature of theory and experiment that we employ today. The big questions are in some respects the same ones that scientists have been asking for 400 years, but because of technological and theoretical advances, the little questions we now ask have evolved tremendously.

GALILEO'S CONTRIBUTIONS TO SCIENCE

Scientists knock on heaven's door in an attempt to cross the threshold separating the known from the unknown. At any moment we start with a set of rules and equations that predict phenomena we can currently measure. But we are always trying to move into regimes that we haven't yet been able to explore with experiments. With technology and mathematics we systematically approach questions that in the past were the subject of mere speculation or faith. With better and more numerous observations and with improved theoretical frameworks that encompass newer measurements, scientists develop a more comprehensive understanding of the world.

I better understood the key role Galileo played in developing this way of thinking as I explored Padua and its historical landmarks. The Scrovegni Chapel is one of its most famous sites, housing Giotto's frescoes from the early fourteenth century. These paintings are notable for many reasons, but to scientists the extremely realistic image of the 1301 passing of Halley's comet (over the *Adoration of the Magi*) is a marvel. (See Figure 6.) The comet had been clearly visible to the naked eye at the time the painting was made.

But the images weren't yet scientific. My tour guide pointed to an

[**FIGURE 6**] Giotto painted this scene, which appears in the Scrovegni Chapel, in the early fourteenth century when Halley's comet was visible to the naked eye.

astral image in the Palazzo della Ragione that she had initially been told was the Milky Way. She remarked that a more expert guide had afterward explained to her the anachronistic nature of the interpretation. At the time the painting was made, people were just illustrating what they saw. It might have been a starry sky, but it was not anything so well defined as our galaxy. Science, as we understand it today, was yet to arrive.

Before Galileo, science relied on unmediated observations and pure thought. Aristotelian science was the model for the way people had tried to understand the world. Math could be used to make deductions, but the underlying assumptions were taken on faith or in accordance with direct observations.

Galileo explicitly refused to base his research on a "mondo di carta" (a world of paper)—he wanted to read and study the "libro della natura" (the book of nature). In achieving this goal, he changed the methodology of observation and, furthermore, recognized the power of experiments. Galileo understood how to construct and use these artificial situations to make deductions about the nature of physical law. With experiments,

Galileo could test hypotheses about the laws of nature that he could prove—and, as importantly, disprove.

Some of his experiments involved inclined planes: the tilted flat surfaces that feature so prominently—and somewhat annoyingly—in every introductory physics text. For Galileo, inclined planes weren't just some made-up classroom problem, as they sometimes appear to introductory physics students. They were a way to study the velocity of falling bodies by spreading out the descent of objects over a horizontal distance so that he could make careful measurements of how they "fell." He measured time with a water chronometer, but he also cleverly added bells at specific points so that he could use his gifted musical ear to listen and establish speed as a ball rolled down, as illustrated in Figure 7. Through these and other experiments dealing with motion and gravity, Galileo, along with Johannes Kepler and René Descartes, laid the foundation for the classical mechanical laws that Isaac Newton so famously developed.

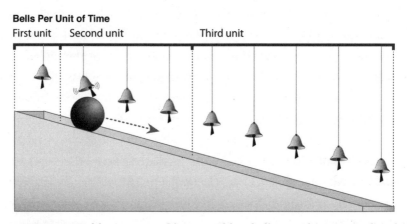

[FIGURE 7] Galileo measured how quickly a ball went down an inclined plane, using bells to register their passage.

Galileo's science also went beyond what he could observe. He created thought experiments—abstractions based on what he did see—in order to make predictions that would apply to experiments no one at the time could actually perform. Perhaps most famous is his prediction that objects—in the absence of resistance—all fall at the same rate. Even though he couldn't set up the idealized situation, he predicted what would occur. Galileo understood gravity's role in objects falling toward the Earth, but he also knew that air resistance slows them down. Good science involves understanding all the factors that might enter into a measurement. Thought experiments and actual physical experiments helped him to better understand the nature of gravity.

In an interesting historical coincidence, Newton, one of the greatest physicists to continue this scientific tradition, was born the year that Galileo died. (At a talk Stephen Hawking gave, he expressed his pleasure that his own birth came precisely three centuries later.) The tradition of designing physical or thought experiments, interpreting them, and understanding their limitations is one that scientists today continue, whatever their year of birth. Current experiments are more subtle and rely on far more advanced technology, but the idea of creating an apparatus to confirm or rule out the predictions from hypotheses continues to define science and its methods in research today.

In addition to experiments—the artificial situations he created to test hypotheses—another of Galileo's game-changing contributions to science was understanding and believing in technology's potential for advancing our observations of the universe as it presents itself. With experiments, he moved beyond pure intellect and reason, and with new devices, he moved beyond unfiltered observations.

Much of earlier science relied on direct unmediated observations. People touched or saw objects with their own senses, not through an intervening device that in some way altered the images. Tycho Brahe, who among other things discovered a supernova and accurately measured the orbits of the planets, made the last famous astronomical observations before Galileo entered the scene. Tycho did use precise measuring instruments, such as large quadrants, sextants, and armillary spheres. He in fact designed and paid for the construction of instruments of greater precision than anyone had used before, leading to measurements that were sufficiently accurate to allow Kepler to deduce elliptical orbits. Yet Tycho made all his measurements through careful observations with his naked eye, with no intermediary lens or other device.

Notably, Galileo had an artistically trained eye and an astute musical ear—he was, after all, the son of a music theorist and lutenist—but he nonetheless recognized how observations that employed technology as a mediator to his observations could improve on his already formidable faculties. Galileo trusted that the indirect measurements he could make with observational tools at both large and small scales would go far beyond those made purely with his unassisted faculties.

Galileo's best-known application of technology was the use of telescopes to explore the stars. His use of this instrument changed the way we do science, the way we think about the universe, and the way we see ourselves. Galileo didn't invent the telescope. It was invented in 1608 by Hans Lippershey in the Netherlands—but the Dutch used telescopes to spy on others, hence the alternative name of spyglass. Yet Galileo was among the first to realize that the device was a potentially potent tool to make observations of the cosmos not possible with the naked eye. He updated the spyglass invented in the Netherlands by developing a telescope capable of magnifying sizes by a factor of 20. Within a year of being presented with a carnival toy, he turned it into a scientific instrument.

Galileo's act of observing through intermediate devices was a radical departure from previous ways of measuring and represented a major advance essential to all modern science. People were initially suspicious of such indirect observations. Even today, some are skeptical about the reality of the observations made with big proton colliders or the data that computers on satellites or telescopes record. But the digital data cataloged by these devices are every bit as real as—and in many respects more accurate than—anything we can observe directly. After all, our hearing comes from oscillations of air hitting our eardrums and our vision from electromagnetic waves hitting our retinas and being processed by our brains. This means that we too are a sort of technology—and not a highly reliable one at that, as anyone who has experienced an optical illusion can attest. (See Figure 8 for an example.) The beauty of scientific measurements is that we can unambiguously deduce aspects of physical reality, including the nature of elementary particles and their properties, from experiments such as those physicists perform today with large and precise detectors.

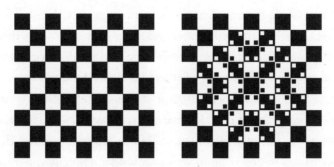

[**FIGURE 8**] Our eyes are not always the most reliable means of ascertaining external reality. Here the two checkerboards are the same, but the dots on the one to the right make the squares appear very different.

Although our instinct might be that observations made unaided with our eyes are the most reliable and that we should be suspicious of abstraction, science teaches us to transcend this all too human inclination. The measurements we make with the instruments we design are more trustworthy than our naked eyes, and can be improved and verified through repetition.

In 1611, the church accepted the radical proposition that indirect measurements are valid. As Tom Levenson relates in his book *Measure for Measure*,[7] the scientific establishment of the church had to decide whether observations from a telescope were trustworthy. Cardinal Robert Bellarmine pressed the church scholars to decide this issue, and on March 24, 1611, the four leading church mathematicians concluded that Galileo's discoveries were all valid: the telescope had indeed produced accurate and reliable observations.

Another commemorative brass medallion that the Paduans shared with me beautifully summed up the pivotal nature of Galileo's achievement. On one side is a picture of the 1609 presentation of the telescope to the Signoria of the Republic of Venice and to the Doge, Leonardo Dona. The other side has an inscription noting that the act "marks the true birth of the modern astronomical telescope" and begins the "revolution in man's perception of the world beyond planet Earth," "a historic moment that crosses the boundaries of Astronomy, making [it] one of the starting points of modern Science."

Galileo's observational advantages led to an explosion of further discoveries. Repeatedly, as he stared up into the cosmos, he found new objects that were beyond the range of the naked eye. He found stars in the Pleiades and throughout the sky that no one had seen before, sprinkled among the brighter ones that were already known. He publicized his discoveries in his famous 1610 book, *Sidereus Nuncius (Starry Messenger)*, that he raced to complete in about six weeks. He hastily performed his research while the printer worked on the manuscript, eager to impress and gain the support of Cosimo II de' Medici, the Grand Duke of Tuscany—and a member of one of Italy's richest families—before someone else with a telescope might manage to publish first.

Because of Galileo's insightful observations, an explosion of understanding occurred. He asked a different type of question: *how* rather than *why*. The detailed discoveries that were possible only with his telescope naturally led him to the conclusions that were to anger the Vatican. Specific observations convinced Galileo that Copernicus had been correct. For him, the only worldview that could consistently explain all of his observations relied on a cosmology in which the Sun, and not the Earth, was the center of the galaxy around which all planets orbited.

The moons of Jupiter were among the most critical of these observations. Galileo could see the moons as they appeared and disappeared and moved in accordance with their orbit around the giant planet. Before this discovery, a stationary Earth seemed the obvious and only way to explain the Moon's fixed orbit. The discovery of Jupiter's moons meant that it too had satellites in tow despite its motion. This lent credence to the possibility that the Earth could also be moving and even orbiting about a separate central body—a phenomenon that was explained only later when Newton developed his theory of gravity and its prediction of the mutual attraction of celestial objects.

Galileo named Jupiter's moons Medicean stars, in honor of Cosimo II de' Medici—further demonstrating his understanding of funding—another key aspect of modern science. The Medicis indeed decided to support Galileo's research. Later on however, after Galileo had been granted funding for life from the city of Florence, the moons were to be renamed Galilean satellites in honor of their discoverer.

Galileo also used his telescope to observe the hills and valleys of the Moon. Before his discoveries, the heavens were thought to be perfectly unchanging, ruled by absolute regularity and constancy. The prevailing Aristotelian view maintained that while everything between the Moon and the Earth was imperfect and inconstant, celestial objects beyond our planet were supposed to be spherical and invariant—of divine essence. Comets and meteors were considered weather phenomena like clouds and winds, and our term *meteorology* harks back to this classification. Galileo's detailed observations implied that imperfection extended beyond the human and sublunar domain. The Moon was not a perfectly smooth sphere and was in fact more similar to the Earth than anyone had dared to suppose. With the discovery of the textured topography of the Moon, the dichotomy between terrestrial and celestial objects was called into question. The Earth was no longer unique, but seemed to be a celestial object like any other.

The art historian Joseph Koerner explained to me that Galileo could use light and shadows to identify craters in part because of his artistic background. Galileo's perspectival training helped him understand the projections he saw. He immediately recognized the implications of these images, even though they weren't fully three-dimensional. He wasn't interested in mapping the Moon, but in understanding its texture. And he understood right away what he saw.

The third significant set of observations that validated the Copernican point of view related to the phases of Venus—illustrated in Figure 9. These observations were particularly significant in establishing that celestial bodies orbited around the Sun. The Earth clearly was not unique in any obvious way, and Venus clearly didn't rotate around it.

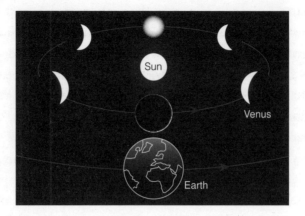

[**FIGURE 9**] Galileo's observation of the phases of Venus demonstrated that it too must orbit the Sun, invalidating the Ptolemaic system.

From an astronomical perspective, the Earth was not so special. The other planets behaved like ours, orbiting the Sun with satellites orbiting them. Furthermore, even beyond the Earth—evidently sullied by human beings—not everything was unblemished perfection. Even the Sun was besmirched by sunspots that Galileo had also identified.

Armed with these observations, Galileo famously concluded that we are not the center of the universe and that the Earth revolves around the Sun. The Earth is not the focal point. Galileo wrote up these radical conclusions. In doing so, he defied the church—although he later professed to reject Copernicanism in order to reduce his punishment to house arrest.

As if his observations and theorizing about the large scales of the cosmos were not enough, Galileo also radically altered our ability to perceive small scales. He recognized that intermediate devices could reveal phenomena at small scales, just as they did at large ones, and he advanced scientific knowledge at both frontiers. In addition to his (in)famous astronomical investigations, he turned technology inward—to investigate the microscopic world.

I was a little surprised when a young Italian physicist, Michele Doro, who was my guide to the San Gaetano exhibit in Padua, said without hesitation that Galileo had invented the microscope. I'd say that outside Italy at least the consensus is that it was invented in the Netherlands, but whether it was Hans Lippershey or Zacharias Janssen (or his father) is anyone's guess. Whether or not Galileo invented the telescope (and he almost certainly didn't), the fact is that he built a microscope and used it to observe smaller scales. It could be used to observe insects with accuracies never before possible. In his letter to friends and other scientists, Galileo was the first we know of to write about the microscope and its potential. The exhibition displayed the first publication to display the systematic observations that could be made with a Galilean microscope: dating from 1630, it illustrated Francesco Stelluti's detailed studies of bees.

The exhibit also showed how Galileo had studied bones—exploring how their structural properties would need to change with size. Apparently, in addition to his many other insights, Galileo was acutely aware of the significance of scale.

The exhibit left no doubt that Galileo fully understood the methods and goals of science—the quantitative, predictive, and conceptual framework that tries to describe definite objects, which act according to the dictates of precise rules. Once these rules have provided well-tested predictions about the world, they can be used to anticipate future phenomena. Science searches for the most economical interpretation that can explain and predict all observations.

The story of the Copernican revolution nicely illustrates this point too. In Galileo's era, Tycho Brahe, the great observational astronomer, came to a different—and wrong—conclusion about the nature of the solar system. He supported an odd hybrid of the Ptolemaic system, with the Earth at the center, and the Copernican system, where planets orbited the Sun. (See Figure 10 for a comparison.) The *Tychonic* universe agreed with observations, but it wasn't the most elegant interpretation. It was, however, more satisfactory to the Jesuits than Galileo's view, since according to Tycho's premises—as with the Ptolemaic theory that Galileo's observations contradicted—the Earth didn't move.[8]

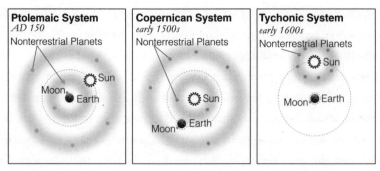

[**FIGURE 10**] Three proposals to describe the cosmos: Ptolemy postulated that the Sun, along with the Moon and other planets, circled the Earth. Copernicus (correctly) suggested that all the planets orbit the Sun. Tycho Brahe postulated that nonterrestrial planets orbited the Sun, which in turn orbited the Earth at the center.

Galileo rightly recognized the jury-rigged nature of the Tychonic interpretation and came to the correct and most economical conclusion. Newton's rival Robert Hooke later noted that both the Copernican and Tychonic theories agreed with Galileo's data, but one was more elegant, saying "but from the proportion and harmony of the World, [one] cannot help but embrace the Copernican Arguments."[9] Galileo's instincts about the truth of the more beautiful theory turned out to be correct, and his interpretation ultimately prevailed when Newton's theory of gravity explained the consistency of the Copernican setup and predicted planetary orbits. Tycho Brahe's theory, as was true for Ptolemy's, was a dead end. It was wrong. It wasn't absorbed in later theories because it couldn't be. Unlike the situation with an effective theory, no approximation of the true theory leads to these non-Copernican interpretations.

As the failure of the original Tychonic theory showed, and as Newtonian physics verified, the subjective criterion of the more economical explanation can also play an important role in the initial scientific interpretation. Research involves the search for underlying laws and principles that will encompass the structures and interactions being observed. Once a sufficient number of observations exist, a theory that economically incorporates the results while providing a predictive underlying framework ultimately wins out. At any point in time, logic takes you only so far—something particle physicists are painfully aware of as we await the data that will ultimately determine what we believe about the underlying nature of the universe.

Galileo helped lay the groundwork for how all scientists work today. Understanding the progression that he and others initiated helps us to better understand the nature of science—in particular, how indirect observations and experiments help us ascertain the correct physical description—as well as some of the major questions that physicists ask today. Modern science builds on all his insights—the usefulness of technology, experiment, theory, and mathematical formulation—in its attempts to match observations to theory. Critically, Galileo recognized the interplay of all these elements in formulating physical descriptions of the world.

Today we can be more free in our thinking, allowing the Copernican revolution to continue as we explore the outer reaches of the cosmos, and theorize about possible extra dimensions or alternative universes. New ideas continue to make human beings less and less central, both literally and figuratively. And observations and experiments will either confirm or reject our proposals.

The indirect methods of observation that Galileo employed currently find dramatic expression in the Large Hadron Collider's elaborate detectors. A final display in the Paduan exhibit showed the evolution of science up to modern times, and even presented pieces of LHC experiments. Our guide confessed he had been confused by this until he recognized that the LHC is the ultimate microscope to date, probing shorter distances than have ever been observed.

As we enter new regimes of precision in measurement and theory, Galileo's understanding of how to design and interpret experiments continues to reverberate. His legacy lives on as we use devices to create images far from visible to the naked eye and apply his insights into how the scientific method works, using experiments to confirm or refute scientific ideas. The conference participants in Padua were thinking about what might be found soon and what it could mean, in the hope we will soon once again cross new thresholds of knowledge. In the interim, we'll keep knocking.

CHAPTER THREE

LIVING IN A MATERIAL WORLD

In February 2008, the poet Katherine Coles and the biologist and mathematician Fred Adler, both from the University of Utah in Salt Lake City, organized an interdisciplinary conference entitled "A Universe in a Grain of Sand." The meeting's topic was the role of scale in various disciplines—a theme that could capitalize on the wide-ranging interests of the diverse group of speakers and attendees. Dividing up our observations into different-sized categories so that we can make sense of and organize them and piece them back together was a subject to which our panel—consisting of a physicist, an architectural critic, and an English professor—could all contribute in interesting ways.

In her opening talk, the literary critic and poet Linda Gregerson described the universe as "sublime." The word precisely captures what makes the universe so wonderful and so frustrating at the same time. A great deal seems beyond our reach and our comprehension, while still appearing to be close enough to tantalize us—to dare us to enter and understand. The challenge for all approaches to knowledge is to make those less accessible aspects of the universe more immediate, more understandable, and ultimately less foreign. People want to learn to read and understand the book of nature and accommodate those lessons into the comprehensible world.

Humanity employs different methods and strives toward contrasting goals in the attempt to unravel the mysteries of life and the world. Art, science, and religion—though they might involve common creative impulses—offer distinct means and methods of approach toward bridging the gaps in our understanding.

So before returning to the world of modern physics, the remainder of this part of the book contrasts these various ways of thinking, introduces some historical context for the science-religion debate, and presents at least one aspect of that debate that won't ever be resolved. In examining these issues, we'll explore science's materialist and mechanistic premises—an essential feature of a scientific approach to knowledge. In all likelihood, those who are at extreme ends of the spectrum won't change their minds, but this discussion might nonetheless help in more precisely identifying the roots of the differences.

THE SCALE OF THE UNKNOWN

The German poet Rainer Maria Rilke rather dramatically captured the paradox at the heart of our feelings when faced with the sublime when he wrote: "For beauty is nothing but the beginning of terror, which we are still just able to endure, and we are so awed because it serenely disdains to annihilate us."[10] In her Salt Lake City talk, Linda Gregerson addressed the sublime in subtle, illuminating, and somewhat less intimidating words. She elaborated on Immanuel Kant's distinction between the beautiful, which "would have us believe we are made for this universe and it for us" and the sublime, which is far more scary. Gregerson described how people feel "apprehension in beholding the sublime" because it seems to be "a poorer fit"—less suited to human interactions and perceptions.

The word "sublime" reemerged in 2009 in discussions of music, art, and science with my collaborators on a physics-based opera about these themes. For our conductor, Clement Power, particular pieces of music occasionally achieved the epitome of simultaneous terror and beauty with which others had defined it. Sublime music for Clement was at a pinnacle beyond his usual powers of comprehension—resisting ready interpretation or explanation.

The sublime proffers scales and poses questions that just might lie beyond our intellectual reach. It is for these reasons both terrifying and compelling. The range of the sublime changes over time as the scales we are comfortable with cover an increasingly large domain. But at any given moment, we still want to gain insights about behavior or events at scales far too small or far too large for us to readily comprehend.

Our universe is in many respects sublime. It prompts wonder but can be daunting—even frightening—in its complexity. Nonetheless, the components fit together in marvelous ways. Art, science, and religion all aim to channel people's curiosity and enlighten us by pushing the frontiers of our understanding. They promise, in their different ways, to help transcend the narrow confines of individual experience and allow us to enter into—and comprehend—the realm of the sublime. (See Figure 11.)

Art allows us to explore the universe through a filter of human perceptions and emotions. It examines how our senses access the world

[FIGURE 11] Caspar David Friedrich's *Wanderer Above the Sea of Fog* (1818), an iconic painting of the sublime—a recurring theme in art and music.

and what we can learn from this interaction—highlighting how people participate in and observe the universe around us. Art is very much a function of human beings, giving us a clearer view of our intuitions and how we as people perceive the world. Unlike science, it is not seeking objective truths that transcend human interactions. Art has to do with our physical and emotional responses to the external world, bearing directly on internal experiences, needs, and capacities that science might never reach.

Science, on the other hand, seeks objective and verifiable truth about the world. It is interested in the elements of which the universe is composed and how those elements interact. Although referring to his trade of forensic investigation, Sherlock Holmes admirably described science's methodology in his inimitable style when he advised Dr. Watson: "Detection is, or ought to be, an exact science and should be treated in the same cold and unemotional manner. You have attempted to tinge it with romanticism, which produces much the same effect as if you worked a love-story or an elopement into the fifth proposition of Euclid . . . The only point in the case which deserved mention was the curious analytical reasoning from effects to causes, by which I succeeded in unravelling it."[11]

No doubt Sir Arthur Conan Doyle would have had Holmes express similar methodology for unraveling the secrets of the universe. Practitioners of science attempt to keep human limitations or prejudices from clouding the picture so that they can trust themselves to obtain an unbiased understanding of reality. They do so with logic and collective observations. Scientists try to objectively figure out how things happen and what underlying physical framework could account for what they observe.

As a sidebar, however, someone should let Sherlock know that he's using inductive, not deductive logic, as do most detectives and scientists when they are trying to piece together the evidence. Scientists and detectives inductively work from observations to try to establish a consistent framework that matches all the measured phenomena. Once the theory is in place, scientists and detectives make deductions, too, in order to predict other phenomena and relationships in the world. But by then—for detectives at least—the work is done.

Religion is yet another approach that many use to respond to the challenge that Gregerson described of relating to the hard-to-access aspects of the universe. The seventeenth-century British author Sir Thomas Browne wrote in his *Religio Medici,* "I love to lose myself in a mystery, to pursue my reason to an O altitudo."[12] For Browne and others like him, logic and the scientific method are believed to be insufficient to access all truth—which they trust religion alone to address. The key distinction between science and religion might well be the character of the questions they choose to ask. Religion includes questions that fall outside the domain of science. Religion asks "why," in the sense of the presumption of an underlying purpose, whereas science asks "how." Science doesn't rely on any sense of an underlying goal for nature. That is a line of inquiry we leave to religion or philosophy, or abandon altogether.

During our Los Angeles conversation, the screenwriter Scott Derrickson told me that there was originally a line in *The Day the Earth Stood Still* (he directed a remake of the 1951 version in 2008) which troubled him so much that he thought about it for days afterward. The Jennifer Connolly character, when talking about her husband's death, was supposed to have commented that "the universe is random."

Scott was disturbed by those words. Underlying physical laws do include randomness, but their whole point is to encapsulate order so that at least some aspects of the universe can be regarded as predictable phenomena. Scott told me that it took several weeks after the line was removed for him to identify the word he had been looking for—"indifferent." My ears perked up when I heard that exact line in the TV show *Mad Men,* enunciated by the lead character, Don Draper, in a way that made it sound distasteful.

But an unconcerned universe is not a bad thing—or a good one for that matter. Scientists don't look for underlying intention in the way that religion often does. Objective science simply requires that we treat the universe as indifferent. Indeed, science in its neutral stance sometimes removes the stigma of evil from human conditions by pointing to their physical, as opposed to moral, origins. We now know, for example, that mental disease and addiction have "innocent" genetic and physical sources that can shift them into the category of diseases exempt from the moral sphere.

Even so, science doesn't address all moral issues (though it doesn't disown them either as is sometimes alleged). Nor does science ask about the reasons for the universe's behavior or inquire into the morality of human affairs. Though logical thinking certainly helps in dealing with the modern world and some scientists today do search for physiological bases for moral actions, science's purpose, broadly speaking, is not to resolve the status of humans' moral standing.

The dividing line isn't always precise, and theologians can sometimes ask scientific questions while scientists might get their initial ideas or directions from a worldview that inspires them—sometimes even a religious one. Moreover, because science is done by human beings, intermediate stages during which scientists are formulating their theories will frequently involve unscientific human instincts such as faith in the existence of answers or emotions about particular beliefs. And, needless to say, this works the other way too: artists and theologians can be guided by observations and a scientific understanding of the world.

But these sometimes blurry divisions don't eliminate the distinctions in ultimate goals. Science aims for a predictive physical picture that can explain how things work. The methods and goals of science and religion are intrinsically different, with science addressing physical reality, and religion addressing psychological or social human desires or needs.

The separate aims shouldn't be a source of conflict—in fact they seem in principle to create a nice division of labor. However, religions don't always stick to questions of purpose or comfort. Many religions attempt to address the external reality of the universe as well, as can be seen even in the definition of the word: *The American Heritage Dictionary* tells us that religion is "belief in a divine or superhuman power or powers to be obeyed and worshiped as the creator(s) and ruler(s) of the universe." Dictionary.com says that religion is "A set of beliefs concerning the causes, nature, and purpose of the universe, especially when considered as the creation of a superhuman agency or agencies, usually involving devotional and ritual observations, and of constructing a moral code governing the morality of human affairs." Religion in these definitions is not only about people's relationship to the world—be it moral or emotional or spiritual—but it's about the world itself. This leaves religious views open to falsification. When science encroaches on domains of knowledge that religion attempts to explain, disagreements are bound to arise.

Despite humanity's shared desire for wisdom, people using different methods to ask questions and find answers or people with different goals haven't always gotten along and the pursuit of truth hasn't always neatly separated along lines that would avoid controversy. When people apply religious beliefs to the natural world, observations of nature can push back, and religion has to accommodate these findings. This was as true for the early church—which had, for example, to reconcile free will with God's infinite powers—as it is for religious thinkers today.

ARE SCIENCE AND RELIGION COMPATIBLE?

Science and religion didn't always face this quandary. Before the scientific revolution, religion and science peacefully coexisted. In the Middle Ages, the Roman Catholic Church was content to allow a generous interpretation of scripture, which lasted until the Reformation threatened the church's dominance. Galileo's evidence for the Copernican heliocentric theory, which contradicted the church's claims about the heavens, was particularly troubling in this context—the publication of his results not only defied church orders, but explicitly questioned the church's sole authority in interpreting scripture. The clergy were therefore none too fond of Galileo and his claims.

More recent history has provided numerous instances of conflict between science and religion. The second law of thermodynamics, which says that the world is moving toward increasing disorder, can dismay people who believe that God created an ideal world. The theory of evolution of course creates similar problems, erupting most recently with "debates" over intelligent design. Even the expanding universe can be disturbing to those who want to believe that we live in a perfect universe, notwithstanding that it was Georges Lemaître, a Catholic priest, who first proposed the Big Bang theory.

One of the more amusing examples of a scientist confronting his faith concerned the English naturalist Philip Gosse. He faced a quandary when—in the early nineteenth century—he realized that the Earth's strata, which hold fossils of extinct animals, contradicted the idea that the Earth could be only 6,000 years old. In his book *Omphalos,* he resolved his conflict by deciding the Earth was created recently—but included specially created "bones" and "fossils" from animals that had never existed and other misleading signs of its (nonexistent) history. Gosse posited that a world in working order should show marks of change, even if they had never actually occurred. This interpretation might sound silly, but technically it does work. However, no one else has ever seemed to take this interpretation very seriously. Gosse himself switched to marine biology to avoid the annoying tests of faith that the dinosaur bones posed.

Happily, most correct scientific ideas become less radical-seeming and more acceptable over time. In the end, scientific discoveries generally prevail. Today no one questions the heliocentric point of view or the universe's expansion. But literal interpretations do still cause problems like Gosse's for believers who take them too seriously.

Less literal readings of scripture helped avoid such conflicts prior to the seventeenth century. In a conversation over lunch, the scholar and historian of religion Karen Armstrong explained how the current conflict between religion and science didn't really exist early on. Religious texts were then read on many levels, so interpretation was less literal and dogmatic and consequently less confrontational.

In the fifth century, Augustine made this viewpoint explicit: "Often a non-Christian knows something about the earth, the heavens, and the other parts of the world, about the motions and orbits of the stars and even their sizes and distances, and this knowledge he holds with certainty from reason and experience. It is thus offensive and disgraceful for an unbeliever to hear a Christian talk nonsense about such things, claiming that what he is saying is based in Scripture. We should do all that we can to avoid such an embarrassing situation, lest the unbeliever see only ignorance in the Christian and laugh to scorn."[13]

Augustine, in his subtlety, went even further. He explained that God deliberately introduced riddles into scripture to give people the pleasure of figuring them out.[14] This referred as much to obscure words as to passages that required metaphorical interpretation. Augustine seems to have had some fun with the logic and illogic of it all, and tried to interpret basic paradoxes. How could anyone completely understand or appreciate God's plan, for example—at least in the absence of time travel?[15]

Galileo himself adhered closely to the Augustinian stance. In a 1615 letter to Madame Christina of Lorraine, the Grand Duchess of Tuscany, he wrote, "I think in the first place that it is very pious to say and prudent to affirm that the Holy Bible can never speak untruth—whenever its true meaning is understood."[16] He even claimed that Copernicus felt similarly, asserting that Copernicus "did not ignore the Bible, but he knew very well that if his doctrine were proved, then it could not contradict the Scripture when they were rightly understood."[17]

In his zeal, Galileo also wrote, quoting Augustine, "If anyone shall set the authority of Holy Writ against clear and manifest reason, he who does this knows not what he has undertaken; for he opposes to the truth not the meaning of the Bible, which is beyond his comprehension, but rather his own interpretation; not what is in the Bible, but what he has found in himself and imagines to be there."[18]

Augustine's less dogmatic approach to scripture assumed the text always had a rational meaning. Any apparent contradiction with observations of the external world necessarily represented the reader's misunderstanding, even if the explanation wasn't manifest. Augustine viewed the Bible as the product of human formulation of divine revelation.

Construing the Bible, at least in part, as a reflection of the writers' subjective experiences, Augustine's interpretation of scripture comes close in some respects to our definition of art. The church wouldn't need to backtrack in the face of scientific discoveries with the Augustinian cast of mind.

Galileo realized this. For he and others who thought similarly, science and the Bible couldn't possibly be in conflict if the words were properly interpreted. Any apparent conflict lies not with the scientific facts, but with human understanding. The Bible might be incomprehensible to humans at times and might superficially appear to contradict our observations, but according to the Augustinian interpretation, the Bible is never wrong. Galileo was devout and didn't think he had the authority to contradict scripture, even when logic would tell him to do so. Many years later, Pope John Paul II went so far as to declare Galileo a better theologian than those who had opposed him.

But Galileo also believed in his discoveries. In a bit of religious trash talking, he presciently advised: "Take note, theologians, that in your desire to make matters of faith out of propositions relating to the fixity of sun and earth you run the risk of eventually having to condemn as heretics those who would declare the earth to stand still and the sun to change position—eventually, I say, at such a time as it might be physically or logically proved that the earth moves and the sun stands still."[19]

Clearly Christian religions didn't always stick to such a philosophy, or Galileo wouldn't have been imprisoned and newspapers today wouldn't be reporting controversies over intelligent design. Though many practitioners of religion have flexible beliefs, a rigid interpretation of physical phenomena is likely to prove problematic. A literal reading of scripture is a risky point of view to uphold. Over time, as technology permits us to scale new regimes, science and religion will have more overlapping domains and potential contradictions can only increase.

Today, a significant proportion of the world's religious population aims to avoid such conflicts through a more liberal interpretation of their faith. They don't necessarily rely on a strict interpretation of scripture or the dogma of any particular faith. They believe they maintain the tenets of their spiritual life while accepting the findings of rigorous science.

PHYSICAL CORRELATES

The intrinsic problem is that the contradictions between science and religion run deeper than any specific words or phrasing. Even without worrying about a literal interpretation of any particular text, religion and science rely on incompatible logical tenets when we consider that religion addresses issues in our world and existence through the intervention of an external deity. Divine actions—whether applied to mountains or your conscience—don't happen within the framework of science.

The crucial contrast is between religion as a social or psychological experience and religion that is based on a God who actively influences us or our world through external intervention. After all, religion is a purely personal enterprise for some. Those who feel this way might relish the social connections that come from being part of a like-minded religious organization or the psychological benefits that come from viewing themselves in the context of a larger world. Faith for people in this category has to do with its practice and the way they choose to live their lives. It is a source of comfort, with a shared set of goals.

Many such people regard themselves as spiritual. Religion enhances their existence—it provides context, meaning, and purpose, as well as a sense of community. They don't see religion's role as explaining the mechanics of the universe. Religion addresses their personal sense of awe and wonder, and it might help in their interactions with others and the world. Many such people would argue that religion and science can perfectly readily coexist.

But religion is usually more than a way of life or a philosophy. Most religions involve a deity who can intervene in mysterious ways that go beyond what people can describe or science involve. Such a belief, even for more open-minded religious people who welcome scientific advances, inevitably introduces a quandary about how to reconcile such activity with the dictates of science. Even allowing for a God or some spiritual force that might have exerted influence earlier on as a prime mover, it is inconceivable from a scientific perspective that God could continue to intervene without introducing some material trace of his actions.

To understand the conflict—and better appreciate the nature of science—we need to more fully understand science's materialist viewpoint, which tells us that science applies to a material universe and that active influences have physical correlates. Built into the scientific view is the idea—introduced in Chapter 1—that we can identify the components of matter at each level of structure. What exists at larger scales is built from material at smaller scales. Even though we can't necessarily explain everything about bigger scales by knowing all the underlying physical elements, those components are nonetheless essential. The material makeup of phenomena that interest us won't always suffice to explain them, but the physical correlates are instrumental to their existence.

Some people turn to religion to answer difficult questions that they don't think science will ever get to. Indeed, the materialist scientific view doesn't mean we are guaranteed to understand everything—certainly not by simply understanding just the basic components. In dividing the universe by scales, scientists recognize that we are unlikely to answer all questions at once and that even though fundamental structure might be essential, it won't necessarily answer all our questions directly. Even when we know quantum mechanics, we still use Newton's laws since they tell us how a ball travels through the Earth's gravitational field in a way that would be very difficult to derive from an atomic picture. The ball needs atoms to exist, but the atomic picture doesn't help explain the ball's trajectory, though it is of course compatible with it.

This lesson generalizes to many phenomena we all encounter in our daily lives. We can often ignore underlying details or composition, even though the material is essential. We don't need to know the inner workings of a car in order to drive it. When we cook food, we evaluate if fish is flaky, if the center of a cake is dry, if oatmeal is mushy, or if a soufflé has risen. But unless we practice molecular gastronomy, we rarely pay attention to the buried atomic structure responsible for these changes. However, that doesn't change the fact that food without substance is not very satisfying. The ingredients in a soufflé look nothing like the final product (see Figure 12). Nonetheless, the constituents and molecules in your food that you are happy to ignore are essential to its existence.

[FIGURE 12] A soufflé is very different from the ingredients that comprise it. In a similar manner, matter might have very different properties—or even appear to obey very different physical laws—from the more fundamental matter of which it is composed.

Similarly, anyone would be hard-pressed to say decisively what music is. But any attempt to describe the phenomenon and our emotional response to it would almost certainly involve viewing music on a level apart from atoms or neurons. Even though we apprehend music when our ears register the sound waves produced by a particularly well-tuned instrument, music is much more than the individual oscillating atoms of air that generate the sound or the physical response of our ears and our brains.

Yet the materialist view still stands, and the substrate is essential. Music arises from those molecules of air. Get rid of the ear's mechanical response to material phenomena and you have no more music. (And in space no one will hear you scream.) It's just that somehow our perception and understanding of music goes beyond that materialistic description. Questions about how we as human beings perceive music won't be addressed if we simply focus on oscillating molecules. Understanding music involves weighing chords and harmonies and lack of harmony in ways that never mention molecules or oscillations. But music nonetheless requires those oscillations, or at least the sensory impression they leave in our brains.

Similarly, understanding an animal's basic components is only one step to understanding the processes that make up life. We almost certainly won't understand everything without a better knowledge of how those components aggregate to produce the phenomena with which we are familiar. Life is an *emergent phenomenon* that goes beyond the basic ingredients.

Most likely consciousness will also turn out to be in this category. Though we don't have a comprehensive theory of consciousness, thoughts and feelings are ultimately rooted in electrical, chemical, and physical properties of the brain. Scientists can observe material mechanistic phenomena in the brain associated with thoughts and feelings, even if they can't put it together to see how it works. This material view is essential but not necessarily sufficient for understanding all the phenomena in our world.

We aren't guaranteed to understand consciousness in terms of the most fundamental units, but we might ultimately figure out principles that apply on some larger, more composite or emergent scale. With future scientific advances, scientists will better understand the fundamental chemistry and electrical channels of the brain and thereby understand the basic functioning units. Consciousness will probably be explained as a phenomenon that scientists will only fully understand by identifying and studying the correct composite pieces.

This means that not only neuroscientists, who study basic brain chemistry, stand a shot at making progress. Developmental psychologists, who ask how a baby's thought process differs from our own,[20] or others who might ask how human thought differs from that of a dog, stand a good chance of making progress as well. Just as music is not one thing but has many levels and many layers, my guess is that so too is consciousness. And by asking questions at a larger level we might gain insights both about consciousness itself and about what are the right questions to ask when we do go ahead and study the building blocks— namely, the chemistry and physics of the brain. As with a lovely soufflé, we will have to understand emergent systems that arise as well. Nonetheless, no human thought or action will occur without affecting some physical component of our body.

Though perhaps less mysterious than the theory of consciousness, physics advances by studying phenomena on various scales. Physicists ask different questions when studying disparate sizes and different aggregates. The questions we ask about sending a spaceship to Mars are very different from the questions we ask about how quarks interact. Both are legitimate questions to study, but we won't readily extrapolate one from the other. Nonetheless, the matter that gets sent out into space is made up of the fundamental components that we ultimately hope to understand.

I've occasionally heard people mock as reductionist the materialist view that particle physicists employ and point out all the phenomena we won't—or don't—address. Sometimes these are physical or biological processes such as brain function or hurricanes, and sometimes they are spiritual phenomena—where I in turn become a little perplexed about what people mean, but which I would have to agree we never address. Physical theories address structure from the largest to the smallest scales that we can hypothesize about or study with experiments. Over time, we build a consistent picture of how one layer of reality proceeds from the next. The basic elements are essential to reality, but good scientists don't assert that knowledge of them in itself explains everything. Explanations call for further research.

Even if string theory turns out to explain quantum gravity, the "theory of everything" will remain a horrible misnomer. In the unlikely event that physicists arrive at such an all-embracing fundamental theory, we would still have to face lots of questions about phenomena on larger scales that won't be answered simply by knowing the basic components. Only when scientists understand collective phenomena that arise on larger scales than those described by elementary strings will we hope to explain superconducting materials, monster waves in the ocean, and life. In the process of doing science, we'll address phenomena scale by scale. We will investigate objects and processes at larger distance scales than we would ever be able to handle if we tried to keep track of each component.

Though we focus on different layers of reality to address different questions, the materialist view is nonetheless essential. Physics and other sciences rely on studying the matter that exists in the world. Science at its core relies on objects interacting through mechanical causes and their effects. Something moves because a force acted on it. An engine functions through its consumption of energy. Planets orbit the Sun through its gravitational influence. According to a scientific perspective, human behavior too ultimately requires chemical and physical processes, even if we are still far from understanding how this works. Our moral choices must also ultimately relate at least in part to our genes and hence our evolutionary history. The physical makeup plays a role in our actions.

We might not address all the vital questions at once, but the underlying substrate is always necessary to a scientific description. For a scientist, material mechanistic elements underlie the description of reality. The associated physical correlates are essential to any phenomenon in the world. Even if not sufficient to explain everything, they are required.

This materialist viewpoint works well for science. But it inevitably leads to logical conflicts when religion invokes a God or some other external entity to explain how people or the world behave. The problem is that in order to subscribe both to science and to a God—or any external spirit—who controls the universe or human activity, one has to address the question of at what point does the deity intervene and how does He do it. According to the materialist, mechanistic point of view of science, if genes that influence our behavior are a result of random mutations that allowed a species to evolve, God can be responsible for our behavior only if He physically intervened by producing that apparently random mutation. To guide our activities today, God had to influence the ostensibly random mutation that was critical to our development. If He did, how did He do that? Did He apply a force or transfer energy? Is God manipulating electrical processes in our brains? Is He pushing us to act in a certain way or creating a thunderstorm for any particular individual so he or she can't get to their destination? On a larger level, if God gives purpose to the universe, how does He apply His will?

The problem is that not only does much of this seem silly, but that these questions seem to have no sensible answer that is consistent with science as we understand it. How could this "God magic" possibly work?

Clearly people who want to believe that God can intervene to help them or alter the world at some point have to invoke nonscientific thinking. Even if science doesn't necessarily tell us why things happen, we do know how things move and interact. If God has no physical influence, things won't move. Even our thoughts, which ultimately rely on electrical signals moving in our brains, won't be affected.

If such external influences are intrinsic to religion, then logic and scientific thought dictate that there must be a mechanism by which this influence is transmitted. A religious or spiritual belief that involves an invisible undetectable force that nonetheless influences human actions and behavior or that of the world itself produces a situation in which a believer has no choice but to have faith and abandon logic—or simply not care.

This incompatibility strikes me as a critical logical impasse in methods and understanding. Stephen Jay Gould's purportedly "nonoverlapping magisteria"—those of science, covering the empirical universe, and religion, extending into moral inquiry—do overlap and face this intractable paradox too. Though believers might relegate the latter to religion, and even though science has yet to answer some deep and fundamental questions of interest to humanity, once we talk about substance and activity—be it in and of the brain or in reference to celestial objects—we are in the domain of science.

RATIONAL CONFLICTS AND IRRATIONAL ESCAPE CLAUSES

However, the incompatibility doesn't necessarily trouble all believers. It so happened that when I was on a plane ride from Boston to Los Angeles, I was seated next to a young actor who had trained as a molecular biologist, but who had some surprising views about evolution. Before embarking on his acting career, he had coordinated science teaching for three years in urban schools. When I met him, he was returning from the inauguration of President Obama, and he was brimming over with enthusiasm and optimism, and wanting to leave the world a better place. Along with continuing his successful acting career, his ambition was to open schools worldwide to teach science and scientific methodology.

But our conversation took a surprising turn. The curriculum he planned would include at least one course on religion. Religion had been a big part of his own life, and he trusted people to make their own judgments. But that wasn't the biggest surprise. He then went on to explain his belief that man descended from Adam as opposed to ascending from apes. I didn't get how someone trained as a biologist could not believe in evolution. This inconsistency goes further than any violation of the materialist universe through God's intermediate intervention of the sort I've just discussed. He told me how he could learn the science and understand the logic but that these are simply how man—whatever that means—puts things together. In his mind the logical conclusions of "man" are just not the way it is.

This exchange reinforced to me why we will have a tough time answering questions about the compatibility of science and religion. Empirically based logic-derived science and the revelatory nature of faith are entirely different methods for trying to arrive at truth. You can derive a contradiction only if your rules are logic. Logic tries to resolve paradoxes, whereas much of religious thought thrives on them. If you believe in revelatory truth, you've gone outside the rules of science so there is no contradiction to be had. A believer can interpret the world in a non-rational way that from his perspective is compatible with science, which is to say accept "God magic." Or—like my neighbor on the plane—he can simply decide that he's willing to live with the contradiction.

But although God might have a way of avoiding the logical contradictions, science does not. Religious adherents who want to accept religious explanations for how the world works as well as scientific thinking are obliged to confront a tremendous chasm between scientific discoveries and unseen, imperceptible influences—a gap that is basically unbridgeable by means of logical thought. They have no choice but to temporarily abandon logical (or at least literal) interpretations in matters of faith—or simply not to care about the contradiction.

Either way, it is still possible to be an accomplished scientist. And indeed, religion might well yield valuable psychological benefits. But any religious scientist has to face daily the scientific challenge to his belief. The religious part of your brain cannot act at the same time as the scientific one. They are simply incompatible.

CHAPTER FOUR

LOOKING FOR ANSWERS

I first heard the phrase "knockin' on heaven's door" when listening to the Bob Dylan song at his 1987 concert with the Grateful Dead in Oakland, California. Needless to say, the title of my book is intended differently than the song's lyrics, which I still hear Dylan and Jerry Garcia singing in my head. The phrase differs from its biblical origin as well, though my title does toy with this interpretation. In Matthew, the Bible says, "Ask, and it shall be given you; seek, and ye shall find; knock, and it shall be opened unto you: For every one that asketh receiveth; and he that seeketh findeth; and to him that knocketh it shall be opened."[21]

According to these words, people can search for knowledge, but the ultimate object is to gain access to God. People's curiosity about the world and active inquiries are mere stepping-stones to the Divine—the universe itself is secondary. Answers might be forthcoming or a believer might be spurred to more actively seek truth, but without God, knowledge is inaccessible or not worth pursuing. People can't do it on their own—they are not the final arbiters.

The title of my book refers to science's different philosophy and goals. Science is not about passive comprehension and belief. And truth about the universe is an end in itself. Scientists actively approach the door to knowledge—the boundary of the domain of what we know. We question and explore and we change our views when facts and logic force us to do so. We are confident only in what we can verify through experiments or in what we can deduce from experimentally confirmed hypotheses.

Scientists know a remarkable amount about the universe, but we also know that much more remains to be understood. A great deal remains beyond the reach of current experiments—or even any experiment we can dream of. Yet despite our limitations, each new discovery lets us advance another rung in our ascent toward truth. Sometimes a single step can have a revolutionary impact on the way we see the world. While acknowledging that our ambitious aspirations are not always satisfied, scientists steadfastly seek access to a richer understanding as advancements in technology make more of the world's ingredients accessible to our gaze. We then search for more comprehensive theories that can accommodate any newly acquired information.

The key question then: who has the capacity—or the right—to look for answers? Do people investigate on their own or trust higher authorities? Before entering the world of physics, this part of the book concludes by contrasting the scientific and religious perspectives.

WHO'S IN CHARGE?

We've seen that in the seventeenth century, the ascent of scientific thinking splintered the Christian attitude to knowledge—leading to conflicts between different conceptual frameworks that continue to this day. But a second source of division between science and religion was about authority. In the eyes of the church, Galileo's claim to be able to think for himself and presume the capacity to independently understand the universe deviated too far from Christian religious belief.

When Galileo pioneered the scientific method, he rejected a blind allegiance to authority in favor of making and interpreting observations on his own. He would change his views in accordance with observations. In doing so, Galileo unleashed a whole new way of approaching knowledge about the world—one that would lead to much greater understanding of and influence over nature. Yet despite (or more accurately because of) the publication of his major advances, Galileo was imprisoned. His openness in his conclusion about the solar system saying that the Earth is not the center was too threatening to the religious powers of the time and their strict interpretation of scripture. With Galileo and other independent thinkers who precipitated the scientific revolution, any literal biblical interpretation of the nature, origin, and behavior of the universe had become subject to refutation.

Galileo's timing was especially poor since his radical claims coincided with the heyday of the Counter-Reformation, the Catholic Church's response to its Protestant offshoots. Catholicism felt itself seriously threatened then by Martin Luther's advocacy of independent thought and interpreting scripture by looking directly to the text, rather than through an unquestioning acceptance of the church's interpretation. Galileo supported Luther's views and went even a step further. He rejected authority and furthermore explicitly contradicted the Catholic interpretation of religious texts.[22] His modern scientific methods were based on direct observations of nature that he then tried to interpret with the most economical hypotheses that could account for the results. Despite Galileo's devotion to the Catholic Church, his inquisitive ideas and methods were too similar to Protestant thinking in the clergy's eyes. Galileo had inadvertently entered into a religious turf war.

Ironically, the Counter-Reformation might nonetheless have inadvertently precipitated Copernicus's espousal of a heliocentric universe. The Catholic Church had wanted to ensure that its calendar was reliable so that celebrations would occur at the right time of year and its rituals would be properly maintained. Copernicus was one of the astronomers asked by the church to attempt to reform the Julian calendar to make it more compatible with the motion of the planets and the stars. It was this very research that led him to his observations and ultimately to his radical claims.

Luther himself did not accept Copernicus's theory. But neither did most anyone else until Galileo's advanced observations and ultimately Newton's theory of gravity validated it later on. Luther did, however, accept other advances made in astronomy and medicine, which he found consistent with an open-minded appreciation of nature. He wasn't necessarily a great scientific advocate, but the Reformation created a way of thinking—an atmosphere where new ideas were discussed and accepted—that encouraged modern scientific methods. Thanks also in part to the development of printing, scientific as well as religious ideas could rapidly spread and diminish the authority of the Catholic Church.

Luther held that secular scientific pursuits were potentially as valuable as religious ones. Scientists such as the great astronomer Johannes Kepler felt similarly. Kepler wrote to Michael Maestlin, his former professor at Tübingen, "I wanted to become a theologian, and for a long time I was restless. Now, however, observe how through my effort God is being celebrated in astronomy."[23]

In this view, science was a way of acknowledging the spectacular nature of God and what he created and the fact that explanations for how things worked were rich and varied. Science became a means of better understanding God's rational and orderly universe, and furthermore helping humankind. Notably, early modern scientists, far from rejecting religion, construed their inquiry as a form of praise for God's creation. They viewed both the Book of Nature and the Book of God as paths to edification and revelation. The study of nature in this view was a form of gratitude and acknowledgment to their creator.

We occasionally hear this viewpoint in more recent times as well. The Pakistani physicist Abdus Salam, during the speech he gave when receiving the 1979 Nobel Prize for his role in creating the Standard Model of particle physics, asserted, "The Holy Prophet of Islam emphasized that the quest for knowledge and sciences is obligatory upon every Muslim, man and woman. He enjoined his followers to seek knowledge even if they had to travel to China in its search. Here clearly he had scientific rather than religious knowledge in mind, as well as an emphasis on the internationalism of the scientific quest."

WHY DO PEOPLE CARE?

Despite the essential differences the last chapter described, some religious believers are happy to apply the scientific and religious parts of their brains separately and continue to view understanding nature as a way of understanding God. Many who don't actively pursue science too are happy to allow scientific progress to proceed unfettered. Still, the rift between science and religion nonetheless persists for many in the United States and other parts of the world. It occasionally expands to the point where it causes violence or at the very least interferes with education.

From the point of view of religious authority, challenges to religion such as science can be suspect for many reasons, including some that have nothing to do with truth or logic. For those in charge, God can always be invoked as the trump card that justifies their point of view. Independent inquiry of any kind is clearly a potential threat. Prying into God's secrets might furthermore undermine the moral power of the church and the secular authority of the rulers on Earth. Such questioning could also interfere with humility and community loyalty, and might even lead one to forget God's importance. No wonder religious authorities are sometimes worried.

But why do individuals align themselves with this point of view? The real question for me is not what the differences are between science and religion. Those can be reasonably well delineated as we argued in the previous chapter. The important questions to answer are these: Why do people care so much? Why are so many people suspicious of scientists and scientific progress? And why does this conflict over authority erupt so often and even continue to this day?

It so happened that I was on a mailing list for the Cambridge Roundtable on Science, Art and Religion, a series of discussions among Harvard and MIT affiliates. The first one I attended, on the topic of the seventeenth-century poet George Herbert and the New Atheists, helped shed some light on some of these questions.

Stanley Fish, the literary scholar turned law professor, was the principal speaker at the event. He began his remarks by summarizing the views of the New Atheists and their antagonism toward religious faith. The New Atheists are those authors, including Christopher Hitchens, Richard Dawkins, Sam Harris, and Daniel Dennett, who have countered religion with harsh and critical words in bestselling books.

After his brief report of their views, Fish proceeded to criticize their lack of understanding of religion, a perspective that seemed to fall on a receptive audience since I think as a nonbeliever I was in the minority at the discussion. Fish argued that the New Atheists would have a stronger case if they had considered the challenges to self-reliance that religious faithful have to contend with.

Faith requires active questioning, and many religions demand it of the observant. Yet at the same time, many religions, some branches of Protestantism among them, call for a rejection or suppression of independent will. In Calvin's words: "Man by nature inclines to deluded self-admiration. Here, then, is what God's truth requires us to seek in examining ourselves: it requires the kind of knowledge that will strip us of all confidence in our own ability, deprive us of all occasion for boasting, and lead us to submission."[24]

These particular words applied primarily to moral questions. But the belief in the necessity for external guidance is unscientific, and it can be difficult to know where to draw the line.

The struggle between the desire for knowledge and the mistrust of human pride reverberates throughout religious literature, including the Herbert poems that Fish and the Roundtable participants discussed. The Cambridge conversation elaborated on Herbert's inner conflicts about his relationships with knowledge and with God. For Herbert, self-generated understanding was a sign of sinful pride. Similar warnings appear in the writings of John Milton. Although he firmly believed in the necessity for robust intellectual inquiry, he nonetheless has Raphael tell Adam in *Paradise Lost* that he should not inquire too curiously into the motion of the stars, for "they need not thy belief."

Surprisingly (at least to me), notable representatives of our group of Harvard and MIT professors in attendance at the roundtable event approved of Herbert's attempts at self-renunciation, believing it was a good thing to suppress one's individuality and align oneself with this greater force. (Anyone who knows Harvard and MIT professors would also be surprised at this alleged denial of ego.)

Maybe the question of whether people can access truth on their own is the real issue at the heart of the religion/science debate. Is it possible that the negative attitudes toward science we hear today are partially rooted in the admittedly extreme beliefs expressed by Herbert and Milton? I'm not sure we are arguing so much about how the world came to be as about who has a right to figure things out and whose conclusions we should trust.

The universe is humbling. Nature hides many of its most interesting mysteries. Yet scientists are arrogant enough to believe we can solve them. Is it blasphemous to search for answers or is it merely presumptuous? Einstein as well as the Nobel Prize–winning physicist David Gross described scientists as thinking they are wrestling with God in order to learn the answers to the big questions about how nature works. David certainly didn't mean this literally (and certainly not humbly)—he was recognizing our miraculous ability to intuit the world around us.

This legacy of not trusting our ability to figure things out for ourselves continues in other respects as well, when we see it in humor, movies, and a good deal of today's politics. Sincerity and respect for facts have become somewhat unfashionable in our ironic and often anti-intellectual era. The degree to which some people will go to deny the successes of science can be amazing. I was once at a party where I met someone who boldly insisted to me that she didn't believe in science. So I asked her whether she had taken the same elevator to the eleventh floor that I had. Did her phone work? How did her electronic invitation reach her?

Many people still consider it embarrassing or at best quaint to be earnest about facts or logic. One source of anti-intellectual antiscientific sentiment might be resentment at the act of egotism in a person feeling powerful enough to tackle the world. Those who have an underlying sense that we don't have the right to take on enormous intellectual challenges believe these are the domain of higher powers than we possess. This peculiar anti-ego, anti-progress trend can still be heard in the playground and the country club.

For some individuals, the idea that you can decipher the world is a source of optimism and leads to a sense of greater understanding and influence. But for others, science and scientific authorities who know more and have greater skill in these technical areas are a source of fear. People divide themselves according to who feels qualified to engage in scientific activities and to evaluate scientific conclusions, and who feels left out and powerless in the face of scientific thought and therefore views such pursuits as acts of ego.

Most people want to feel empowered and to experience a sense of belonging. The question each individual faces is whether religion or science offers a greater sense of control over the world. Where do you find trust, comfort, and understanding? Do you prefer to believe that you can figure things out for yourself or at least trust fellow humans to do so? People want answers and guidance that science can't yet provide.

Nonetheless, science has told us much about what the universe is made of and how it works. When you put together all of what we know, the picture scientists have deduced over time fits together miraculously well. Scientific ideas lead to correct predictions. So some of us trust in its authority, and many recognize the remarkable lessons of science through the ages.

We constantly move beyond human intuition as we explore regions to which we don't have immediate access, and we have yet to make discoveries that bring back the centrality of humans in our description of the world. The Copernican revolution consistently repeats itself as we realize how we are just one of many sets of objects of a random size in a random place in what appears—in the scientific viewpoint—to be a randomly operating universe.

People's curiosity and the ability to make progress toward satisfying this hunger for information make humanity very special indeed. We are the one species equipped to ask questions and systematically chip away to find the answers. We question, we interact, we communicate, we hypothesize, we make abstractions, and in all of this we end up with a richer view of the universe and our place within.

This doesn't mean that science necessarily will answer all questions. People who think science will solve all human problems are probably on the wrong track as well. But it does mean that the pursuit of science has been and will continue to be a worthwhile endeavor. We don't yet know all the answers. But scientifically inclined people, whether or not they have religious faith, try to pry open the universe and find them. Part II explores what they've found so far and what's now on the horizon.

Part II: SCALING MATTER

CHAPTER FIVE

THE MAGICAL MYSTERY TOUR

Though the ancient Greek philosopher Democritus might have started off on the right track when he posited the existence of atoms 2,500 years ago, no one could have accurately guessed what the true elementary components of matter would turn out to be. Some of the physical theories that apply at small distances are so counterintuitive that even the most creative and open-minded people would never have imagined them if experiments hadn't forced scientists to accept their new and confounding premises. Once scientists of the last century had the technology to probe atomic scales, they found that the inner structure of matter repeatedly defied expectations. The pieces fit together in a way that is far more magical than anything we will see on a stage.

Any human being will have difficulty creating an accurate visual image of what's going on at the minuscule scales that particle physicists study today. The elementary components that combine to form the stuff we recognize as matter are very different from what we access immediately through our senses. Those components operate according to unfamiliar physical laws. As scales decrease, matter seems to be governed by properties so different that they appear to be part of entirely different universes.

Many confusions in trying to comprehend this strange inner structure arise from lack of familiarity with the variety of ingredients that emerge at different scales and the range of sizes at which different theories most readily apply. We need to know what exists and to have a sense of the sizes and scales that different theories describe in order to fully understand the physical world.

Later on we will explore the different sizes relevant to space, the final frontier. This chapter first looks inward, starting with familiar scales and ending deep in the interior of matter—the other final frontier. From commonly encountered length scales to the innards of an atom (where quantum mechanics is essential) to the *Planck scale* (where gravity would be as powerful as the other known forces), we'll explore what we know and how it all fits together. Let's now take a tour of this remarkable inner landscape that enterprising physicists and others have deciphered over time.

SCALING THE UNIVERSE

Our journey begins at human scales—the ones we see and touch in our daily lives. It's no coincidence that a meter—not one-millionth of a meter and not ten thousand meters—is, roughly speaking, the size of a person. It's about twice the size of a baby and half the size of a fully grown man. It would be rather strange to find that the basic unit we use for common measurements was one-hundredth the size of the Milky Way or the length of an ant's leg.

Nonetheless, a standard physical unit defined in terms of any particular human wouldn't be all that useful since a measuring stick should be a length we all agree on and understand.[25] So in 1791, the French Academy of Sciences established a standard. A meter was to be defined either as the length of a pendulum with a half period of one second or one ten-millionth of the length of the Earth's meridian along a quadrant (that is the distance from the Equator to the North Pole).

Neither definition has much to do with us humans. The French were simply trying to find an objective measure that we could all agree on and be comfortable with. They converged on the latter choice of definition to avoid the uncertainties introduced by the slightly varying force of gravity over the surface of the Earth.

The definition was arbitrary. It was designed to make the measure of a meter precise and standard so that everyone could agree on what it was.

But one ten-millionth was no coincidence. With the official French definition, a meter stick is something you can comfortably hold in your hands.

Most of us are better approximated by two meters, but none of us
are 10, or even three meters in height. A meter is a human scale, and
when objects are this size, we're pretty comfortable with them—at least
insofar as our ability to observe and interact with them (we'll stay away
from meter-long crocodiles). We know the rules of physics that apply
since they are the ones we witness in our daily existence. Our intuition
is based on a lifetime of observing objects and people and animals whose
size can be reasonably described in terms of meters.

I sometimes find it remarkable how constrained our comfort zone can
be. The NBA basketball player Joakim Noah is a friend of my cousin.
My family and I never tire of commenting on his height. We can look at
photos or marks on a door frame charting his height at various ages and
marvel at him blocking a smaller guy's shot. Joakim is mesmerizingly tall.
But the fact is, he is only about 15 percent taller than the average human
being, and his body works pretty much like everyone else's. The exact
proportions might be different, sometimes giving a mechanical advantage and sometimes not. But the rules his bones and muscles follow are
pretty much the same that yours do.

Newton's laws of motion, written down in 1687, still tell us what happens when we apply force to a given mass. They apply to the bones in
our body and they apply to the ball Joakim throws. With these laws we
can calculate the trajectory of a ball he tosses here on Earth and predict
the path the planet Mercury takes when orbiting the Sun. In all cases,
Newton's laws tell us that motion will continue at the same speed unless a
force acts on the object. That force will accelerate an object in accordance
with its mass. An action will induce an equal and opposite reaction.

Newton's laws work admirably for a well-understood range of lengths,
speeds, and densities. Disparities appear only at the very small distances
where quantum mechanics changes the rules, at extremely high speeds
where relativity applies, or at enormous densities such as those in a black
hole where general relativity takes over.

The effects of any of the new theories that supersede Newton's laws are too small to ever be observed at ordinary distances, speeds, or densities. But with determination and technology we can reach the regimes where we encounter these limitations.

JOURNEY INSIDE

We have to travel a ways down before we encounter new physics components and new physical laws. But a lot goes on in the range of scales between a meter and the size of an atom. Many of the objects we encounter in our daily existence as well as in life itself have important features we can notice only when we explore smaller systems where different behaviors or substructures become prominent. (See Figure 13 for some scales that we refer to in this chapter.)

Of course, a lot of objects we're familiar with are made by simply putting together a single fundamental unit many times, with few details or any internal structure of interest. These *extensive systems* grow like walls of bricks. We can make walls bigger or smaller by adding more or fewer bricks, but the basic functional unit is always the same. A large wall is in many respects just like a small wall. This type of scaling is exemplified in many large systems that grow with the number of repeated elementary components. This applies, for example, to many large organizations as well as computer memory chips that are composed of large numbers of identical transistors.

A different type of scaling that applies to other types of large systems is exponential growth, which occurs when the connections, rather than the fundamental elements, determine a system's behavior. Although such systems too grow by adding many similar units, the behavior depends on the number of connections—not just the number of basic units. These connections don't extend just to an adjacent part, as with bricks, but can extend to other units across the system. Neural systems composed of many synaptic connections, cells with many interacting proteins, and the Internet with a large number of connected computers are all examples. This is a worthy subject of study in itself, and some forms of physics also deal with related emergent macroscopic behavior.

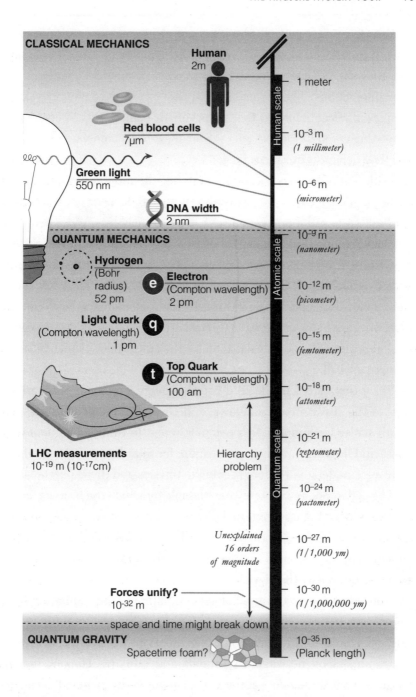

[FIGURE 13] A tour of small scales, and the length units that are used to describe them.

But elementary particle physics is not about complex multi-unit systems. It focuses on identifying elementary components and the physical laws they obey. Particle physics zones in on basic physical quantities and their interactions. These smaller components are of course relevant to complex physical behaviors that involve many components interacting in interesting ways. But identifying the smallest basic components and the way they behave is our focus here.

With technology and biological systems, the individual components of the larger systems have internal structure too. After all, computers are built from microprocessors built from transistors. And when doctors look inside human beings, they find organs and blood vessels and everything else that one encounters upon dissection that are in turn built from cells and DNA that one can see only with more advanced technology. The operation of those internal elements is nothing like what we see when we observe only the surface. The elements change at smaller scales. The best description for the rules those elements follow changes as well.

Since the history of the study of physiology is in some ways analogous to the study of physical laws, and covers some of the interesting length scales for humans, let's take a moment to think a bit about ourselves and how some aspects of the more familiar inner workings of the body were understood before turning to physics and the external world.

The collarbone is an interesting example for which the function could only be understood upon internal dissection. It has its name because on the surface it seems like a collar. But when scientists probed inside the human body they found a key-like piece to the bone that gave it another name we often use: the clavicle.

Nor did anyone understand blood circulation or the capillary system connecting arteries and veins until the early seventeenth century when William Harvey did meticulous experiments to explore the details of hearts and blood networks in animals and humans. Harvey, though English, studied medicine at the University of Padua, where he learned quite a lot from his mentor Hieronymus Fabricius, who was interested in blood flow as well but misunderstood the role of veins and their valves.

Not only did Harvey change our picture of the actual objects involved—here we have networks of arteries and veins carrying blood in a branching network to capillaries working on smaller and smaller scales—but Harvey also discovered a process. Blood is transferred back and forth to cells in ways that no one anticipated until they actually looked. Harvey discovered more than a catalog—he discovered a whole new system.

However, Harvey did not yet have the tools to physically discover the capillary system, which Marcello Malpighi succeeded in doing only in 1661. Harvey's suggestions had included hypotheses based on theoretical arguments that were only later validated by experiments. Although Harvey made detailed illustrations, he couldn't achieve the same level of resolution that users of the microscope such as Leeuwenhoek would subsequently attain.

Our circulatory system contains red blood cells. Those internal elements are only seven micrometers long—roughly one hundred thousandth the size of a meter stick. That's 100 times smaller than the thickness of a credit card—about the size of a fog droplet and about 10 times smaller than what we see with the naked eye (which is in turn a bit smaller than a human hair).

Blood flow and circulation is certainly not the only human process doctors have deciphered over time. Nor has the exploration of inner structure in human beings stopped at the micrometer scale. The discovery of entirely new elements and systems has since been repeated at successively smaller scales, in humans as much as in inanimate physical systems.

Coming down in size to about a tenth of a micron—10 million times smaller than a meter—we find DNA, the fundamental building block of living beings that encodes genetic information. That size is still about 1,000 times bigger than an atom, but is nonetheless a scale where molecular physics (that is, chemistry) plays an important role. Although still not fully understood, the molecular processes occurring within DNA underlie the abundantly broad spectrum of life that covers the globe. DNA molecules contain millions of nucleotides, so the significant role of quantum mechanical atomic bonds should not be surprising.

DNA can itself be categorized on different scales. With its twisty convoluted molecular structure, the total length of human DNA can be measured in meters. But DNA strands are only about two thousandths of a micron—two nanometers wide. That's a little smaller than the current smallest transistor gate of a microprocessor, which is about 30 nanometers in size. A single nucleotide is only 0.33 nm long, comparable in size to a water molecule. A gene is about 1,000–100,000 nucleotides long. The most useful description of a gene will involve different types of questions than those we would confer on individual nucleotides. DNA therefore operates in different ways on different length scales. With DNA, scientists ask different questions and use different descriptions on different scales.

Biology resembles physics in the way that smaller units give rise to the structure that we see at large scales. But biology involves far more than understanding the individual elements of living systems. Biology's goals are far more ambitious. Although ultimately we believe the laws of physics underlie the processes at work in the human body, functional biological systems are complex and intricate and often have difficult-to-anticipate consequences. Disentangling the basic units and the complicated feedback mechanisms is enormously difficult—complicated further by the combinatorics of the genetic code. Even with knowledge of the basic units, we still have the formidable task of resolving more complicated emergent science, notably that responsible for life.

Physicists too can't always understand processes at larger scales through understanding the structure of individual subunits, but most physics systems are simpler in this respect than biological ones. Although composite structure is complex and can have very different properties than the smaller units, feedback mechanisms and evolving structure usually play less of a role. For physicists, finding the simplest, most elementary component is an important goal.

ATOMIC SCALES

As we move away from the mechanics of living systems and descend further in scale to understand basic physical elements themselves, the next length at which we will momentarily pause is the atomic scale, 100 picometers, which is about 10,000 million (10^{10}) times smaller than a meter. The precise scale of an atom is difficult to pin down since it involves electrons that circulate around a nucleus but are never static. However, it is customary to categorize the average distance of the electron from the nucleus and label that as an atom's size.

People conjure up pictures to explain physical processes on these small scales, but they are necessarily based on analogies. We have no choice but to apply descriptions we're familiar with from our experiences at ordinary length scales in order to describe a completely different structure that exhibits strange and unintuitive behavior.

Faithfully drawing the interior of an atom is impossible with the physiology most readily at our disposal—namely, our senses and our human-sized manual dexterity. Our vision, for example, relies on phenomena made visible by light composed of electromagnetic waves. These light waves—the ones in the optical spectrum—have a wavelength that varies between about 380 and 750 nanometers. That is far larger than the size of an atom, which is only about a tenth of a nanometer. (See Figure 14.)

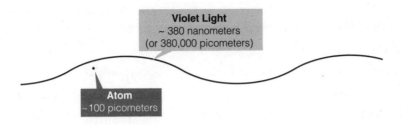

[FIGURE 14] An individual atom is a mere speck relative to even the smallest wavelength of visible light.

This means that probing within the atom with visual light to try to see directly with our eyes is as impossible as threading a needle with mittens on. The wavelengths involved force us to implicitly smear over the smaller sizes that these overly extended waves could never resolve. So when we want to literally "see" quarks or even a proton, we're asking for something intrinsically impossible. We simply don't have the capacity to accurately visualize what is there.

But confusing our ability to picture phenomena with our confidence in their reality is a mistake that scientists cannot afford to make. Not seeing or even having a mental image doesn't mean that we can't deduce the physical elements or processes that are happening at these scales.

From our hypothetical vantage point on the scale of an atom, the world would appear incredible because the rules of physics are extremely different from those that apply to the scales we tick off on our measuring sticks at familiar lengths. The world of an atom looks nothing like what we think of when we visualize matter. (See Figure 15.)

Parts of the Atom

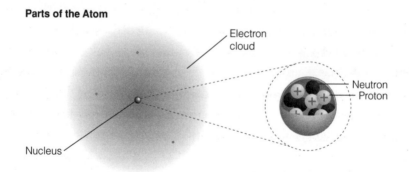

[**FIGURE 15**] An atom consists of electrons orbiting a central nucleus, which consists of positively charged protons, each of charge one, and neutral neutrons, which have zero charge.

Perhaps the first and most striking observation one might make would be that the atom consists primarily of empty space.[26] The nucleus, the center of an atom, is about 10,000 times smaller in radius than the electron orbits. An average nucleus is roughly 10^{-14} meters, 10 femtometers, in size. A hydrogen nucleus is about 10 times smaller than that. The nucleus is as small compared to the radius of an atom as the radius of the Sun is when compared to the size of the solar system. An atom is mostly empty. The volume of a nucleus is a mere trillionth of the volume of an atom.

That's not what we observe or touch when we pound our fist on a door or drink cool liquid through a straw. Our senses lead us to think of matter as continuous. Yet on atomic scales we find that matter is mostly devoid of anything substantial. It is only because our senses average over smaller sizes that matter appears to be solid and continuous. On atomic scales, it is not.

Near emptiness is not all that is surprising about matter on the scale of an atom. What took the physics world by storm and still mystifies physicists and nonphysicists alike is that even the most basic premises of Newtonian physics break down at this tiny distance. The wave nature of matter and the uncertainty principle—key elements of quantum mechanics—are critical to understanding atomic electrons. They don't follow simple curves describing the definite paths that we often see drawn. According to quantum mechanics, no one can measure both the location and the momentum of a particle with infinite precision, a necessary prerequisite for following an object's path through time. Heisenberg's uncertainty principle, developed by Werner Heisenberg in 1926, tell us that the accuracy with which position is known limits the maximum precision with which one can measure momentum.[27] If electrons were to follow classical trajectories, we would know at any given time exactly where the electron is and how fast and in what direction it is moving so that we could know where it will be at any later time, contradicting Heisenberg's principle.

Quantum mechanics tells us that electrons don't occupy fixed locations in the atoms as the classical picture would assert. Instead, probability distributions tell us how likely electrons are to be found in any particular point in space, and all we know are these probabilities. We can predict the average position of an electron as a function of time, but any particular measurement is subject to the uncertainty principle.

Bear in mind that these distributions are not arbitrary. The electrons can't have just any old energy or probability distribution. There is no good classical way to describe an electron's orbit—it can only be described in probabilistic terms. But the probability distributions are in fact precise functions. With quantum mechanics, we can write down an equation describing the wave solution for an electron, and this tells us the probability for it to be at any given point in space.

Another property of an atom that is remarkable from the perspective of a classical Newtonian physicist is that the electrons in an atom can occupy only fixed quantized energy levels. Electron orbits depend on their energies, and those particular energy levels and the associated probabilities must be consistent with quantum mechanical rules.

The electrons' quantized levels are essential to understanding the atom. In the early twentieth century, an important clue that the classical rules had to radically change was that classically, electrons circling a nucleus are not stable. They would radiate energy and quickly fall into the center. Not only would this be nothing like an atom, it wouldn't permit the structure of matter that follows from stable atoms as we know them.

Niels Bohr in 1912 was faced with a challenging choice—abandon classical physics or abandon his belief in observed reality. Bohr wisely chose the former and assumed classical laws don't apply at the small distances occupied by electrons in an atom. This was one of the key insights that led to the development of quantum physics.

Once Bohr ceded Newton's laws, at least in this limited regime, he could postulate that electrons occupied fixed energy levels—according to a quantization condition that he proposed involving a quantity called *orbital angular momentum*. According to Bohr, his quantization rule applied on an atomic scale. The rules were different from those we use at macroscopic scales, such as for the Earth circulating around the Sun.

Technically, quantum mechanics still applies to these larger systems as well. But the effects are far too small to ever measure or notice. When you observe the orbit of the Earth or any macroscopic object for that matter, quantum mechanics can be ignored. The effects average out in all such measurements so that any prediction you make agrees with its classical counterpart. As discussed in the first chapter, for measurements on macroscopic scales, classical predictions generally remain extremely good approximations—so good that you can't distinguish that quantum mechanics is in fact the deeper underlying structure. Classical predictions are analogous to the words and images on an extremely high-resolution computer screen. Underlying them are the many pixels that are like the quantum mechanical atomic substructure. But the images or words are all we generally need (or want) to see.

Quantum mechanics constitutes a change in paradigm that becomes apparent only at the atomic scale. Despite Bohr's radical assumption, he didn't have to abandon what was known before. He didn't assume classical Newtonian physics was wrong. He simply assumed that classical laws cease to apply for electrons in an atom. Macroscopic matter, which consists of so many atoms that quantum effects can't be isolated, obeys Newton's laws, at least at the level at which anyone could measure the success of its predictions. Newton's laws are not wrong. We don't abandon them in the regime in which they apply. But at the atomic scale, Newton's laws had to fail. And they failed in an observable and spectacular fashion that led to the development of the new rules of quantum mechanics.

NUCLEAR PHYSICS

As we continue our journey down in scale into the atomic nucleus itself, we will continue to see the emergence of different descriptions, different basic components, and even different physical laws. But the basic quantum mechanical paradigm will remain intact.

Inside the atom, we'll now explore inner structure with size of about 10 femtometers, the nuclear size of a hundred thousandth of a nanometer. So far as we have measured to date, electrons are fundamental—that is, there don't seem to be any smaller components of electrons. The nucleus, on the other hand, is not a fundamental object. It is composed of smaller elements, known as nucleons. Nucleons are either protons or neutrons. Protons have positive electric charge and neutrons are neutral, with neither a positive nor negative charge.

One way to understand the nature of protons and neutrons is to recognize that they are not fundamental either. George Gamow, the great nuclear physicist and science popularizer, was so excited about the discovery of protons and neutrons that he thought it was the final "other frontier": he didn't think any further substructure existed. In his words:

"Instead of a rather large number of 'indivisible' atoms of classical physics, we are left with only three essentially different entities; protons, electrons, and neutrons . . . Thus it seems we have actually hit the bottom in our search for the basic elements of which matter is formed."[28]

That was a little shortsighted. More precisely, it was not shortsighted enough. There does exist further substructure—more elementary components to the proton and neutron—but the more fundamental elements were challenging to find. One had to be able to study length scales smaller than the size of the proton and neutron, which required higher energies or smaller probes than existed when Gamow made his inaccurate prediction.

If we were to now enter inside the nucleus to see nucleons and protons with size about a fermi—about ten times smaller than the nucleus itself—we would encounter objects Murray Gell-Mann and George Zweig suspected existed inside nucleons. Gell-Mann creatively named these units of substructure *quarks,* in his telling inspired by a line from James Joyce's *Finnegans Wake* ("three quarks for Muster Mark"). The up and down quarks inside a nucleon are the more fundamental objects of smaller size (the two *up* and one *down* quarks inside are shown in Figure 16) that a force called the *strong nuclear force* binds together to form protons and neutrons. Despite its generic name, the strong force is a specific force of nature—one that complements the other known forces of electromagnetism, gravity, and the weak nuclear force that we'll discuss later.

The strong force is called the strong force because it is strong—that's

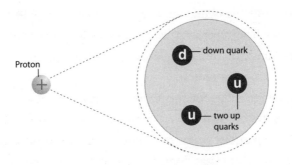

[**FIGURE 16**] The charge of a proton is carried by three valence quarks—two up quarks and a down quark.

an actual quote from a fellow physicist. Even though it sounds pretty silly, it's in fact true. That's why quarks are always found bound together into objects such as protons and neutrons for which the direct influence of the strong nuclear force cancels. The force is so strong that in the absence of other influences the strongly interacting components won't ever be found far apart.

One can never isolate a single quark. It's as if all quarks carry a sort of glue that becomes sticky at long distances (the particles that communicate the strong force are for this reason known as *gluons*). You might think of an elastic band whose restoring force comes into play only when you stretch it. Inside a proton or neutron, quarks are free to move around. But trying to remove one of the quarks any significant distance away would require additional energy.

Though this description is entirely correct and fair, one should be careful in its interpretation. One can't help but think of quarks as all bound together in a sack with some tangible barrier from which they cannot escape. In fact, one model of nuclear systems essentially treats the protons and neutrons in precisely this way. But that model, unlike others we will later encounter, is not a hypothesis for what is really going on. Its purpose was solely to make calculations in a range of distances and energies where forces were so strong our familiar methods don't apply.

Protons and neutrons are not sausages. There is no synthetic casing that surrounds the quarks in a proton. Protons are stable collections of three quarks held together through the strong force. Because of the strong interactions, three light quarks concertedly act as one single object, either a neutron or proton.

Another significant consequence of the strong force—and quantum mechanics—is the ready creation of additional *virtual* particles inside a proton or neutron—particles permitted by quantum mechanics that don't last forever but at any given time contribute energy. The mass—and hence, à la Einstein's $E = mc^2$, the energy—in a proton or neutron is not carried just by the quarks themselves but also by the bonds that tie them together. The strong force is like the elastic band tying together two balls that itself carries energy. "Plucking" the stored energy allows new particles to be created.

So long as the net charge of the new particles is zero, this particle creation from the energy in the proton doesn't violate any known physical laws. For example, a positively charged proton cannot suddenly change into a neutral object when virtual particles are created.

This means that every time a quark—which is a particle that carries nonzero charge—is created, an *antiquark*—which is a particle identical in mass to a quark but with opposite charge—must also be formed. In fact, quark-antiquark pairs can both be created and destroyed. For example, a quark and antiquark can produce a photon (the particle that communicates the electromagnetic force), which in turn produces another particle/antiparticle pair. (See Figure 17.) Their total charge is zero, so even with pair creation and destruction, the charge inside the proton will never change.

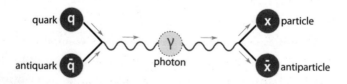

[FIGURE 17] Sufficiently energetic quarks and antiquarks can annihilate into energy that can, in turn, create other charged particles and their antiparticles.

In addition to quarks and antiquarks, the *proton sea* (that's the technical term)—consisting of the virtual particles that are created—contains *gluons* as well. Gluons are the particles that communicate the strong force. They are analogous to the photon that is exchanged between electrically charged particles to create electromagnetic interactions. Gluons (there are eight different ones) act in a similar manner to communicate the strong nuclear force. They are exchanged between particles that carry the charge that the strong force acts on, and their exchange binds or repels the quarks to or from each other.

However, unlike photons, which carry no electric charge and therefore don't directly experience the electromagnetic force, gluons themselves are subject to the strong force. So whereas photons transmit forces over enormous distances—so we can turn on a TV and get a signal generated miles away—gluons, like quarks, cannot travel far before they interact. Gluons bind objects on small scales comparable in size to a proton.

If we take a course-grained view of the proton and focus just on the elements carrying the proton charge, we would say that a proton is primarily composed of three quarks. However, the proton contains a lot more than the three *valence* quarks—the two up quarks and the lone down quark—that contribute to its charge. In addition to the three quarks responsible for a proton's charge, inside a proton is a sea of virtual particles—that is, quark/antiquark pairs and gluons. The closer we examine a proton, the more virtual quark-antiquark pairs and gluons we would find. The exact distribution depends on the energy with which we probe it. At energies with which protons are colliding together today, we find a substantial amount of their energy is carried by virtual gluons and quarks and antiquarks of different types. They are not important for determining electric charge—the sum of the charges of all this virtual stuff is zero—but as we will see later on, they are important for predictions about proton collisions when we need to know exactly what is inside a proton and what carries its energy. (See Figure 18 for the more complicated structure inside a proton.)

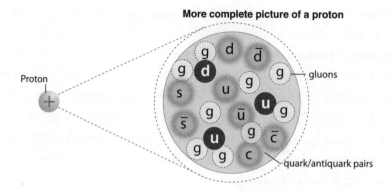

[**FIGURE 18**] The LHC collides protons together at high energy, each of which contains three valence quarks plus many virtual quarks and gluons that can also participate in the collisions.

Now that we have descended to the scale of quarks, held together by the strong nuclear force, I would like to be able to tell you what happens at yet smaller scales. Is there structure inside a quark? Or inside an electron for that matter? As of now, we have no evidence for such a thing. No experiment to date has given any evidence of further substructure. In terms of our journey inside matter, quarks and electrons are the end of the line—so far.

However, the LHC is now exploring an energy scale more than 1,000 times higher—and hence a distance more than 1,000 times smaller—than the scales associated with the proton mass. The LHC achieves its milestones by colliding together two proton beams that have been accelerated to extremely high energy—higher energy than has ever been achieved before here on Earth. The beams of protons at the LHC consist of a few thousand bunches of 100 billion highly lined-up, or collimated, protons concentrated in tiny packets that circulate in the underground tunnel. There are 1,232 superconducting magnets located around the ring to keep the protons inside the beam pipe while electric fields accelerate them to high energies. Other magnets (392 to be exact) reorient the beams so that the two beams stop streaming by each other and collide.

Then—and here's where all the action happens—magnets guide the two proton beams around the ring in a precise path so that they collide in a region smaller across than the width of a human hair. When this collision occurs, some of the energy of the accelerated protons will be converted to mass—as Einstein's famous formula, $E = mc^2$, tells us. And with these collisions and the energy they release, new elementary particles, heavier than any seen before, could be created.

When the protons meet, quarks and gluons occasionally collide with a great deal of energy in a very concentrated region—much as if you had pebbles hidden inside balloons that were smashed together. The LHC provides such high energy that in the events of interest, individual components of the colliding protons crash together. These include the two up quarks and the down quark responsible for the proton's charge. But at LHC energies, virtual particles carry a sizable fraction of the proton's energy as well. At the LHC, along with the three quarks contributing to the proton's charge, the virtual "sea" of particles also collide.

And when that happens—and here is the key to all of particle physics —the numbers and types of particles can change. New results from the LHC should teach us more about smaller distances and sizes. In addition to telling us about possible substructure, it should tell us about other aspects of physical processes that could be relevant at smaller distances. LHC energies are the final short-distance experimental frontier, at least for quite some time.

BEYOND TECHNOLOGY

We've now finished our introductory journey to the smaller scales accessible with current or even imagined technology. However, current human limitations on our ability to explore do not constrain the nature of reality. Even if it seems that we will have a tough time developing technology to explore much smaller scales, we can still try to deduce structure and interactions at those distances through theoretical and mathematical arguments.

We've come a long way since the time of the Greeks. We now recognize that without experimental evidence it is impossible to be certain of what exists at these minuscule scales we would also like to understand. Nonetheless, even in the absence of measurements, theoretical clues can guide our explorations and suggest how matter and forces could behave at tinier length scales. We can investigate possibilities that could help explain and relate the phenomena that occur at measurable scales, even if the fundamental components are not accessible directly.

We don't yet know which, if any, of our theoretical speculative ideas will turn out to be right. Yet even without direct experimental access to very small distances, the scales we have observed constrain what can consistently exist—since it is the underlying theory that has to ultimately account for what we see. That is, experimental results, even on larger distance scales, limit the possibilities and motivate us to speculate in certain specific directions.

Because we haven't yet explored these energies, we don't know much about them. People even speculate the existence of a *desert,* a paucity of interesting lengths or energies, between those of the LHC and those applying to much shorter distances or higher energies. Probably this is lack of imagination or data at work. But for many, the next interesting scale has to do with *unification.*

One of the most intriguing speculations about shorter distances concerns the unification of forces at short distances. It is a concept that sparks both the scientific and the popular imagination. According to such a scenario, the world we see around us fails to reveal the fundamental underlying theory that incorporates all known forces (or, at least, all forces aside from gravity) together with its beauty and simplicity. Many physicists have earnestly searched for such unification from the time the existence of more than one force was first understood.

One of the most interesting such speculations was made by Howard Georgi and Sheldon Glashow in 1974. They suggested that even though we observe three distinct nongravitational forces with different strengths (the electromagnetic and the weak and strong nuclear forces) at low energies, only one force with a single strength will exist at much higher energies. (See Figure 19.)[29] This one force was called a unified force because it encompasses the three known forces. The speculation was called a *Grand Unified Theory* (*GUT*) because Georgi and Glashow thought that was funny.

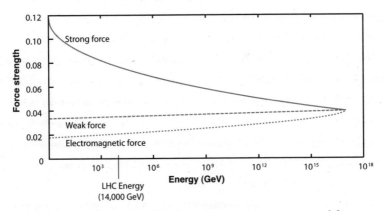

Strength of Standard Model Forces as a Function of Energy

[**FIGURE 19**] At high energy, the three known nongravitational forces might have the same strength and, therefore, could possibly unify into a single force.

This possibility of the strength of forces converging seems to be more than idle speculation. Calculations using quantum mechanics and special relativity indicate it might well be the case.[30] But the energy scale at which it would occur is far above the energies we can study with collider experiments. The distances where the unified force would operate is about 10^{-30} cm. Even though such a size is far removed from anything we can directly observe, we can look for indirect consequences of unification.

One such possibility is proton decay. According to Georgi and Glashow's theory—which introduces new interactions between quarks and leptons—protons should decay. Given the rather specific nature of their proposal, physicists could calculate the rate at which this should occur. So far, no experimental evidence for unification has been found, ruling out their specific suggestion. That doesn't mean that unification is necessarily incorrect. The theory may be more subtle than the one they proposed.

The study of unification demonstrates how we can extend our knowledge beyond scales we directly observe. Using theory, we can try to extrapolate what we have experimentally verified to as yet inaccessible energies. Sometimes we're lucky and clever experiments suggest themselves that allow us to test whether the extrapolation agrees with data or was somehow too naive. In the case of Grand Unified Theories, proton-decay experiments permitted scientists to indirectly study interactions at distances far too tiny for direct observation. These experiments allowed them to test the proposal. One lesson from this example is that we occasionally gain interesting insights into matter and forces and even come up with ways to extend the implications of our experiments to much higher energies and more general phenomena by speculating about distance scales that at first seem to be too remote to be relevant.

The next (and last) stop on our theoretical journey is a distance known as the *Planck length*, namely, 10^{-33} cm. To give a sense of just how minuscule this length is, its size is about as small relative to a proton as a proton is relative to the width of Rhode Island. At this scale, even something as fundamental as our basic notions of space and time will probably fail. We don't even know how to imagine a hypothetical experiment to probe distances smaller than the Planck length. It is the smallest possible scale we can imagine.

This lack of experimental probes of the Planck length could be more than a symptom of our limited imagination, technology, or even funding. The inaccessibility of shorter distances could be a true restriction imposed by the laws of physics. As we will see in the following chapter, quantum mechanics tells us that small probes require high energies. But once the energy trapped in a small region is too big, matter collapses into a black hole. At this point, gravity takes over. More energy then makes the black holes bigger—not smaller—much as we are accustomed to from more familiar macroscopic situations where quantum mechanics plays only a limited role. We just don't know how to explore any distance tinier than the Planck length. More energy doesn't help. Very likely, traditional ideas about space no longer apply at this tiny size.

I recently gave a lecture where, after explaining the current state of particle physics and our suggestions for the possible nature of extra dimensions, someone quoted back to me a statement I had forgotten I'd made about the possible limitations of our notion of spacetime. I was asked how I could reconcile speculations about extra dimensions with the idea of spacetime breaking down.

The speculations for the breakdown of space and possibly time apply only at the unobservably small Planck length. Since no one has observed scales smaller than 10^{-17} cm, the requirement of a nice smooth geometry at measurable distances is not violated. Even if the notion of space itself breaks down at the Planck scale, this is still much smaller than the lengths we explore. There is no inconsistency so long as a smooth recognizable structure emerges when we average over larger, observable scales. After all, different scales often exhibit very different behaviors. Einstein can talk about smooth geometries of space on large scales. But his ideas might break down at smaller scales—so long as they're so tiny and yield such negligible effects on measurable scales that the new more fundamental ingredients have no discernible impact we can observe.

Independently of whether or not spacetime breaks down, a critical feature of the Planck length that our equations certainly tell us would be true is that at this distance, gravity, whose strength is minuscule when acting on fundamental particles at the distances we can measure, would become a strong force—comparable in strength to the other forces we know. At the Planck length, our standard formulation of gravity according to Einstein's theory of relativity would cease to apply. Unlike larger distances where we know how to make predictions that agree well with measurements, quantum mechanics and relativity are inconsistent when we apply the theories we generally use in this tiny regime. We don't even know how to try to make predictions. General relativity is based on smooth classical spatial geometry. At the Planck length, quantum fluctuations can make a spacetime foam with too much structure for our conventional formulation of gravity to apply.

To address physical predictions at the Planck scale, we need a new conceptual framework that combines quantum mechanics and gravity into a single more comprehensive theory known as *quantum gravity*. The physical laws that work most effectively at the Planck scale must be very different from the ones that have proven successful on observable scales. The understanding of this scale could conceivably involve a paradigm shift as fundamental as the transition from classical to quantum mechanics. Even if we can't make measurements at the tiniest distances, we have a chance of learning about the fundamental theory of gravity, space, and time through increasingly advanced theoretical speculations.

The most popular candidate for such a theory is known as *string theory*. Originally string theory was formulated as a theory that replaces fundamental particles with fundamental strings. We now know that string theory also involves fundamental objects other than strings (which we'll learn a little more about in Chapter 17), and the name is sometimes replaced with a broader (but less well-defined) term, *M-theory*. This theory is currently the most promising suggestion for addressing the problem of quantum gravity.

However, string theory poses enormous conceptual and mathematical challenges. No one yet knows how to formulate string theory to answer all the questions we would want a theory of quantum gravity to address. Furthermore, the string scale of 10^{-33} cm is likely to be beyond the reach of any experiment we can think about.

So a reasonable question is whether investigating string theory is a reasonable expenditure of time and resources. I am often asked this question. Why would anyone study a theory so unlikely to yield experimental consequences? Some physicists find mathematical and theoretical consistency reason enough. Those people think they can repeat the type of success Einstein had when he developed his general theory of relativity, based in large part on purely theoretical and mathematical investigations.

But another motivation for studying string theory—one that I think is very important—is that it can and has provided new ways of thinking about ideas that apply on measurable scales. Two of those ideas are *supersymmetry* and theories of *extra dimensions,* ideas that we will address in Chapter 17. These theories do have experimental consequences if they address problems in particle physics. In fact, if certain extra-dimensional theories prove correct and explain phenomena at LHC energies, even evidence of string theory could possibly appear at much lower energies. A discovery of supersymmetry or extra dimensions won't be proof of string theory. But it will be a validation of the utility of working on abstract ideas, even those without direct experimental consequences. It will of course also be a testimony to the utility of experiments in probing even initially abstract-seeming ideas.

CHAPTER SIX

"SEEING" IS BELIEVING

Scientists could decipher what matter is made of only when tools were developed that let them look inside. The word "look" refers not to direct observations but to the indirect techniques that people use to probe the tiny sizes inaccessible to the naked eye.

It's rarely easy. Yet despite the challenges and the counterintuitive results that experiments sometimes display, reality is real. Physical laws, even at tiny scales, can give rise to measurable consequences that eventually become accessible to cleverer investigations. Our current knowledge about matter and how it interacts is the culmination of many years of insight and innovation and theoretical development that permit us to consistently interpret a variety of experimental results. Through indirect observations, pioneered by Galileo centuries ago, physicists have deduced what is present at matter's core.

We'll now explore the current state of particle physics and the theoretical insights and experimental discoveries that have led us to where we are today. Inevitably, the description will have a rather list-like aspect to it as I enumerate the ingredients that compose the matter we know and how they were discovered. The list is a lot more interesting when we remember the very different behaviors of these diverse ingredients on different scales. The chair you are sitting on is ultimately reducible to these elements, but it's quite a train of discoveries to get from here to there.

As Richard Feynman mischievously explained when talking about one of his theories, "If you don't like it, go somewhere else—perhaps to another universe where the rules are simpler . . . I'm going to tell you what it looks like to human beings who have struggled as hard as they can to understand. If you don't like it, that's too bad."[31] You may think that some of what we believe to be true is so crazy or cumbersome that you won't want to accept it. But that won't change the fact that it's the way nature works.

SMALL WAVELENGTHS

Small distances seem strange because they are unfamiliar. We need tiny probes to observe what is happening on the smallest scales. The page (or screen) you are currently reading looks very different from what resides at matter's core. That's because the very act of seeing has to do with observing visible light. That light is emitted from electrons in orbits around nuclei at the center of atoms. As Figure 14 illustrated, the wavelength of that light is never small enough to let us probe inside nuclei.

We need to be more clever—or more ruthless, depending on how you look at it, to detect what is happening on the tiny scale of a nucleus. Small wavelengths are required. That shouldn't be so hard to believe. Imagine a fictional wave with wavelength equal to the size of the universe. No interaction of this wave could possibly have sufficient information to locate anything in space. Unless there are smaller oscillations in this wave that can resolve structure in the universe, we would have no way, with only this enormous wavelength wave as our guide, to determine that anything is in any particular place. It would be like covering a pile of stuff with a net and asking where your wallet is located in the mess underneath. You can't find it unless you have enough resolution to look inside on smaller scales.

With waves, you need peaks and troughs with the right spacing— variations on the scale of whatever it is we are trying to resolve—to be able to identify where something is or what its size or shape might be. You can think of a wavelength the size of the net. If all I know is that something is inside it, I can say with certainty only that something is within a region whose size is that of the net with which I caught it. To say anything more requires either a smaller net or some other way of searching for variations on a more sensitive scale.

Quantum mechanics tells us that waves characterize the probability of finding a particle in any given location. Those waves might be waves associated with light. Or they might be the waves that quantum mechanics tells us are secretly carried by any individual particle. The wavelength of those waves tells us the possible resolution one can hope to attain when we use a particle or radiation to probe small distances.

Quantum mechanics also tells us that short wavelengths require high energies. That's because it relates frequencies to energies, and the waves with the highest frequencies and shortest wavelengths carry the most energy. Quantum mechanics thereby connects high energies and short distances, telling us that only experiments operating at high energies can probe into the inner workings of matter. That is the fundamental reason we need machines that accelerate particles to high energy if we want to probe matter's fundamental core.

Quantum mechanical wave relations tell us that high energies allow us to probe tiny distances and the interactions that occur there. Only with higher energies, and hence shorter wavelengths, can we study these smaller sizes. The quantum mechanical uncertainty relation that tells us small distances connect to large momenta combined with connections among energy, mass, and momenta provided by special relativity make these connections precise.

On top of that, Einstein taught us that energy and mass are interconvertible. When particles collide, their mass can turn into energy. So at higher energies, heavier matter can be produced, since $E = mc^2$. This equation means that larger energy—E—permits the creation of heavier particles with bigger mass—m. And that energy is ecumenical—capable of creating any type of particle that is kinematically accessible (which is to say light enough).

This tells us that the higher energies we currently explore are taking us to smaller sizes, and the particles that get created are our key to understanding the fundamental laws of physics that apply at these scales. Any new high-energy particles and interactions that emerge at short distances hold the clues to decoding the underpinnings of the so-called Standard Model of particle physics, which describes our current understanding of matter's most basic elements and their interactions. We'll now consider a few key Standard Model discoveries, and the methods we now use to advance our knowledge some more.

THE DISCOVERIES OF ELECTRONS AND QUARKS

Each of the destinations on our initial tour of the atom—the electrons circulating around a nucleus and the quarks held together by gluons inside the protons and neutrons—were experimentally discovered with successively higher-energy and hence shorter-distance probes. We've seen that the electrons in an atom are bound to a nucleus through the mutual attraction due to their opposite charges. The attractive force gives the bound system—the atom—lower energy than the charged ingredients in isolation. Therefore, to isolate and study electrons, someone had to add enough energy to *ionize* them, which is to say to free the electrons by ripping them off. Once isolated, physicists could learn more about the electron by studying its properties, such as its charge and its mass.

The discovery of the nucleus, the other part of the atom, was more surprising still. In an experiment analogous to particle experiments today, Ernest Rutherford and his students discovered the nucleus by shooting Helium nuclei (then called alpha particles since nuclei hadn't been discovered) at a thin gold foil. The alpha particles turned out to have enough energy for Rutherford to identify the structure inside the nucleus. He and his colleagues found that the alpha particles they shot at the foil sometimes scattered at much greater angles than they would have anticipated. (See Figure 20.) They expected scatterings like those from tissue paper and instead discovered ones seeming more like they were ricocheting off marbles inside. In Rutherford's own words:

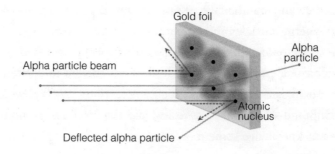

[**FIGURE 20**] Rutherford's experiment scattered alpha particles (which we now know to be Helium nuclei) off gold foil. The unexpectedly large deflections of some of the alpha particles demonstrated the existence of concentrated masses at the centers of the atoms—atomic nuclei.

"It was quite the most incredible event that has ever happened to me in my life. It was almost as incredible as if you fired a 15-inch shell at a piece of tissue paper and it came back and hit you. On consideration, I realized that this scattering backward must be the result of a single collision, and when I made calculations I saw that it was impossible to get anything of that order of magnitude unless you took a system in which the greater part of the mass of the atom was concentrated in a minute nucleus. It was then that I had the idea of an atom with a minute massive centre carrying a charge."[32]

The experimental discovery of quarks inside protons and neutrons used methods in some respects similar to Rutherford's but required even higher energies than that of the alpha particles he had used. Those higher energies required a particle accelerator that could accelerate electrons and the photons they radiated to sufficiently high energies.

The first circular particle accelerator was named a *cyclotron*, due to the circular paths along which the particles were accelerated. Ernest Lawrence built the first cyclotron at the University of California in 1932. It was less than a foot in diameter and was very feeble by modern standards. It produced nowhere near the energy needed to discover quarks. That milestone could happen only with a number of improvements in accelerator technology (that nicely gave rise to a couple of important discoveries along the way).

Well before quarks and the inner structure of the nucleus could be explored, Emilio Segrè and Owen Chamberlain received the 1959 Nobel Prize for their discovery of antiprotons at Lawrence Berkeley Laboratory's *Bevatron* in 1955. The Bevatron was a more sophisticated accelerator than a cyclotron and could raise the protons to energy more than six times their rest mass—more than enough to create proton-antiproton pairs. The proton beam at the Bevatron bombarded targets and (via the magic of $E = mc^2$) produced exotic matter, which includes antiprotons and antineutrons.

Antimatter plays a big role in particle physics, so let's take a brief detour to explore this remarkable counterpart to the matter we observe. Because the charges of matter and antimatter particles add up to zero, matter can annihilate with its associated antimatter when they meet. For example, antiprotons—one form of antimatter—can combine with protons to produce pure energy according to Einstein's equation $E = mc^2$.

The British physicist Paul Dirac first "discovered" antimatter mathematically in 1927 when he tried to find the equation that describes the electron. The only equation he could write down consistent with known symmetry principles implied the existence of a particle with the same mass and opposite charge—a particle that no one had ever seen before.

Dirac racked his brain before capitulating to the equation and admitting this mysterious particle had to exist. The American physicist Carl Anderson discovered the positron in 1932, verifying Dirac's assertion that "The equation was smarter than I was." Antiprotons, which are significantly heavier, were not discovered until more than twenty years later.

The discovery of antiprotons was important not only for establishing their existence, but also for demonstrating a matter-antimatter symmetry in the laws of physics essential to the workings of the universe. The world is, after all, made of matter, not antimatter. Most of the mass of ordinary matter is carried by protons and neutrons, not by their antiparticles. This asymmetry in matter and antimatter is critical to the world as we know it. Yet we don't yet know how it arose.

DISCOVERY OF QUARKS

Between 1967 and 1973, Jerome Friedman, Henry Kendall, and Richard Taylor led a series of experiments that established the existence of quarks inside protons and neutrons. They did their work at a linear accelerator, which—unlike the circular cyclotrons and Bevatrons before it—accelerated electrons along a straight line. The accelerator center was named SLAC, the Stanford Linear Accelerator Center, located in Palo Alto. The electrons that SLAC accelerated radiated photons. These energetic—and hence short-wavelength—photons interacted with quarks inside the nuclei. Friedman, Kendall, and Taylor measured the change in interaction rate as the energy of the collision increased. Without structure, the rate would have gone down. With structure, the rate still decreased, but much more slowly. As with Rutherford's discovery of the nucleus many years before, the projectile (the photon in this case) scattered differently than if the proton was a blob that lacked structure.

Nonetheless, even with experiments performed at the requisite energy, identifying quarks wasn't entirely straightforward. Technology and theory both had to progress to the point that the experimental signatures could be anticipated and understood. Insightful experiments and theoretical analyses performed by the theoretical physicists James Bjorken and Richard Feynman showed that the rates agreed with the predictions based on structure inside the nucleus, thereby demonstrating that structure in protons and neutrons—namely, quarks—had been discovered. Friedman, Kendall, and Taylor were awarded the Nobel Prize in 1990 for their discovery.

No one could have hoped to use their eyes to directly observe a quark or its properties. The methods were necessarily indirect. Nonetheless, measurements confirmed quarks' existence. Agreement between predictions and measured properties, as well as the explanatory nature of the quark hypothesis in the first place, established their existence.

Physicists and engineers have over time developed different and better types of accelerators that operated on increasingly larger scales, accelerating particles to ever higher energy. Bigger and better accelerators produced increasingly energetic particles that were used to probe structure at smaller and smaller distances. The discoveries they made established the Standard Model, as each of its elements was discovered.

FIXED-TARGET EXPERIMENTS
VERSUS PARTICLE COLLIDERS

The type of experiment that discovered quarks, in which a beam of accelerated electrons was aimed at stationary matter, is known as a *fixed-target* experiment. It involves a single beam of electrons that is directed toward matter. The matter target is a sitting duck.

The current highest-energy accelerators are different. They involve collisions of two particle beams, both of which have been accelerated to high energy. (See Figure 21 for a comparison.) As one can imagine, those beams have to be highly focused into a small region to guarantee that any collisions can take place. This significantly reduces the number of collisions you can expect, since a beam is much more likely to interact with a chunk of matter than with another beam.

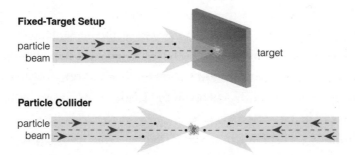

[FIGURE 21] Some particle accelerators generate interactions between a beam of particles and a fixed target. Others collide together two particle beams.

However, beam-beam collisions have one big advantage. These collisions can achieve far higher energy. Einstein could have told you the reason that colliders are now favored over fixed-target experiments. It has to do with what is known as the *invariant mass* of the system. Although Einstein is famous for his theory of "relativity," he thought a better name would have been "Invariantentheorie." The real point of his quest was to find a way to avoid being misled by a particular frame of reference—to find the invariant quantities that characterize a system.

This idea is probably more familiar to you for spatial quantities such as length. Length of a stationary object doesn't depend on how it is oriented in space. An object has a fixed size that has nothing to do with you or your observations, unlike its coordinates, which depend on an arbitrary set of axes and directions you impose.

Similarly, Einstein showed how to characterize events in a way that doesn't depend on an observer's orientation or motion. Invariant mass is a measure of total energy. It tells you how massive an object can be created with the energy in your system.

To determine the amount of invariant mass, one could ask this instead: if your system were sitting still—that is, if it had no overall velocity or momentum—how much energy would it contain? If a system has no momentum, Einstein's equation $E = mc^2$ applies. Therefore, knowing the energy for a system at rest is equivalent to knowing its invariant mass. When the system is not at rest, we need to use a more complicated version of his formula that depends on the value of momentum as well as energy.

Suppose we collide together two beams with the same energy and equal and opposite momentum. When they collide, the momenta add up to zero. That means that the total system is already at rest. Therefore, all the energy—the sum of the energy of the particles in the two individual beams—can be converted to mass.

A fixed-target experiment is very different. One beam has large momentum, but the target itself has none. Not all the energy is available to make new particles because the combined system of the target and the beam particle that hit it is still moving. Because of this motion, not all the energy from the collision can be transferred into making new particles, since some of the energy remains as kinetic energy associated with the motion. It turns out that the available energy scales only with the square root of the product of the energy of the beam and the target. That means, for example, that if we were to increase the energy of a proton beam by 100 and collide it with a proton at rest, the energy available to make new particles would increase by only a factor of 10.

This tells us there is a big difference between fixed-target and beam-beam collisions. The energy of a beam-beam collision is far greater—much bigger than twice as big as a beam-target collision, which is perhaps what you might assume. But that guess would be based on Newtonian thinking, which doesn't apply for the relativistic particles in that beam that travel at nearly the speed of light. The difference in net energy of fixed-target compared with beam-beam collisions is much bigger than the simple guess because at near the speed of light, relativity comes into play. When we want to achieve high energies, we have no choice but to turn to particle colliders, which accelerate two beams of particles to high energy before colliding them together. Accelerating two beams together allows for much higher energy, and hence much richer collisions.

The LHC is an example of a collider. It bangs together two beams of particles that magnets deflect so that they will be aimed toward each other. The principal parameters that determine the capabilities of a collider such as the LHC are the type of particles that collide, their energy after acceleration, and the machine's *luminosity* (the intensity of the combined beams and hence the number of events that occur).

TYPES OF COLLIDERS

Once we have decided that two beams colliding can provide higher energy (and hence explore shorter distances) than fixed-target experiments, the next question is what to collide. This leads to some interesting choices. In particular, we have to decide which particles to accelerate so that they participate in the collision.

It's a good idea to use matter that's readily available here on Earth. In principle, we could try to collide together unstable particles, such as particles called *muons* that rapidly decay into electrons, or heavy quarks such as *top quarks* that decay into other lighter matter.

In that case, we would first have to make these particles in a laboratory since they are not readily available. But even if we could make them and accelerate them before they decayed, we'd have to ensure that the radiation from the decay could be safely diverted. None of these problems are necessarily insurmountable, particularly in the case of muons, whose feasibility as particle beams is currently under investigation. But they certainly pose additional challenges that we don't face with stable particles.

So let's go with the more straightforward option: stable particles available here on Earth that don't decay. This means light particles or at least bound stable configurations of light particles such as protons. We also would want the particles to be charged, so that we can readily accelerate them with an electric field. This leaves protons and electrons as options—particles that are conveniently situated in abundance.

Which should we choose? Both have their advantages and their downsides. Electrons have the advantage that they yield nice clean collisions. After all, electrons are fundamental particles. When you collide an electron into something, the electron doesn't partition its energy into lots of substructure. So far as we know, the electron is all there is. Because the electron doesn't divide, we can follow very precisely what happens when it collides with anything else.

That's not true for protons. Recall that protons are composed of three quarks bound together by the strong nuclear force with gluons exchanged among the quarks that "glue" the whole thing together, as was discussed in Chapter 5. When a proton collides at high energy, the interaction you are interested in—that could produce some heavy particle—generally involves only one individual particle inside the proton, such as a single quark.

That quark certainly won't carry all the energy of the proton. So even though the proton might be very energetic, the quark will generally have much less energy. It can still have quite a bit of energy, just not as much as if the proton could impart all its energy into that single quark.

On top of that, collisions involving protons are very messy. That's because the other stuff in the proton still hangs around, even if it's not involved in the super-high-energy collision we care about. All the remaining particles still interact through strong interactions (aptly named), which means there is a flurry of activity surrounding (and obscuring) the interaction you are interested in.

So why would anyone ever want to collide a proton in that case? The reason is that the proton is heavier than an electron. In fact, the proton mass is about 2,000 times greater than that of an electron. It turns out that's a very good thing when we try to accelerate a proton to high energy. To get to these enormous energies, electric fields accelerate particles around a ring so that they can be accelerated more and more in each successive go-round. But accelerated particles radiate, and the lighter they are, the more they do so.

This means that even though we'd love to collide together super-high-energy electrons, this won't happen any time soon. We can accelerate electrons to very high energies, but high-energy electrons radiate away a significant fraction of their energy when they are accelerated around a circle. (That's why the Stanford Linear Accelerator Center [SLAC] in Palo Alto, California, which accelerated electrons, was a linear collider.) So in terms of pure energy and discovery potential, protons win out. Protons can be accelerated to sufficiently high energy that even their quark and gluon subcomponents can carry more energy than an accelerated electron.

In truth, physicists have learned a lot about particles from both types of colliders—those colliding protons and those colliding electrons. Colliders with an electron beam don't operate at the lofty energies that the highest-energy proton accelerators have attained. But the experiments at colliders with electron beams have achieved measurements more precise than proton collider people could even dream about. In particular, in the 1990s, experiments performed at SLAC and also the Large Electron-Positron collider (LEP) (the blandness of the names never ceases to amuse me) at CERN achieved spectacular precision in verifying the predictions of the Standard Model of particle physics.

These *precision electroweak measurement* experiments exploited the many different processes that can be predicted with knowledge of the electroweak interactions. For example, they measured the weak force carriers' masses, the rates of decay into different types of particles, and asymmetries in the forward and backward parts of the detectors that tell even more about the nature of the weak interactions.

Precision electroweak measurements explicitly apply the effective theory idea. Once physicists perform enough experiments to pin down the few parameters of the Standard Model such as the interaction strengths of each of the forces, everything else can be predicted. Physicists check for consistency of all the measurements and look for deviations that would tell us whether something is missing. All told so far, measurements indicate that the Standard Model works extraordinarily well—so well that we still don't have the clues we need to know what lies beyond except that whatever it is, its effects at LEP energies must be small.

That tells us that getting more information about heavier particles and higher-energy interactions requires directly investigating processes at energies that are considerably higher than those that were achieved at LEP and SLAC. Electron collisions simply won't achieve the energies we think we'll need to pin down the question of what gives particles mass and why they are the masses they are—at least not in the near future. That will require proton collisions.

That's why physicists decided to accelerate protons rather than electrons inside the tunnel that had been built in the 1980s to house LEP. CERN ultimately shut down LEP operations to make way for preparations for its new colossal enterprise, the LHC. Because protons don't radiate nearly as much energy away, the LHC far more efficiently boosts them to higher energies. Its collisions are messier than those involving electrons, and experimental challenges abound. But with protons in the beam, we have a chance to attain energies high enough to directly tell us the answers we've been seeking for several decades.

PARTICLES OR ANTIPARTICLES?

But we still have one more question to answer before we can decide what to collide. After all, collisions involve two beams. We've decided that high energies mandate that one beam consist of protons. But will the other beam be made of particles—that is, protons—or their antiparticles—namely, antiprotons? Protons and antiprotons have the same mass and therefore radiate at the same rate. Other criteria must be used to decide between them.

Clearly protons are more plentiful. We don't see too many antiprotons lying around since they would annihilate with the abundant protons in our surroundings, turning into energy or other, more elementary particles. So why would anyone even consider making beams of antiparticles? What is to be gained?

The answer could be quite a bit. First of all, acceleration is simpler since the same magnetic field can be used to direct protons and antiprotons in opposite directions. But the most important reason has to do with the particles that could be produced.

Particles and antiparticles have equal masses but opposite charges. This means that the incoming particle and antiparticle together carry exactly the same charge as pure energy carries—namely, nothing. According to $E = mc^2$, this means that a particle and its antiparticle can turn into energy, which can in turn create any other particle and antiparticle together, so long as they are not too heavy and have a strong enough interaction with the initial particle-antiparticle pair.

These particles that are created could in principle be new and exotic particles whose charges are different from those of particles in the Standard Model. A colliding particle and antiparticle have no net charge, and neither does an exotic particle plus its antiparticle. So even though the exotic particle's charges can be different from those in the Standard Model, a particle and antiparticle together have zero charge and can in principle be produced.

Let's apply this reasoning to electrons. Were we to collide together two particles with equal charges such as two electrons, we could make only objects that carry the same charge as whatever went in. It could produce either a single object with net charge two or two different objects like electrons that each carry a charge of one. That's rather restrictive.

Colliding two particles with the same charge is very limiting. On the other hand, colliding together particles and antiparticles opens many new doors that wouldn't be possible were we to collide only particles. Because of the greater number of possible new final states, electron-positron collisions have much more potential than electron-electron collisions. For example, collisions involving electrons and their antiparticles—namely, positrons—have produced uncharged particles like the Z *gauge boson* (that's how LEP worked) as well as any particle-antiparticle pair light enough to be produced. Although we pay a steep price when we use antiparticles in the collisions—since they are so difficult to store—we win big when the new exotic particles we hope to discover have different charges than the particles we collide.

Most recently, the highest-energy colliders used one beam of protons and one beam of antiprotons. That of course required a way to make and store antiprotons. Efficiently stored antiprotons were one of CERN's major accomplishments. Earlier on, before CERN constructed the electron-positron collider, LEP, the lab produced high-energy proton and antiproton beams.

The most important discoveries from the collision of protons and antiprotons at CERN were the electroweak gauge bosons that communicate the electroweak force for which Carlo Rubbia and Simon van der Meer received the Nobel Prize in 1984. As with the other forces, the weak force is communicated by particles. In this case they are known as the *weak gauge bosons*—the positively and negatively charged W and neutral Z *vector bosons*—and these three particles are responsible for the weak nuclear force. I still think of the Ws and the Z as the "bloody vector bosons" due to a drunken exclamation of a British physicist who lumbered into the dormitories where visiting physicists and summer students—including me—resided at the time. He was concerned about America's dominance and was looking forward to Europe's first major discovery. When the Ws and the Z vector bosons were discovered at CERN in the 1980s, the Standard Model of particle physics, for which the weak force was an essential component, was experimentally verified.

Critical to the success of these experiments was the method that Van der Meer developed to store antiprotons, which is clearly a difficult task since antiprotons want nothing better than to find protons with which to annihilate. In Van der Meer's process, known as *stochastic cooling,* the electric signals of a bunch of particles drove a device that "kicked" any particle with particularly high momentum, eventually cooling the entire bunch so that they didn't move as rapidly and therefore didn't immediately escape or hit the container so that even antiprotons could be stored.

The idea of a proton-antiproton collider wasn't restricted to Europe. The highest-energy collider of this type was the *Tevatron,* built in Batavia, Illinois. The Tevatron reached an energy of 2 TeV (an energy equivalent to about 2,000 times the proton's rest energy).[33] Protons and antiprotons collided together to make other particles that we could study in detail. The most important Tevatron discovery was the *top* quark, the heaviest and the last Standard Model particle to be found.

However, the LHC is different from either CERN's first collider or the Tevatron. (See Figure 22 for a summary of the collider types.) Rather than protons and antiprotons, the LHC collides together two proton beams. The reason the LHC chooses two proton beams over a beam of protons and another of antiprotons is subtle but worth understanding. The most opportunistic collisions are those where the net charge of the incoming particles adds up to zero. That's the type of collision we already discussed. You can produce anything plus its antiparticle (assuming you have enough energy) when your net charge is zero. If two electrons come in, the net charge of whatever is produced would have to be minus two, which rules out a lot of possibilities. You might think colliding together two protons is an equally bad idea. After all, the net charge of two protons is two, which doesn't seem to be a big improvement.

If protons were fundamental particles, this would be absolutely right. However, as we explored in Chapter 5, protons are made up of subunits.

A COMPARISON OF DIFFERENT COLLIDERS

ACCELERATOR/ YEAR INAUGURATED LAB @ LOCATION	COLLIDING PARTICLES	SHAPE	ENERGY SIZE
Stanford Linear Collider **(SLC)** 1989 *SLAC @ Menlo Park, CA*	electron & positron	Linear $e^- \cdots \blacktriangleright \blacktriangleleft \cdots e^+$	**100 GeV** 3.2 km
Tevatron 1983 *Fermilab @ Batavia, Illinois*	proton & antiproton	Circular $p \blacktriangleright \blacktriangleleft \bar{p}$	**1,960 GeV** 6.3 km
Large Electron-Positron Collider(s) **(LEP/LEP2)*** 1989/2000 *CERN @ Geneva, Switzerland*	electron & positron	Circular $e^- \blacktriangleright \blacktriangleleft e^+$	**90GeV/** **209 GeV** 26.6 km
Large Hadron Collider (LHC) 2008 *CERN @ Geneva, Switzerland*	proton & proton	Circular $p \blacktriangleright \blacktriangleleft p$	**7,000 GeV-** **14,000 GeV** 26.6 km

*LEP was upgraded to LEP2

[**FIGURE 22**] A comparison of different colliders showing their energies, what collides, and the accelerator shape.

Protons contain quarks that are bound together through gluons. Even so, if the three *valence* quarks—two up quarks and a down—that carry its charge were all there were inside a proton, that still wouldn't be very good: the charges of two valence quarks never add to zero either.

However, most of the mass of the proton isn't coming from the mass of the quarks it contains. Its mass is primarily due to the energy involved in binding the proton together. A proton traveling at high momentum contains a lot of energy. With all this energy, protons contain a sea of quarks and antiquarks and gluons in addition to the three valence quarks responsible for the protons' charge. That is, if you were to poke a high-energy proton, you would find not only the three valence quarks, but also a sea of quarks and antiquarks and gluons whose charge adds up to zero.

Therefore, when we consider proton collisions, we have to be a little more careful in our logic than we were with electrons. The interesting events are the result of subunits colliding. The collisions involve the charges of the subunits and not the protons. Even though the sea quarks and gluons don't contribute to the net proton charge, they do contribute to its composition. When protons collide together, it could be that one of the three valence quarks in the proton hits another valence quark and the net charge in the collision doesn't add to zero. When the net charge of the event doesn't vanish, interesting events involving the correct sum of charges might occasionally occur, but the collision won't have the broad capacities that net-charge-zero collisions do.

But a lot of interesting collisions will happen because of the virtual sea, which allows a quark to meet an antiquark or a gluon to hit a gluon, yielding collisions that carry no net charge. When protons bang together, a quark inside one proton might hit an antiquark inside the other, even if that is not what happens most of the time. All of the possible processes that can happen, including those from the collision of the sea particles, play a role when we ask what happens at the LHC. These sea collisions in fact become more and more likely as the protons are accelerated to higher energy.

The total proton charge doesn't determine the particles that get made, since the rest of the proton just goes forward, avoiding the collision. The pieces of the protons that don't collide carry away the rest of the net proton charges, which just disappear down the beam pipe. This was the subtle answer to the question the Paduan mayor asked, which was where the proton charges go during an LHC collision. It has to do with the composite nature of the proton and the high energy that guarantees that only the smallest elements we know of—quarks and gluons—directly collide.

Because only pieces of the proton collide and those pieces can be virtual particles that collide with net zero charge, the choice of proton-proton versus proton-antiproton collider is not so obvious. Whereas in the past, it was worth the sacrifice at lower-energy colliders to make antiprotons in order to guarantee interesting events, at LHC energies that's not such an obvious choice. At the high energies the LHC will achieve, a significant fraction of the energy of the proton is carried by sea quarks, antiquarks, and gluons.

LHC physicists and engineers made the design choice to collide together two proton beams, rather than a proton and an antiproton beam.[34]

This makes generating high luminosity—that is, a higher number of events—a far more accessible goal. It's considerably easier to make proton beams than antiproton beams.

So—rather than a proton-antiproton collider—the LHC is a proton-proton collider. With its many collisions—more readily achievable with protons colliding with protons—it has enormous potential.

CHAPTER SEVEN

THE EDGE OF THE UNIVERSE

On December 1, 2009, I reluctantly woke up at 6:00 A.M. at the Marriott near the Barcelona airport in order to catch a plane. I was visiting to attend the Spanish premiere of a small opera—for which I'd written a libretto—about physics and discovery. The weekend had been enormously satisfying, but I was exhausted and eager to get home. However, I was briefly delayed by a lovely surprise.

The lead story in the newspaper that the hotel provided at my door that morning was "Atom-smasher Sets Record Levels." Rather than the usual headline reporting a horrible disaster or some temporary curiosity, a story about the record energies that the Large Hadron Collider had achieved a couple of days before was the most important news of the day. The excitement in the article about the milestone for the LHC was palpable.

A couple of weeks later, when the two high-energy beams of protons actually collided with each other, the *New York Times* ran a front-page news article titled "Collider Sets Record, and Europe Takes U.S.'s Lead."[35] The record energy reported by the earlier news was now on track to be only the first of a series of milestones to be set by the LHC during this decade.

The LHC is now probing the tiniest distances ever studied. At the same time, satellite and telescope observations are exploring the largest scales in the cosmos, studying the rate at which its expansion accelerates and investigating details of the relic cosmic microwave background radiation left over from the time of the Big Bang.

We currently understand a lot about the makeup of the universe. Yet as with most progress, further questions have emerged as our knowledge has grown. Some have exposed crucial gaps in our theoretical frameworks. Nonetheless, in many cases, we understand the nature of the missing links well enough to know what we need to look for and how.

So let's take a closer look at what's on the horizon—what experiments are out there and what we anticipate they might find. This chapter is about some of the chief questions and physics investigations that the rest of the book will explore.

REACHING BEYOND
THE STANDARD MODEL AT THE LHC

The Standard Model of particle physics tells us how to make predictions about the light particles we're made of. It also describes other heavier particles with similar interactions. These heavy particles interact with light and nuclei through the same forces the particles that constitute our bodies and our solar system experience.

Physicists know about the electron, and heavier similarly charged particles called the *muon* and the *tau*. We know that these particles—called *leptons*—are paired with neutral particles (particles with no charge that don't directly experience electromagnetic interactions) called *neutrinos,* which interact only via the prosaically named *weak force*. The weak force is responsible for radioactive beta decay of neutrons into protons (and beta decay of nuclei in general) and to some of the nuclear processes that occur in the Sun. All Standard Model matter experiences the weak force.

We also know about quarks, which are found inside protons and neutrons. Quarks experience both the weak and electromagnetic forces, as well as the strong nuclear force, which holds light quarks together inside protons and neutrons. The strong force poses calculational challenges, but we understand its basic structure.

The quarks and leptons, together with the strong, weak, and electromagnetic forces, form the essence of the Standard Model. (See Figure 23 for a summary of the particle physics Standard Model.) With these ingredients, physicists have been able to successfully predict the results of all particle physics experiments to date. We understand the Standard Model's particles and how its forces act very well.

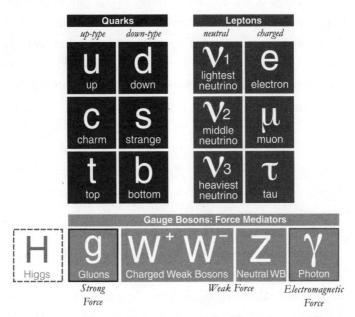

[**FIGURE 23**] The elements of the Standard Model of particle physics, which describe matter's most basic known elements and their inter-actions. Up- and down-type quarks experience the strong, weak, and electromagnetic forces. Charged leptons experience the weak and electromagnetic forces, while neutrinos experience only the weak force. Gluons, weak gauge bosons, and the photon communicate these forces. The Higgs boson is yet to be found.

However, some big puzzles remain.

Chief among these challenges is how gravity fits in. That's a big ques-tion that the LHC has some chance to explore but is far from guaranteed to resolve. The LHC's energy—though high from the perspective of what has been previously achieved here on Earth and from the requirement of what it will take to address some of the big puzzles that come next on this list—is much too low to definitively answer the questions relating to quantum gravity. To do so, we would need to study the infinitesimally tiny lengths where both quantum mechanical and gravitational effects can emerge—and that is far beyond the reach of the LHC. If we're lucky, and gravity plays a big role in addressing the particle problems that we'll soon consider related to mass, then we will be in a much better position to answer this question and the LHC might reveal important information about gravity and space itself. Otherwise, experimental tests of any quantum theory of gravity—including string theory—are most likely a long way off.

However, gravity's relation to the other forces isn't the only major question left unanswered at this point. Another critical gap in our understanding—one that the LHC is definitively poised to resolve—is the way in which the masses of the fundamental particles arise. That probably sounds like a pretty strange question (unless of course you read my first book) since we tend to think of the mass of something as a given—an intrinsic inalienable property of the particle.

And in some sense that is correct. Mass is one of the properties—along with charge and interactions—that define a particle. Particles always carry nonzero energy, but mass is an intrinsic property that can take many possible values including zero. One of Einstein's major insights was to recognize that the value of a particle's mass tells how much energy it has when it's at rest. But particles don't always have a nonvanishing value for their masses. And those that have zero mass, like the photon, are never at rest.

However, the nonzero masses of elementary particles, which are an intrinsic property they possess, are an enormous mystery. Not only quarks and leptons, but also weak gauge bosons—the particles that communicate the weak force—have nonzero mass. Experimenters have measured these masses, but the simplest physics rules simply don't allow them. Standard Model predictions work if we just assume particles have these masses. But we don't know where they came from in the first place. Clearly the simplest rules don't apply and something more subtle is afoot.

Particle physicists believe these nonvanishing masses arise only because something very dramatic occurred in the early universe in a process that is most commonly called the *Higgs mechanism* in honor of the Scottish physicist Peter Higgs who was among the first to show how masses could arise. At least six authors contributed similar ideas, however, so you might also hear about the Englert-Brout-Higgs-Guralnik-Hagen-Kibble mechanism, though I will stick with the name Higgs.[36] The idea—whatever we call it—is that a phase transition (perhaps like the phase transition of liquid water bubbling into gaseous steam) took place that actually changed the nature of the universe. Whereas early on, particles had no mass and zipped around at the speed of light, later on— after this phase transition involving the so-called Higgs field—particles had masses and traveled more slowly. The Higgs mechanism tells how elementary particles go from having zero mass in the absence of the Higgs field to the nonzero masses we have measured in experiments.

If particle physicists are correct and the Higgs mechanism is at work in the universe, the LHC will reveal telltale signs that betray the universe's history. In its simplest implementation, the evidence is a particle—the eponymous *Higgs boson*. In more elaborate physical theories in which the Higgs mechanism is nonetheless at work, the Higgs boson might be accompanied by other particles with about the same mass, or the Higgs might be replaced by some other particle altogether.

Independently of how the Higgs mechanism is implemented, we expect the LHC to produce something interesting. It might be a Higgs boson. It might be evidence of a more exotic theory such as *technicolor* that we will discuss later on. Or it could be something completely unforeseen. If all goes as planned, experiments at the LHC will discern what it was that implemented the Higgs mechanism. No matter what is found, the discovery will tell us something interesting about how particles acquire their masses.

The Standard Model of particle physics, which describes matter's most basic elements and their interactions, works beautifully. Its predictions have been confirmed many times at a high level of precision. This Higgs particle is the last remaining piece of the Standard Model puzzle.[37] We now assume particles have masses. But when we understand the

Higgs mechanism, we'll know how those masses came about. The Higgs mechanism, which is explored further in Chapter 16, is essential to a more satisfactory understanding of mass.

And there is another, even bigger, puzzle in particle physics where the LHC should help. Experiments at the Large Hadron Collider are likely to illuminate the solution to a question known as the *hierarchy problem of particle physics*. The Higgs mechanism addresses the question of why fundamental particles have mass. The hierarchy problem asks the question why those masses are what they are.

Not only do particle physicists believe that masses arose because of a so-called Higgs field that permeates the universe, we also believe we know the energy at which the transition from massless to massive particles occurred. That's because the Higgs mechanism gives masses to some particles in a predictable manner that depends only on the strength of the weak nuclear force and the energy at which the transition occurs.

The peculiar thing is that this transition energy doesn't really make sense from an underlying theoretical perspective. If you put together what we know from quantum mechanics and special relativity, you can actually calculate contributions to particle masses, and they are far bigger than what is measured. Calculations based on quantum mechanics and special relativity tell us that without a richer theory, masses should be much greater—in fact, 10 quadrillion, or 10^{16}, times as big. The theory only hangs together with an enormous fudge physicists unabashedly call "fine-tuning."

The hierarchy problem of particle physics poses one of the biggest challenges to the underlying description of matter. We want to know why the masses are so different from what we would have expected. Quantum mechanical calculations would lead us to believe they should be much bigger than the *weak energy scale* that determines their masses. Our inability to understand the weak energy scale in the superficially simplest version of the Standard Model is a real stumbling block to a fully complete theory.

The likely possibility is that a more interesting, more subtle theory subsumes the most naive model—a possibility we physicists find much more compelling than a fine-tuned theory of nature. Despite the ambitious scope of the question of what theory solves the hierarchy problem, the Large Hadron Collider is likely to shed light on it. Quantum mechanics and relativity dictate not only contributions to masses, but also the energy at which new phenomena must appear. That energy scale is the one the LHC will probe.

We anticipate that at the LHC a more interesting theory will emerge. This theory, which will address these mysteries about masses, should reveal itself when new particles and forces or symmetries show up. It's one of the big secrets we hope LHC experiments will unmask.

The answer is interesting in itself. But it is likely to be the key to deep insights into other aspects of nature as well. Two of the most compelling suggested answers to the problem involve either extensions of symmetries of space and time, or revisions of our notion of space itself.

Scenarios that are further explained in Chapter 17 tell us that space might contain more than the three dimensions we know about: up-down, forward-backward, and left-right. In particular, it could contain entirely unseen dimensions that hold the key to understanding particle properties and masses. If that's the case, the LHC will provide evidence of these dimensions in the form of particles known as *Kaluza-Klein* particles that travel throughout the full higher-dimensional spacetime.

No matter what theory solves the hierarchy problem, it should provide experimentally accessible evidence at the weak energy scale. A train of theoretical logic will connect what we find at the LHC to whatever ultimately resolves this problem. It might be something we anticipate or it might be unforeseen, but it should be spectacular either way.

DARK MATTER

In addition to these particle physics issues, the LHC could also help illuminate the nature of the *dark matter* of the universe, the matter that exerts gravitational influence but does not absorb or emit light. Everything we see—the Earth, the chair you're sitting on, your pet parakeet—is made up of Standard Model particles that interact with light. But visible stuff that interacts with light and whose interactions we understand constitutes only about four percent of the energy density of the universe. About 23 percent of its energy is carried by something known as dark matter that has yet to be positively ID'd.

Dark matter is indeed matter. That is, it clumps together through gravity's influence and thereby (along with ordinary matter) contributes to structures—galaxies, for example. However, unlike familiar matter such as the stuff we're made of and the stars in the sky, it doesn't emit or absorb light. Because we generally see things through light that is emitted or absorbed, dark matter is hard to "see."

Really, the term "dark matter" is a misnomer. So-called dark matter isn't exactly dark. Dark stuff absorbs light. We can actually see dark stuff where light is absorbed. Dark matter, on the other hand, doesn't interact with light of any kind in any observable way. Technically speaking, "dark" matter is transparent. But I'll continue to use conventional terminology and refer to this elusive substance as dark.

We know dark matter exists because of its gravitational effects. But without seeing it directly, we won't know what it is. Is it composed of many tiny identical particles? If so, what is the particle's mass and how does it interact?

We might, however, soon learn much more. Remarkably, the LHC might in fact have the right energy to make particles that could be the dark matter. The key criterion for dark matter is that the universe contains the right amount to exert the measured gravitational effects. That is, the *relic density*—the amount of stored energy that our cosmological models predict survives to this day—has to agree with that measured value. The surprising fact is that if you have a stable particle whose mass corresponds to the weak energy scale that the LHC will explore (again via $E = mc^2$) and whose interactions also involve particles with that energy, its relic density will be in the right ballpark to be dark matter.

The LHC could therefore not only give us insights into particle physics questions, but also give us clues to what is out there in the universe today and how it all began, questions that are incorporated into the science of cosmology, which tells us how the universe has evolved.

As with the elementary particles and their interactions, we understand a surprising amount about the universe's history. Yet also as with particle physics, some very big questions remain. Chief among these difficult questions are these: What is the dark matter?, What is the even more mysterious entity called *dark energy*?, and What drove a period of exponential expansion of the early universe known as *cosmological inflation*?

Today is a tremendous time for observations that might tell us the answers to these questions. Dark matter investigations are at the forefront of the overlap between particle physics and cosmology. Dark matter's interactions with ordinary matter—matter we can make detectors from—are extremely weak, so weak that we are still looking for any evidence of dark matter aside from its gravitational effects.

Current searches therefore rely on the leap of faith that dark matter, despite its near invisibility, nonetheless interacts weakly—but not impossibly weakly—with matter that we know. This isn't merely a wishful guess. It's based on the calculation mentioned above that shows that stable particles whose interactions are connected to the energy scale that the LHC will explore have the right density to be dark matter. We hope that even though we haven't yet identified dark matter, we have a good chance of detecting it in the near future.

However, most cosmology experiments don't take place at accelerators. Dedicated outward-looking experiments on Earth and out in space are primarily responsible for addressing and advancing our understanding of potential solutions to cosmological questions.

For example, astrophysicists have sent satellites into space to observe the universe from an environment not obscured by dust and physical and chemical processes on or near the Earth's surface. Telescopes and experiments here on Earth give us additional insights in an environment scientists can more directly control. These experiments in space and on Earth are poised to shed light on many aspects of how the universe has come to be.

We're hoping that a sufficiently strong signal in any of these experiments (which we will describe in Chapter 21) will let us decipher the mysteries of dark matter. These experiments could tell us the nature of dark matter and illuminate its interactions and mass. In the meantime, theorists are thinking hard about all possible models of dark matter and how to use all these detection strategies to learn what dark matter really is.

DARK ENERGY

Ordinary matter and dark matter still do not provide the sum total of the energy in the universe—together they constitute only about 27 percent. Even more mysterious than dark matter is the substance that constitutes the remaining 73 percent and that has become known as dark energy.

The discovery of dark energy was the most profound physics wake-up call of the late twentieth century. Although there is much we don't yet know about the evolution of the universe, we have a spectacularly successful understanding of the universe's evolution based on the so-called Big Bang theory supplemented by a period of exponential expansion of the universe known as cosmological inflation.

This theory has agreed with a range of observations, including observations of the microwave radiation in the sky—the microwave background radiation left over from the time of the Big Bang. Originally the universe was a hot dense fireball. But during the 13.75 billion years of its existence it has diluted and cooled substantially, leaving this much cooler radiation that is a mere 2.7 degrees kelvin today—only a few degrees Celsius above absolute zero. Other evidence for the Big Bang theory of expansion can be found in detailed studies of the abundances of nuclei that were made during the universe's early evolution and in measurements of the universe's expansion itself.

The underlying equations we use to figure out how the universe evolves are the equations Einstein developed in the early twentieth century that tell us how to derive the gravitational field from a given distribution of matter or energy. These equations apply to the gravitational field between the Earth and the Sun but they also apply to the universe as a whole. In all cases, in order to derive the consequences of these equations, we need to know the matter and energy that surround us.

The shocking observation was that measurements of the characteristics of the universe required the presence of this new form of energy that is not carried by matter. This energy is not carried by particles or other stuff, and it doesn't clump like conventional matter. It doesn't dilute as the universe expands but maintains a constant density. The expansion of the universe is slowly accelerating as a consequence of this mysterious energy, which resides throughout the universe, even if it were empty of matter.

Einstein had originally proposed such a form of energy in what he called the *universal constant,* which later became known to physicists as the *cosmological constant.* Shortly after, he thought it a mistake and, indeed, that his use of it to try to explain why the universe was static was misguided. The universe does in fact expand, as Edwin Hubble showed soon after Einstein proposed the idea. The expansion is not only real, but it now seems that its current acceleration is due to the funny type of energy that Einstein had introduced and quickly dismissed in the 1930s.

We want to understand this mysterious dark energy better. Observations at this point are designed to determine whether it is just the sort of background energy that Einstein first proposed or whether it is a new form of energy that changes with time. Or is it something entirely unanticipated that we don't yet even know how to think about?

OTHER COSMOLOGICAL INVESTIGATIONS

This is only a sampling—albeit an important one—of what we are now investigating. In addition to what I have already described, many more cosmological investigations are in store. Gravity wave detectors will look for gravitational radiation from merging black holes and other exciting phenomena involving large amounts of mass and energy. Cosmic microwave experiments will tell us more about inflation. Cosmic ray searches will tell us new details about the content of the universe. And infrared radiation detectors could find new exotic objects in the sky.

In some cases, we will understand the observations sufficiently well to know what they imply about the underlying nature of matter and physical laws. In other cases, we'll spend a lot of time unraveling the implications. Regardless of what happens, the interplay between theory and data will lead us to loftier interpretations of the universe around us and expand our knowledge into currently inaccessible domains.

Some experiments might yield results soon. Others could take many years. As data come in, theorists will be forced to revisit and sometimes even abandon suggested explanations so we can improve our theories and apply them correctly. That might sound discouraging, but it's not as bad as you might think. We eagerly anticipate the clues that will help us answer our questions as experimental results guide our investigations and ensure that we make progress—even when new results might require abandoning old ideas. Our hypotheses are initially rooted in theoretical consistency and elegance, but, as we will see throughout this book, ultimately it is experiment—not rigid belief—that determines what is correct.

Part III:
MACHINERY, MEASUREMENTS, AND PROBABILITY

CHAPTER EIGHT

ONE RING TO RULE THEM ALL

I am not one prone to overstatement, since I usually find that great events or achievements speak for themselves. This reluctance to embellish can get me into trouble in America, where people overuse superlatives so much that mere praise without an "est" at the end is sometimes misinterpreted as slander by faint praise. I'm frequently encouraged to add a few buzzwords or adverbs to my statements of support to avoid any misunderstanding. But in the case of the LHC I'll go out on a limb and say there is no question that it's a stupendous achievement. The LHC has an uncanny authority and beauty. The technology overwhelms.

In this chapter, we'll embark on our exploration of this incredible machine. In the chapter that follows, we'll enter the roller coaster construction adventure and a few chapters later, the world of the experiments that record what the LHC creates. But for the time being, we'll focus on the machine itself, which isolates, accelerates, and collides together the energetic protons that we hope will reveal new inner worlds.

THE LARGE HADRON COLLIDER

The first time I visited the LHC, I was surprised at the sense of awe it inspired—this in spite of my having visited particle colliders and detectors many times before. Its scale was simply different. We entered, put on our helmets, walked down into and through the LHC tunnel, stopped at an enormous pit into which the ATLAS (A Toroidal LHC

ApparatuS) detector would ultimately be lowered, and finally arrived at the experimental apparatus itself. It was still under construction, which meant ATLAS was not yet covered up as it would be when running—but was instead on display in full view.

Although the scientist in me recoils at first in thinking of this incredibly precise technological miracle as an art project—even a major one—I couldn't help taking out my camera and snapping away. The complexity, coherence, and magnitude, as well as the crisscrossing lines and colors, are hard to convey in words. The impression is simply awe-inspiring.

People from the art world have had similar reactions. When the art collector Francesca von Habsburg toured the site, she took along a professional photographer whose pictures were so beautiful they were published in the magazine *Vanity Fair*. When the filmmaker Jesse Dylan, who grew up in a world of culture, first visited the LHC, he viewed it as a remarkable art project—a "culminating achievement" whose beauty he wanted to share. Jesse embarked on a video to convey his impressions of the grandeur of the experiments and the machine.

The actor and science enthusiast Alan Alda, when moderating a panel about the LHC, likened it to one of the wonders of the ancient world. The physicist David Gross compared it to the pyramids. The engineer and entrepreneur Elon Musk—who cofounded PayPal, runs Tesla (the company that makes electric cars), and developed and operates SpaceX (which constructs rockets that will deliver machinery and products to the International Space Station)—said about the LHC, "Definitely one of humanity's greatest achievements."

I've heard such statements from people in all walks of life. The Internet, fast cars, green energy, and space travel are among the most exciting and active areas of applied research today. But going out and trying to understand the fundamental laws of the universe is in a category by itself that astounds and impresses. Art lovers and scientists alike want to understand the world and decipher its origins. You might debate the nature of humanity's greatest achievement, but I don't think anyone would question that one of the most remarkable things we do is to contemplate and investigate what lies beyond the easily accessible. Humans alone take on this challenge.

The collisions we'll study at the LHC are akin to those that took place in the first trillionth of a millisecond after the Big Bang. They will teach us about small distances and about the nature of matter and forces at this very early time. You might think of the Large Hadron Collider as a super-microscope that allows us to study particles and forces at incredibly small sizes—on the order of a tenth of a thousandth of a trillionth of a millimeter.

The LHC achieves these tiny probes by creating higher energy particle collisions than ever before achieved on Earth—up to seven times the energy of the highest existing collider, the Tevatron in Batavia, Illinois. As explained in Chapter 6, quantum mechanics and its use of waves tells us these energies are essential for investigating such small distances. And—along with the increase in energy—the intensity will be 50 times higher than at the Tevatron, making discovering the rare events that could reveal nature's inner workings that much more likely.

Despite my resistance to hyperbole, the LHC belongs to a world that can only be described with superlatives. It is not merely large: the LHC is the biggest machine ever built. It is not merely cold: the 1.9 kelvin (1.9 degrees Celsius above absolute zero) temperature necessary for the LHC's superconducting magnets to operate is the coldest extended region that we know of in the universe—even colder than outer space. The magnetic field is not merely big: the superconducting dipole magnets generating a magnetic field more than 100,000 times stronger than the Earth's are the strongest magnets in industrial production ever made.

And the extremes don't end there. The vacuum inside the proton-containing tubes, a 10 trillionth of an atmosphere, is the most complete vacuum over the largest region ever produced. The energy of the collisions are the highest ever generated on Earth, allowing us to study the interactions that occurred in the early universe the furthest back in time.

The LHC also stores huge amounts of energy. The magnetic field itself stores an amount equivalent to a couple of tons of TNT, while the beams store about a tenth of that. That energy is stored in one-billionth of a gram of matter, a mere submicroscopic speck of material under ordinary circumstances. When the machine is done with the beam, this enormously concentrated energy is dumped into a cylinder of graphite composite eight meters long and one meter in diameter, which is encased in 1,000 tons of concrete.

The extremes achieved at the LHC push technology to its limits. They don't come cheaply and the superlatives extend to cost. The LHC's $9 billion price tag also makes it the most expensive machine ever built. CERN paid about two-thirds of the cost of the machine, with CERN's 20 member countries contributing to the CERN budget according to their means, ranging from 20 percent from Germany to 0.2 percent from Bulgaria. The remainder was paid for by nonmember states, including the United States, Japan, and Canada. CERN contributes 20 percent to the experiments themselves, which are funded by international collaborations. As of 2008, when the machine was essentially built, the United States had more than 1,000 scientists working on CMS and ATLAS and had contributed $531 million toward the LHC enterprise.

THE BEGINNING OF THE LHC

CERN, which houses the LHC, is a research facility, with many programs operating simultaneously. However, CERN's resources are generally concentrated in a single flagship program. In the 1980s, that program was the *SpbarpS* collider,[38] which found the force carriers essential to the Standard Model of particle physics. The stellar experiments that took place there in 1983 discovered the weak gauge bosons—the two charged *W* bosons and the neutral Z boson, which communicate the weak force. Those were the key missing Standard Model ingredients at the time, and the discovery earned the accelerator project leaders a Nobel Prize.

Even so, while the SpbarpS was operating, scientists and engineers were already planning a collider known as LEP, which would collide together electrons and their antiparticles known as positrons to study the weak interactions and the Standard Model in exquisite detail. This dream came to fruition in the 1990s, when through its very accurate measurements, LEP studied millions of weak gauge bosons that taught physicists a great deal about Standard Model physics interactions.

LEP was a circular collider with a 27 kilometer circumference. Electrons and positrons were repeatedly boosted in this ring as they orbited around. As we saw in Chapter 6, circular colliders can be inefficient when accelerating light particles such as electrons, since such particles radiate when accelerated on a circular path. The electron beams at the LEP energy of about 100 GeV lost about three percent of their energy each time they went around. This wasn't too great a loss, but if anyone had wanted to accelerate electrons around this tunnel at any higher energy, the loss during each rotation would have been a deal breaker. Increasing the energy by a factor of 10 would have increased energy loss by a factor of 10,000, which would have made the accelerator far too inefficient to be acceptable.

For this reason, while LEP was being envisioned, people were already thinking about CERN's next flagship project—which would presumably run at even higher energy. Because of the electron's unacceptable energy losses, if CERN was to ever build a higher-energy machine, it would require proton beams, which are much heavier and therefore radiate much less. The physicists and engineers who developed LEP were aware of this more desirable possibility so they built the LEP tunnel sufficiently wide to accommodate a possible proton collider in the future, after the electron-positron machine would be dismantled.

Finally, some 25 years later, proton beams now race through the tunnel originally excavated for LEP. (See Figure 24.) The Large Hadron Collider is a couple of years behind schedule and about 20 percent over budget. That's a pity, but perhaps not so unreasonable given that the LHC is the biggest, most international, most expensive, most energetic, most ambitious experiment ever built. As the screenwriter and director James L. Brooks jokingly said when hearing about the LHC's setbacks and recovery, "I know people who take approximately the same amount of time to get their wallpaper just so. Understanding the universe just might have a better kick to it. Then again there's some pretty great wallpaper out there."

[FIGURE 24] The setting for the Large Hadron Collider, with the underground tunnel illustrated in white, and Lake Geneva and mountains in the background. (Photo courtesy of CERN)

THE FELLOWSHIP OF THE RINGS

Protons are everywhere around and within us. However, they are generally bound into nuclei surrounded by electrons inside atoms. They aren't isolated from those electrons and they aren't collimated (aligned into columns) inside beams. The LHC first separates and accelerates protons and then steers them to their ultimate destiny. In doing so, they utilize the LHC's many extremes.

The first step in preparing proton beams is to heat hydrogen atoms, which strips off their electrons and leaves the isolated protons that are their nuclei. Magnetic fields divert these protons so that they are channeled into beams. The LHC then accelerates the beams in several stages in distinct regions, with the protons traveling from one accelerator to another, each time increasing their energy before they are diverted from one of the two parallel beams so that they can collide.

The initial acceleration phase takes place in CERN's *linac,* which is a linear stretch of tunnel along which radio waves accelerate protons. When the radio wave is peaked, the associated electric field accelerates the protons. The protons are then made to drift away from the field so they don't decelerate when the field goes down. They subsequently return to the field when it peaks again so that they repeatedly accelerate from one peak to the next. Essentially the radio waves pulse the protons in the way you push a child on a swing. The waves thereby boost the protons, increasing their energy, but only a tiny amount in this first acceleration stage.

In the next stage, the protons are kicked via magnets into a series of rings where they are further accelerated. Each of these accelerators functions similarly to the linear accelerator described above. However, because these next accelerators are ring shaped, they can repeatedly boost the protons' energies as they circle around thousands of times. These circular accelerators thereby transfer quite a bit of energy.

This "fellowship of the rings" that accelerates protons before they enter the large LHC ring consists of the proton synchrotron booster (PSB) that accelerates protons to 1.4 GeV, the proton synchrotron (PS) that brings them up to 26 GeV in energy, and then the super proton synchrotron (SPS) that raises their energy to the so-called injection energy of 450 GeV. (See Figure 25 to see a proton's journey.) This is the energy the protons carry when they enter the last acceleration stage in the large 27 kilometer tunnel.

A couple of these accelerating rings are relics of previous CERN projects. The proton synchrotron, which is the oldest, celebrated its golden anniversary in November 2009, and the proton synchrotron booster was critical to the operation of CERN's last major project—namely, LEP—in the 1980s.

After protons leave the SPS, their 20 minute long *injection phase* begins. At this point the 450 GeV protons that emerged from the SPS are boosted to their full energy inside the large LHC tunnel. The protons in

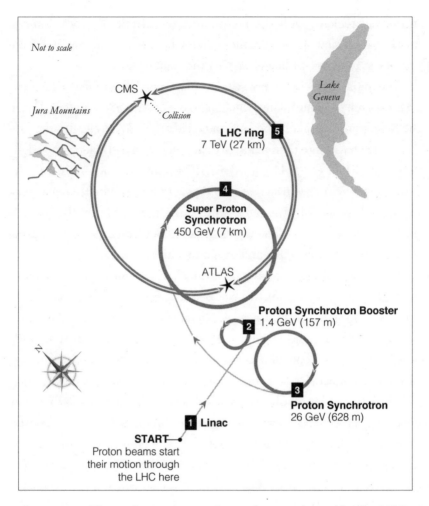

[**FIGURE 25**] The path a proton travels on when accelerated by the LHC.

the tunnel travel along two separate beams going in opposite directions through narrow three-inch pipes that extend on the 27 kilometers of the underground LHC ring.

The 3.8 meter (12 ft.) wide tunnel that was built in the 1980s but that now houses the proton beams in their final acceleration stage is well lit and air conditioned and large enough to comfortably walk around in, as I had the opportunity to do while the LHC was still in the construction phase. I took only a short stroll inside the tunnel on my LHC tour, but it still took me far longer to traverse my few steps than the 89 millionths of a second it takes for the accelerated highly energetic protons traveling at 99.9999991 percent of the speed of light to make it around.

The tunnel sits about 100 meters underground, with the precise depth varying from 50 to 175 meters. This shields the surface from radiation and also means CERN didn't have to buy up (and destroy) all the farmland lying over the tunnel's location during the construction phase. Property rights did, however, delay tunnel excavation back in the 1980s when it was originally constructed for LEP. The problem was that in France, landowners are entitled to the entire region to the Earth's center—not just the farmland they plow. The tunnel could be dug only after the French authorities blessed the operation by signing a "Déclaration d'Utilité Publique," thereby making the underlying rock—and in principle the magma underneath too—public property.

Physicists debate whether the reason for the tilt in the tunnel's depth was geology or if it was done to further deflect radiation, but the fact is the tilt helps with both. The uneven terrain was in fact an interesting constraint on the tunnel's depth and location. The region lying under the CERN site is mostly a type of compact rock known as *molasses,* but underneath the fluvial and marine deposits lie gravel, sand, and loam containing groundwater, and this would not be a good place for a tunnel. The slope keeps the tunnel in the good rock. It also meant that one section of the tunnel at the foot of the beautiful Jura Mountains lying at the edge of CERN could be a little less deep so that getting stuff in and out of vertical shafts in this location was a bit easier (and cheaper).

The final accelerating electric fields in this tunnel are not arranged in a precisely circular fashion. The LHC has eight large arcs alternating with eight 700-meter-long straight sections. Each of these eight sectors can be independently heated up and cooled down, which is important for repairs and instrumentation. After entering the tunnel, protons are accelerated in each of the short straight sections by radio waves, much as they were in the previous acceleration stages that brought them up to injection energy. The acceleration occurs in *radio-frequency (RF) cavities* that contain a 400 MHz radio signal, which is the same frequency you use when you remotely unlock your car door. When this field accelerates a proton bunch that enters such a cavity, it increases the energy of the protons by a mere 485 billionths of a TeV. This doesn't sound like much, but the protons orbit the LHC ring 11,000 times a second. Therefore, it takes only 20 minutes to accelerate the proton beam from its injection energy of 450 GeV to its target energy of 7 TeV, about 15 times higher. Some protons are lost during collisions or stray loose, but most of those protons will continue to circulate for about half a day before the beam is depleted and needs to be dumped into the ground and replaced by fresh newly injected protons.

By design, the protons that circulate in the LHC ring aren't uniformly distributed. They are sent around the ring in bunches—2,808 of them— each containing 115 billion protons. Each bunch starts off 10 centimeters long and one millimeter wide and is separated from the next bunch by about 10 meters. This helps with the acceleration since each bunch is accelerated separately. As a bonus, bundling the protons in this way guarantees that proton bunches interact at intervals of at least 25–75 nanoseconds, which is long enough apart that each bunch collision gets recorded separately. Since so many fewer protons are in a bunch than in a beam, the number of collisions that happen at the same time is under much better control because it is bunches, rather than the full quota of protons in the beam, that will collide at any one time.

CRYODIPOLE MAGNETS

Accelerating the protons to high energy is indeed an impressive achievement. But the real technological tour de force in building the LHC was designing and creating the high-field dipole magnets necessary to keep the protons properly circulating around the ring. Without the dipoles, the protons would go along a straight line. Keeping energetic protons circulating in a ring requires an enormous magnetic field.

Because of the existing tunnel size, the major technical engineering hurdle LHC engineers had to contend with was building magnets as strong as possible on an industrial scale—that is, they could be mass produced.

The strong field is required to keep high-energy protons on track inside the hand-me-down tunnel that LEP had bequeathed. Keeping more energetic protons circulating requires either stronger magnets or a bigger tunnel so that the proton paths curve sufficiently to stay on track. With the LHC, the tunnel size was predetermined, so the target energy was governed by the maximum attainable magnetic field.

The American Superconducting Supercollider, had it been completed, would have resided in a much bigger tunnel (which in fact was partially excavated), 87 kilometer in circumference, and was planned with the goal of achieving 40 TeV—almost three times the LHC's target energy. This vastly greater energy would have been possible because the machine was being designed from scratch, without the constraint in size of an existing tunnel and the consequent requirement of unrealistically large magnetic fields. However, the proposed European plan had the practical advantage that the tunnel and the CERN infrastructure of science, engineering, and logistics already existed.

One of the most impressive objects I saw when I visited CERN was a prototype of LHC's gigantic cylindrical dipole magnets. (See Figure 26 for a cross section.) Even with 1,232 such magnets, each of them is an impressive 15 meters long and weighs 30 tons. The length was determined not by physics considerations but by the relatively narrow LHC tunnel—as well as the imperative of trucking the magnets around on European roads. Each of these magnets cost €700,000, making the net cost of the LHC magnets alone more than a billion dollars.

The narrow pipes that hold the proton beams extend inside the dipoles, which are strung together end to end so that they wind through the extent of the LHC tunnel's interior. They produce a magnetic field that can be as strong as 8.3 tesla, about a thousand times the field of the average refrigerator magnet. As the energy of the proton beams increases from 450 GeV to 7 TeV, the magnetic field increases from 0.54 to 8.3 teslas, in order to keep guiding the increasingly energetic protons around.

The field these magnets produce is so enormous that it would displace the magnets themselves if no restraints were in place. This force

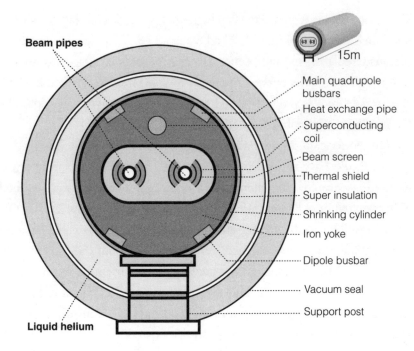

Beam pipes

15m

Main quadrupole busbars

Heat exchange pipe

Superconducting coil

Beam screen

Thermal shield

Super insulation

Shrinking cylinder

Iron yoke

Dipole busbar

Vacuum seal

Support post

Liquid helium

[**FIGURE 26**] Schematic of a cryodipole magnet. Protons are kept circulating around the LHC ring by 1232 such superconducting magnets.

is alleviated through the geometry of the coils, but the magnets are ultimately kept in place through specially constructed collars made of four-centimeter-thick steel.

Superconducting technology is responsible for the LHC's powerful magnets. LHC engineers benefited from the superconducting technology that had been developed for the SSC, as well as for the American Tevatron collider at the Fermilab accelerator center near Chicago, Illinois, and for the German electron-positron collider at the DESY accelerator center in Hamburg.

Ordinary wires such as the copper wires in your home have resistance. This means energy is lost as the current passes through. Superconducting wires, on the other hand, don't dissipate energy. Electrical current passes through unimpeded. Coils of superconducting wire can carry enormous magnetic fields, and, once in place, the field will be maintained.

Each LHC dipole contains coils of niobium-titanium superconducting cables, each of which contains stranded filaments a mere six microns thick—much smaller than a human hair. The LHC contains 1,200 tons of these remarkable filaments. If you unwrapped them, they would be long enough to encircle the orbit of Mars.

When operating, the dipoles need to be extremely cold, since they work only when the temperature is sufficiently low. The superconducting wires are maintained at 1.9 degrees above absolute zero, which is 271 degrees Celsius below the freezing temperature of water. This temperature is even lower than the 2.7-degree cosmic microwave background radiation in outer space. The LHC tunnel houses the coldest extended region in the universe—at least that we know of. The magnets are known as *cryodipoles* to take into account their special refrigerated nature.

In addition to the impressive filament technology used for the magnets, the refrigeration (*cryogenic*) system is also an imposing accomplishment meriting its own superlatives. The system is in fact the world's largest. Flowing helium maintains the extremely low temperature. A casing of approximately 97 metric tons of liquid helium surrounds the magnets to cool the cables. It is not ordinary helium gas, but helium with the necessary pressure to keep it in a *superfluid phase*. Superfluid helium is not subject to the viscosity of ordinary materials, so it can dissipate any heat produced in the dipole system with great efficiency: 10,000 metric tons of liquid nitrogen are first cooled, and this in turn cools the 130 metric tons of helium that circulate in the dipoles.

Not everything at the LHC is beneath the ground. Surface buildings hold equipment, electronics, and refrigeration plants. A conventional refrigerator cools down the helium to 4.5 kelvin and then the final cooling takes place with the pressure reduced. This process (as well as warming up) takes about a month, which means that each time the machine is turned on and off, or any repair is attempted, a good deal of additional time is required to cool.

If something went wrong—for example a tiny amount of heat capable of raising the temperature—the system would *quench,* meaning that superconductivity would be destroyed. Such a quenching would be disastrous if the energy were not properly dissipated, since all the energy stored in the magnets would suddenly be released. Therefore, a special system for detecting quenches and spreading the energy release are in place. The system looks for differences in voltage inconsistent with superconductivity. If detected, the energy is released everywhere, within less than a second, so that the dipole will no longer be superconducting.

Even with superconducting technology, huge currents are needed to achieve the 8.3 tesla magnetic field. The current goes up to almost 12,000 amperes, which is about 40,000 times the current flowing through the lightbulb on your desk.

With the current and the refrigeration, the LHC when running uses an enormous amount of electricity—about the amount required for a small city such as nearby Geneva. To avoid excessive energy expenditures, the accelerator runs only until the cold Swiss winter months when electricity prices go up (with an exception made for the turn-on in 2009). This policy has the extra advantage that it gives the LHC engineers and scientists a nice long Christmas vacation.

THROUGH VACUUM TO COLLISIONS

The final LHC superlative applies to the vacuum inside the pipes where the protons circulate. The system needs to be kept as free as possible of excess matter in order to maintain the cold helium because any stray molecules could transport away heat and energy. Most critically, the proton beam regions have to be as free of gas as possible. If gas were present, protons could collide with it and destroy the nice circulation of the proton beam. The pressure inside the beams is therefore extremely tiny, 10 trillion times smaller than atmospheric pressure—the pressure one million meters above the Earth's surface where the air is extremely rarified. At the LHC, 9,000 cubic meters of air was evacuated to achieve the welcoming space for the proton beam.

Even at this ridiculously low pressure, about three million molecules of gas still reside in every cubic centimeter region in the pipe, so protons do occasionally hit the gas and get deflected. Were enough of these protons to hit a superconducting magnet, they would quench it and destroy the superconductivity. Carbon collimators line the LHC beam in order to remove any stray beam particles that lie outside a three-millimeter aperture, which is plenty large enough to permit the approximately millimeter-wide beam to pass through.

Still, organizing the protons in a millimeter-wide bunch is a tricky task. It is accomplished by other magnets, known as *quadrupole* magnets, that effectively focus and squeeze the beam. The LHC contains 392 such magnets. Quadrupole magnets also divert the proton beams from their independent paths so that they can actually collide.

The beams don't collide precisely or completely head-on, but rather at the infinitesimal angle of about a thousandth of a radian. This is to ensure that only one bunch from each beam collides at a time so that the data are less confusing and the beam stays intact.

When the two bunches from the two circulating beams collide, one hundred billion protons are up against another bunch of 100 billion protons. Quadrupole magnets are also responsible for the especially daunting task of focusing the beams at the regions along the beam where collisions occur and experiments that record the events are situated. At these locations, the magnets squeeze the beams to the tiny size of 16 microns. The beams have to be extremely small and dense so that the hundred billion protons in a bunch are more likely to find one of the hundred billion protons in the other bunch when they pass through.

Most of the protons in a bunch won't find the protons in the other bunch, even when they are directed toward each other so as to collide. Individual protons are only about a millionth of a nanometer in diameter. This means that even though all these protons are kept in bunches of 16 microns, only about 20 protons collide head-on each time the bunches cross.

This is in fact a very good thing. If too many collisions occurred simultaneously, the data would simply be confusing. It would be impossible to tell which particles emerged from which collision. And of course if no collisions occurred, that would be a bad thing as well. By focusing just this number of protons into just this size, the LHC ensures the optimal number of events each time bunches cross.

The individual proton collisions, when they occur, do so almost instantaneously—in a time about 25 orders of magnitude less than a second. This means the time between the sets of proton collisions is set entirely by how frequently the bunches cross, which at full capacity is about every 25 nanoseconds. The beams are crossing more than 10 million times a second. With such frequent collisions, the LHC produces a huge amount of data—about a billion collisions per second. Fortunately, the time between bunch crossing is long enough to let the computers keep track of the interesting individual collisions without confusing collisions that originated in different bunches.

So in the end, the extremes at the LHC are necessary to guarantee both the highest possible energy collisions and the largest number of events that the experiments can handle. Most of the energy just stays in circulation with only the rare proton collision worthy of attention. Despite the massive energy in the beams, the energy of individual bunch collisions involves little more than the kinetic energy of a few mosquitoes in flight. These are protons colliding—not football players or cars. The LHC's extremes concentrate energy in an extremely tiny region, and in elementary particle collisions that experimenters can follow. We'll soon consider some of the hidden ingredients that they might find and the insights into the nature of matter and space that physicists hope those discoveries will provide.

CHAPTER NINE

THE RETURN OF THE RING

I entered graduate school for physics in 1983. The LHC was first officially proposed in 1984. So in some sense I've been waiting for the LHC for the quarter century of my academic career. Now, at long last, my colleagues and I are finally seeing LHC data and realistically anticipating the insights into mass, energy, and matter that the experiments could soon reveal.

The LHC is currently the most important experimental machine for particle physicists. Understandably, as it commenced operation, my physicist colleagues became increasingly anxious and excited. You couldn't enter a seminar room without someone inquiring about what was happening. How much energy would collisions achieve? How many protons will beams contain? Theorists wanted to understand minutiae that had previously been almost an abstraction to those of us engaged in calculations and concepts and not machine or experimental design. The flip side was true as well. Experimenters were as eager as I'd ever seen them to hear about our latest conjectures and learn more about what they might look for and possibly discover.

Even at a conference that took place in December 2009, that was purportedly about dark matter, participants were eagerly commenting on the LHC—which had just completed its incredibly successful debut of acceleration and collisions. At the time, after the near despair of a little more than a year before, everyone was ecstatic. Experimenters were relieved they had data they could study to understand their detectors better. Theorists were happy they might get some answers before too long. Everything was working fabulously well. The beams looked good. Collisions had occurred. And experiments were recording events.

However, reaching this landmark was quite a story, and this chapter tells the tale. So fasten your seat belt. It was a bumpy ride.

A SMALL WORLD AFTER ALL

The story of CERN precedes that of the LHC by several decades. Soon after the end of World War II, a European accelerator center that would host experiments studying elementary particles was first conceived. At that time, many European physicists—some of whom had immigrated to the United States and some of whom were still in France, Italy, and Denmark—wanted to see cutting-edge science restored to their original homelands. Americans and Europeans agreed that it would be best for scientists and science if Europeans joined together in this common enterprise and returned research to Europe so they could repair the residue of devastation and mistrust remaining after the recently ended war.

At a UNESCO conference in Florence in 1950, the American physicist Isidor Rabi recommended the creation of a laboratory that would reestablish a strong scientific community in Europe. In 1952, the Conseil Européen pour la Recherche Nucléaire (hence the acronym CERN) was set up to create such an organization, and on July 1, 1953, representatives from twelve European nations came together to create the institution that became known as "the European Organization for Nuclear Research," and the convention establishing it was ratified the following year. The CERN acronym clearly no longer reflects the name of the research center. And we now study subnuclear, or particle, physics. But as is often true with bureaucracy, the initial legacy remained.

The CERN facility was deliberately built centrally in Europe on a site crossing the Swiss-French border near Geneva. It's wonderful to visit if you like the outdoors. The fabulous setting includes farmland and the Jura Mountains immediately nearby and the Alps readily accessible in the distance. CERN experimenters are on the whole a rather athletic bunch, with their easy access to skiing, climbing, and biking. The CERN site is quite large, covering enough territory for an exhausting run to keep those athletic researchers in shape. The streets are named after famous physicists, so you can drive on Route Curie, Route Pauli, and Route Einstein on a visit to the site. The architecture at CERN was, however, a victim of the time in which it was built, which was the 1950s with bland International Style low-rises, so CERN buildings are rather plain with long hallways and sterile offices. It didn't help the architecture that it was a science complex—look at the science buildings on most any university and you will usually find the ugliest buildings on campus. What enlivens the place (along with the scenery) are the people who work there and their scientific and engineering goals and achievements.

International collaborations would do well to study CERN's evolution and its current operations. It is perhaps the most successful international enterprise ever created. Even in the aftermath of World War II, when the countries had so recently been in conflict, scientists from twelve different nations joined together in this common enterprise.

If competition played any role at all, it was primarily directed against the United States and its burgeoning scientific endeavors. Until experiments at CERN found the W and Z gauge bosons, almost all particle physics discoveries had come from accelerators in America. The drunken physicist who walked into the common area at Fermilab where I was a summer student in 1982 saying how they "had to find the bloody vector bosons" and destroy America's dominance probably expressed the viewpoint of many European physicists at the time—though perhaps somewhat less eloquently and definitely with poorer diction.

CERN scientists did find those bosons. And now, with the LHC, CERN is the undisputed center of experimental particle physics. However, this was by no means predetermined when the LHC was first proposed. The American Superconducting Supercollider (SSC) that President Reagan approved in 1987 would have had almost three times the energy—had Congress continued its support. Although the Clinton administration initially didn't support the project initiated by its Republican predecessors, that changed as President Clinton better understood what was at stake. In June 1993, he tried to prevent the cancellation in a letter to William Natcher, chairman of the House Committee on Appropriations, in which he said, "I want you to know of my continuing support for the Superconducting Super Collider (SSC). . . . Abandoning the SSC at this point would signal that the United States is compromising its position of leadership in basic science—a position unquestioned for generations. These are tough economic times, yet our Administration supports this project as a part of its broad investment package in science and technology. . . . I ask you to support this important and challenging effort." When I met the former president in 2005, he brought up the subject of the SSC and asked what we had lost in abandoning the project. He quickly acknowledged that he too had thought that humanity had forfeited a valuable opportunity.

Around the time that Congress killed the SSC, taxpayers ponied up about $150 billion to pay for the savings and loan crisis, which far exceeded the approximately $10 billion the SSC would have cost. The U.S. annual deficit in comparison amounts to a whopping $600 per American, and the Iraq War to more than $2,000 per citizen. With the SSC we would have had high-energy results already, and we would have reached far higher energies even than the LHC will achieve. With the end of the S&L crisis we left ourselves open to the financial crisis of 2008 and a bailout that was even more expensive to taxpayers.

The LHC's price tag of $9 billion was comparable to the SSC's proposed cost. It amounts to about $15 per European—or as my colleague Luis Álvarez-Gaumé at CERN likes to say, about a beer per European per year during the construction time of the LHC. Assessing the value of fundamental scientific research of the sort taking place at the LHC is always tricky, but fundamental research has spurred electricity, semiconductors, the World Wide Web, and just about all technological advances that have significantly affected our lives. It also inspires technological and scientific thinking, which spreads into all aspects of our economy. The LHC's practical results might be difficult to anticipate, but the science potential is not. I think we can agree that the Europeans in this case are more likely to get their money's worth.

Long-term projects require belief, dedication, and responsibility. Such commitments are becoming increasingly hard to come by in the United States. Our past vision in the U.S. led to tremendous scientific and technological advances. However, this type of essential long-term planning is becoming increasingly rare. You have to hand it to the European Community for their ability to continue to see their projects through. The LHC was first envisioned a quarter century ago and approved in 1994. Yet it was such an ambitious project that only now is it reaching fruition.

Furthermore, CERN has successfully broadened its international appeal to include not only the 20 CERN member states, but also 53 additional nations that have also participated in the design, construction, and testing of instruments—and scientists from 85 countries currently participate. The United States isn't an official member state, but there are more Americans than any other single nationality working on the major experiments.

About 10,000 scientists participate in total—perhaps about half of the total number of particle physicists on Earth. One-fifth of them are full-time employees who live nearby. With the advent of the LHC, the main cafeteria has become so packed that you could barely order food without your tray hitting another physicist—a problem that a new cafeteria extension now helps alleviate.

With its international population, an American arriving at CERN will be struck by the many languages and accents reverberating in the cafeterias, offices, and hallways. The Americans will also notice the cigarettes, cigars, wine, and beer there, which also remind them they're not at home. Some comment as well on the superior quality of the cafeterias, as did one of my freshman students who had worked there over the summer. Europeans, with their more refined palates, tend to find this assessment somewhat questionable.

The many employees and visitors at CERN range from engineers to administrators to the many physicists who actually do the experiments and the more than 100 physicists who participate in the theory division at any given time. CERN is structured hierarchically, with the chief officers and council responsible for all policy matters, including major strategic decisions. The head is known as the director general (DG) which perhaps has the ring of something out of Gilbert and Sullivan, though the many directorships under the DG account for the name. The CERN Council is the ruling body responsible for major strategic decisions such as planning and scheduling projects. It pays special attention to the Scientific Policy Committee, which is the major advisory board that helps evaluate proposals and their scientific merit.

The large experimental collaborations, with thousands of participants, have a structure of their own. Work is distributed according to detector components or types of analyses. A given university group might be responsible for one particular piece of the apparatus or one particular type of potential theoretical interpretation. Theorists at CERN have more freedom than experimenters to work on whatever is of interest to them. Sometimes their work pertains to CERN experiments, but many of them work on more abstract ideas that won't be tested anytime soon.

Nonetheless, all particle physicists at CERN and around the globe are excited about the LHC. They know their future research and the future of the field itself relies on the successful operation and discoveries of the next 10 to 20 years. They understand the challenges, but they also agree in their bones with the superlatives that go with this enterprise.

A BRIEF HISTORY OF THE LHC

Lyn Evans was the LHC's chief architect. Though I'd heard him speak in his lovely lilting Welsh intonation the year before, I finally met him at a conference in California in early January 2010. This was an opportune time since the LHC was finally on track, and even for an understated Welshman, his pleasure was obvious.

Lyn gave a wonderful talk about the roller-coaster ride he'd had since first setting out to build the LHC. He began by telling us about the true inception of the idea in the 1980s, when CERN conducted the first official studies investigating the option of producing a high-energy proton-proton collider. He then told about the 1984 meeting that most people consider the idea's official initiation. Physicists at that time met with machine builders in Lausanne to introduce the idea of colliding together proton beams with 10 TeV of energy—a proposal that was scaled down to 7 TeV beams in the final implementation. Almost a decade later, in December 1993, physicists presented an aggressive plan to the CERN Council, the governing body at CERN over major strategic decisions, to build the LHC during the next 10 years by minimizing all other experimental programs at CERN aside from LEP. At that time, the CERN Council turned it down.

Initially, one argument against the LHC had been the intense competition posed by the SSC. But that disappeared with the project's demise in October 1993, at which time the LHC became the sole candidate for a very high-energy accelerator. Many physicists then became increasingly convinced of the significance of the enterprise. On top of that, machine research was extremely successful. Robert Aymar, who would ultimately head CERN during the LHC construction phase, chaired a review panel in November 1993 that concluded the LHC would be feasible, economical, and safe.

The critical hurdle in planning the LHC was developing strong enough magnets on an industrial scale to keep highly accelerated protons circulating in the ring. As we observed in the previous chapter, the existing tunnel size presented the biggest technical challenge, since its radius was fixed and magnetic fields therefore had to be very big. In his talk, Lyn happily described the "Swiss watch precision" of the first 10-meter-long prototype dipole magnet that engineers and physicists successfully tested in 1994. They reached 8.73 tesla on their first shot, which was their target and a very promising sign.

Unfortunately, however, although European funding is more stable than that of the United States, unforeseen pressures introduced uncertainties for CERN's finances as well. The budget for Germany, which contributes the most to CERN, suffered from the 1990 reunification. Germany therefore reduced its contributions to CERN, and, along with the United Kingdom, didn't want to see any major increase in the CERN budget. Christopher Llewellyn Smith—the British theoretical physicist who succeeded the Nobel Prize–winning physicist Carlo Rubbia as CERN director general—was, like his predecessor, strongly supportive of the LHC. By acquiring funding from Switzerland and France, the two host states that stood to benefit the most from the LHC's construction and operation in their home territory, Llewellyn Smith partially alleviated the serious budget issues.

The CERN Council was appropriately impressed—both with the technology and with the budget resolution—and approved the LHC soon afterward on December 16, 1994. Llewellyn Smith and CERN furthermore convinced nonmember states to join and participate. Japan came on board in 1995, India in 1996, and soon after Russia and Canada, with the United States following in 1997.

With all the contributions from Europe and other nations, the LHC could override a proviso in the original charter that called for construction and operation in two phases, the first of which would involve only two-thirds of the magnets. Both scientifically and in terms of total cost, the reduced magnetic field would have been a poor choice. But the original intention was to allow budgets to balance every year. In 1996, when Germany again reduced its contribution due to its reunification costs, the budget situation again looked grim. However, in 1997, CERN was allowed to compensate for the loss by financing construction with loans for the first time.

After the budget history lowdown, Lyn's talk turned to more happy news. He described the first *test string* of dipoles—a test of magnets combined together in a workable configuration—that took place in December 1998. The successful completion of this test demonstrated the viability and coordination of several of the ultimate LHC components and was a critical milestone in its development.

In 2000, when LEP, the electron-positron collider, had run its course, it was dismantled to pave the way for LHC installation. Yet even though the LHC was ultimately built in a preexisting tunnel and used some of the staff, facilities, and infrastructure that were already in place, a lot of man-hours and resources would be necessary before the transformation from LEP to the LHC could occur.

The five phases of the LHC's development included civil engineering to build caverns and structures for experiments, the installation of general services so that everything could run, the insertion of a cryogenic line to keep the accelerator cold, putting in place all the machine elements including the dipoles and all the associated connections and cables, and ultimately the commissioning of all the hardware to make sure everything worked as anticipated.

The CERN planners started off with a careful schedule to coordinate these construction phases. But as everyone knows, "the best laid plans o' mice an' men gang aft agley." Needless to say, this applied all too well.

Budget issues were a constant nuisance. I remember the frustration and concern of the particle physics community in 2001 as we waited to find out how quickly some serious budget problems at the time could be resolved to allow construction to proceed. CERN management dealt with the cost overruns, but at a price in terms of CERN breadth and infrastructure.

Even after these funding and budget problems were resolved, LHC development still wasn't entirely smooth sailing. Lyn in his talk described how a series of unforeseen events periodically slowed down construction.

Certainly no one involved in excavating the cavern for the CMS (Compact Muon Solenoid) experiment could have foreseen digging into a fourth-century Gallo-Roman villa. The property boundaries were parallel to the farm field boundaries that exist to this day. Excavation was halted while archaeologists studied buried treasure, including some coins from Ostia, Lyon, and London (Ostium, Lugdunum, and Londinium at the time the villa was occupied). Apparently the Romans were better at establishing a common currency than modern Europe, where the euro still hasn't displaced the British pound and the Swiss franc as a means of exchange—particularly annoying for British physicists arriving at CERN who don't have the currency required to pay for a taxi.

Compared to CMS's travails, the 2001 excavation of the ATLAS cavern proceeded relatively uneventfully. Digging the cavern involved removing 300,000 metric tons of rock. The only problem they faced is that once the material was removed, the cavern floor began to rise slightly— at the rate of about a millimeter each year. This might not sound like much, but the movement could in principle interfere with the precise alignment of the detector pieces. So the engineers needed to install sensitive metrology instruments. They are so effective that they not only detect ATLAS movements, but are sufficiently sensitive to have registered the 2004 tsunami and the Sumatra earthquake that triggered it, as well as others that came later.

The procedure for building the ATLAS experiment deep underground was rather impressive. The roof was cast on the surface and suspended by cables while the walls were built up from below until the vault could sit on them. In 2003, the completed excavation was inaugurated with a celebration, notable for the presence of an alpine horn echoing inside, which in Lyn's description was a source of great amusement. Installation and assembly of the experimental apparatus subsequently followed with the components lowered one by one until ultimately the ATLAS experiment was assembled with this "ship in a bottle method" in the excavated cavern belowground.

CMS preparations, on the other hand, continued to face rough seas. It once again got into trouble during excavation since it turned out that the CMS site was infelicitously placed not only over a rare archaeological site but also over an underground river. With the heavy rains that year, the engineers and physicists discovered to their surprise that the 70-meter-long cylinder they inserted into the ground to transport materials down had sunk 30 centimeters. To deal with this unfortunate hindrance, the excavators created walls of ice along the cylinder walls to freeze the ground and stabilize the region. Supporting structures to stabilize the fragile rock around the cavern also had to be installed, including screws up to 40 meters in length. Not surprisingly, the CMS excavation took longer than foreseen.

The only saving grace was that because of CMS's relatively compact size, experimenters and engineers had already been considering constructing and assembling it on the surface. Constructing and installing components is a lot easier aboveground, and everything is faster since there is more room to work in parallel. This aboveground construction had the added critical benefit that the cavern problems wouldn't further delay construction.

However, as you might imagine, it was a rather daunting prospect to lower this enormous apparatus—which is something I had a chance to think about when I first visited CMS in 2007. Indeed, lowering the experiment was no easy task. The largest piece began its 100-meter descent into the CMS pit, carried by a special crane, at the dauntingly low speed of 10 meters per hour. Since there was only a 10-centimeter leeway between the experiment and walls of the shaft, this slow descent and a careful monitoring system were critical. Fifteen large pieces of detector were lowered between November 2006 and January 2008—a brazen piece of timing as the final piece was delivered pretty close to the scheduled LHC start-up date.

Following the CMS water trouble, the next crisis in the construction of the LHC machine itself struck in June 2004, when problems were discovered in the helium distribution line known as the QRL. The CERN engineers who investigated discovered the French firm that had taken on this construction project had replaced the material designated in the original design with what Lyn described as a "five-dollar spacer." The replacement material cracked, allowing thermal contraction of the inner pipes. This faulty component wasn't unique, and all the connections had to be checked.

By this time the cryogenic line had been partially installed and many other pieces had already been produced. To avoid blocking the supply chain and introducing further delays, the CERN engineers decided to repair what had already been produced while leaving industry to correct the problem before delivering the remaining parts. CERN's factory operations and the need to move and reinstall large pieces of the machine cost the LHC a year delay. At least the delay was far less than the decade delay Lyn and others feared had lawyers been involved.

Without pipes and the cryogenic system, no one could install magnets. So 1,000 magnets sat around in the CERN parking lot. Even with the high-end BMWs and Mercedes that grace the lot at times, $1 billion worth of magnets probably exceeded the usual parking lot contents' net worth. No one stole the valuable magnets, but a parking lot isn't a great place to store technology, and further delays associated with restoring the magnets to their initial specification were inevitable.

In 2005, yet another near crisis occurred, having to do with the inner triplet constructed at Fermilab in the United States and in Japan. The inner triplet provides the final focusing of the proton beams before they collide. It combines three quadrupole magnets with cryogenic and power distribution—hence the name. This inner triplet failed during pressure tests. Although the failure was an embarrassment and an annoying delay, the engineers could fix it in the tunnel so the time cost wasn't too severe in the end.

Overall, the year 2005 was more successful than its predecessor. The CMS cavern was inaugurated in February, though no horn graced the day. Another landmark event occurred in February—the lowering of the first cryodipole magnet. Magnet construction had been critical to the LHC enterprise. A close collaboration between CERN and commercial industry facilitated their timely and economical construction. Though designed at CERN, the magnets were produced at companies in France, Germany, and Italy. Initially, CERN engineers, physicists, and technicians placed an order for 30 dipoles in 2000, which they might then carefully examine to ensure quality and cost control before placing the final order for more than 1,000 magnets in 2002. CERN nonetheless maintained responsibility for procuring the main components and raw materials in order to maximize quality and uniformity and minimize cost. To do so, CERN moved 120,000 metric tons of material within Europe, employing an average of 10 big trucks a day for four years. And that was only one piece of the LHC effort.

After delivery, the magnets were all tested and carefully lowered through a vertical shaft into the tunnel near the Jura Mountains that overlook the CERN site. From there, a special vehicle transported them to their destination along the tunnel. Because these magnets are enormous and only a few centimeters separated the wall of the tunnel from the LHC installations, the vehicle was automatically guided by an optically detected line painted on the floor. The vehicle moved forward at a rate of only about a mile an hour in order to limit vibrations. That meant it took seven hours to get a dipole from the lowering point to the opposite end of the ring.

In 2006, after five years of construction, the last of the 1,232 dipole magnets was delivered. In 2007, the big news was the last lowering of a cryodipole and the first successful cooldown of a 3.3-km-long section to the design temperature of −271 degrees Celsius—which allowed the whole thing to be powered up for the first time, with several thousand amps circulating in the superconducting magnets in this section of the tunnel. As often happens at CERN, a champagne celebration marked the occasion.

A continuous cryostat section was closed in November 2007 and everything was looking pretty good until yet another near disaster struck, this time involving the so-called plug-in modules, known as *PIMs*. In the United States, we didn't necessarily follow all the reports about the LHC. But news spread about this one. A CERN colleague told me about the worry that not only had this piece failed, but it could be a ubiquitous problem all around the ring.

The problem is the almost 300-degree differential between a room-temperature LHC and a cool operating one. This difference has an enormous impact on the materials with which it is constructed. Metal parts shrink when cooled and expand when warmed. The dipoles themselves shrink by a few centimeters during the cooldown phase. This might not sound like much for a 15-meter object, but the coils must be accurately positioned to within a tenth of a millimeter to maintain the intense uniform magnetic field required to properly guide the proton beams.

To accommodate the change, dipoles are designed with special *fingers* that straighten out to ensure electrical continuity when the machine is cooled down and that slide back when warmed. However, due to faulty rivets, the fingers collapsed instead of recessing. Worse yet, every interconnection was subject to this failure, and it wasn't clear which ones were problematic. The challenge was to identify and fix each faulty rivet—without introducing a huge delay.

In a tribute to the ingenuity of the CERN engineers, they found a simple method of exploiting the existing electrical pickup located every 53 meters along the beam that was initially installed so that the electronics would be triggered by the beam passage. The engineers installed an oscillator into an object about the size of a Ping-Pong ball, which they could send around the tunnel along the path a beam would take. Each sector was three kilometers long and the ball could blow through, triggering the electronics each time it passed a pickup. When the electronics didn't record a passage, the ball had hit the fingers. The engineers could then go in and fix the problem without having to open every single interconnect along the beam. As one LHC physicist joked, the first LHC collisions were not between protons, but between a Ping-Pong ball and a collapsed finger.

After this last resolution, the LHC seemed to be on track. Once all the hardware was in place, its operation could begin. In 2008, many human fingers crossed when at long last the first test took place.

SEPTEMBER 2008: THE FIRST TESTS

The LHC forms proton beams and after a series of energy boosts injects them into the final circular accelerator. It then sends those beams around the tunnel so that they return to their precise initial position, allowing the protons to circulate many times before being periodically diverted to collide with great efficiency. Each of these steps needs to be tested in turn.

The first milestone was to check whether the beams would actually circulate around the ring. And they could. Amazingly, after its long history of trials and tribulations, in September 2008, CERN fired up its two proton beams with so few hitches that the results exceeded expectations. On that day, for the first time, two proton beams in succession traversed the enormous tunnel in opposite directions. This single step involved commissioning the injection elements, starting the controls and instruments, checking that the magnetic field would keep the protons in the ring, and making sure all the magnets worked to spec and could run simultaneously. The first time this sequence of events was ready was the evening of September 9. Yet everything worked as well as or better than planned when the tests took place the next day.

Everyone involved with the LHC describes September 10, 2008, as a day they will never forget. When I visited a month afterward, I heard many stories of the day's euphoria. People followed the trajectory of two spots of light on a computer screen with unbelievable excitement. The first beam almost returned successfully on its first go-round, and with minor tweaking followed the exact path that was intended within the first hour of its being turned on. The beam at first went around the ring for a few turns. Then each successive burst of protons was adjusted slightly so that the beam was soon circulating hundreds of times. Not long after this, the second beam did the same—taking about one and a half hours to get exactly on track.

Lyn was just as happy that he didn't know about the live video feed at the time from the control room, where the engineers were following the project, to the Internet, where the events were being broadcast for any-one to see. So many people watched those two dots on their screens that the sites were shut down for breaking capacity. People all over Europe—the CERN press office claims a couple of million—sat mesmerized as engineers modified the protons' path to make them successfully circulate around the full circumference of the ring. Meanwhile, inside CERN, the thrill was palpable as physicists and engineers gathered in auditoriums to watch the same thing. At this point, the LHC outlook seemed more than extremely promising. The day was a wonderful success.

But a mere nine days later, euphoria transformed into despair. At the time, two important new features were to be tested. First, the beams were to be accelerated inside the LHC ring to higher energy than they had been during the first test, which used only the beam injection energy that protons have when first entering the LHC ring. The second part of the plan was to collide those beams, which would of course have been a huge milestone in LHC development.

However, at the last moment—on September 19—despite the engi-neers' many considerations and precautions, the test failed. And when it

did, it did so catastrophically. A simple soldering error in the copper casing connecting two magnets combined with too few functioning helium
release valves caused a yearlong delay before protons would first collide.

The problem was that as scientists tried to ramp up the current and
energy of the eighth and final sector, a joint between two magnets along
the busbar that connects them broke. A *busbar* is a superconducting joint
that connects a pair of superconducting magnets. (See Figure 27.) The
splice that holds together a joint between two magnets was the culprit.
The faulty connection created an electrical arc that punctured the helium
enclosure and caused six metric tons of liquid helium—that would ordi-
narily be warmed up slowly—to be suddenly released. Superconductivity
was lost in the quenching that occurred when the liquid helium heated
up and reverted to gas.

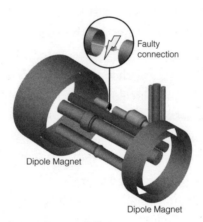

[**FIGURE 27**] A busbar connects different magnets together. A faulty solder in one was responsible for the unfortunate incident in 2008.

The enormous amount of helium released created a huge pressure
wave that effectively caused an explosion. In less than 30 seconds, its
energy displaced some magnets and destroyed the vacuum in the beam
pipe, damaged the insulation, and contaminated 2,000 feet of beam pipe
with soot. Ten dipoles were totally destroyed and 29 more were so damaged they needed to be replaced. Needless to say, this was not exactly
what we had been hoping for. And this was also something no one in

the control rooms had any inkling of until someone noticed that a stop button in the tunnel for one of the computers had been triggered by the escaping helium. Soon afterward, they realized the beam had been lost.

I learned more about the backstory during a visit to CERN a few weeks after the mishap. Keep in mind that the ultimate goal for collisions is a center of mass energy of 14 TeV, or 14 trillion electron volts. The decision was made to keep the energy down to only about 2 TeV for the first run in order to ensure that everything functioned properly. Later the engineers planned to increase it to 10 TeV (5 TeV per beam) for the first actual data runs.

However, the plan became more ambitious following a small delay due to a transformer that broke on September 12. Scientists continued testing the tunnel's eight sectors up to 5.5 TeV during the interval afforded by the short delay and had time to test seven out of the eight sectors. They verified those could run properly at higher energy, but they didn't have the opportunity to test the eighth. They nonetheless decided to charge ahead and attempt higher-energy collisions since there didn't seem to be any problem.

Everything worked fine until the engineers attempted to raise the energy of the last untested sector. The crippling accident occurred when its energy was being raised from about 4 to 5.5 TeV—which required between 7,000 and 9,300 amps of current. This was the last moment for something to go wrong, and it did.

During the year of the delay, everything was repaired at a cost of about $40 million. Although repairing the magnets and the beam took time, they were not impossible tasks. Enough spare magnets were on hand to replace the 39 dipole magnets that were beyond repair. In total, 53 magnets (14 quadrupole and 39 dipole) were replaced in the sector of the tunnel where the incident occurred. In addition, more than four kilometers of the vacuum beam tube were cleaned, a new restraining system for 100 quadrupole magnets was installed, and 900 new helium pressure release ports were added. In addition, 6,500 new detectors were added to the magnet protection system.

The bigger risk was the presence of 10,000 joints between magnets that could potentially cause the same problem. The danger had been identified, but how could anyone trust that this problem would not re-emerge elsewhere in the ring? Mechanisms were needed to detect any similar problem before it could cause any harm. The engineers once again rose to the challenge. Their updated system now looks for minuscule voltage drops that might signal the presence of resistive joints, signaling a break in the closed system that houses the cryogenics that keeps the machine cold. Caution also dictated some delays to improve the helium release valve system and to further study the joints as well as the copper casings of the magnets themselves—which meant a delay in achieving the highest energies at which the LHC is designed to operate. Nonetheless, with all the new systems to monitor and stabilize the LHC, Lyn and others were confident that the kind of pressure buildups that caused the damage will be avoided.

In some sense, we are lucky that engineers and physicists were able to fix things before true operations began and filled the experiments with radiation. The explosion cost the LHC a year before they could even begin to test beams and aim for collisions again. That was a long time, but not so long on the scale of a quest for the underlying theory of matter that we have had for the last 40 years, and in many respects for thousands of years.

On October 21, 2008, the CERN administration did, however, stick to one piece of their initial plan. On that day, I joined 1,500 other physicists and world leaders outside Geneva to celebrate the official LHC inauguration, which had been optimistically planned well in advance—before anyone could have predicted the disastrous events that occurred a mere few weeks before. The day was filled with speeches, music, and—as is important at any European cultural event—good food. It was enjoyable and informative even with the premature timing. Despite anxieties about the September incident, everyone was filled with hope that these experiments would shed light on some of the mysteries surrounding mass, the weakness of gravity, dark matter, and the forces of nature.

Although many CERN scientists were unhappy about the infelicitous timing of the event, I saw the celebration more as a contemplation of this triumph of international cooperation. The day's events did not yet honor discovery but instead recognized the potential of the LHC and the enthusiasm of the many countries participating in its creation. A few of the speeches were truly encouraging and inspirational. The French prime minister, François Fillon, spoke of the importance of basic research and how the world financial crisis should not impede scientific progress. The Swiss president, Pascal Couchepin, spoke of the merit of public service. Professor José Mariano Gago, Portugal's minister for science, technology, and higher education, spoke about valuing science over bureaucracy and the importance of stability for creating important science projects. Many of the foreign partners visited CERN for the first time for the day's celebration. The person seated next to me during the ceremony worked for the European Union in Geneva—but had never set foot inside CERN. Having seen it, he enthusiastically informed me of his intention to return soon with his colleagues and friends.

NOVEMBER 2009: VICTORY AT LAST

The LHC finally came back online on November 20, 2009, and this time, it was a stunning success. Not only did proton beams circulate for the first time in a year, but a few days later, they finally collided, creating sprays of particles that would enter the experiments. Lyn enthusiastically described how the LHC worked better than he had expected—a remark that I found encouraging but a bit peculiar in light of his being in charge of making the machine run as successfully as it had.

What I hadn't understood was how much more quickly all the pieces had fallen into place than would have been anticipated based on the experience with past machines. Maurizio Pierini, a young Italian CMS experimenter, explained to me what Lyn had meant. Tests that took 25 days in the 1980s for the LEP beams of electrons and positrons in the same tunnel were now completed in less than a week. The proton beams were remarkably on target and stable. And the protons stayed in line—very few stray particles were detected. The optics worked, the stability tests worked, realignments worked. The actual beams matched precisely the computer programs that simulated what should occur.

In fact, the experimenters were taken by surprise when they were told Sunday at 5:00 P.M., only a couple of days after the renewed beams began circulation, to expect collisions the next day. They had anticipated a little bit of time between first beams after the shutdown and the first actual collisions they could record and measure. This was now to be their first opportunity to test their experiment with actual proton beams, rather than the cosmic rays they had used while waiting for the machine

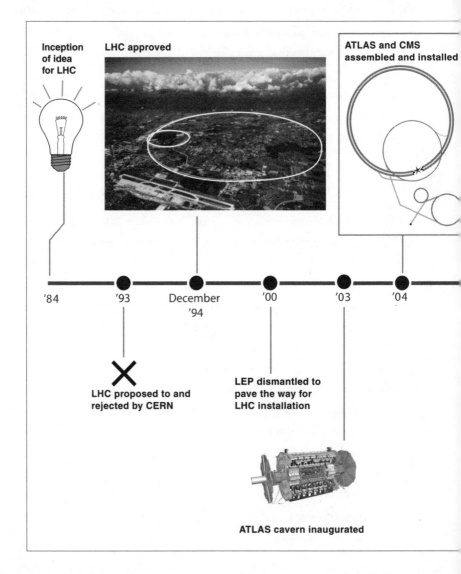

Inception of idea for LHC

LHC approved

ATLAS and CMS assembled and installed

'84   '93   December '94   '00   '03   '04

LHC proposed to and rejected by CERN

LEP dismantled to pave the way for LHC installation

ATLAS cavern inaugurated

to run. The short notice meant, however, that they had very little time to reconfigure their computer *triggers* that tell computers which collisions to record. Maurizio described the anxiety they all felt, since they didn't want to foolishly fumble this opportunity. At the Tevatron, the first test had been mangled by an unfortunate resonance of the beam circulation with the readout system. No one wanted to see this happen again. Of course, in addition to unease, an enormous amount of excitement was shared by everyone involved.

On November 23, the LHC at long last had its first collision. Mil-

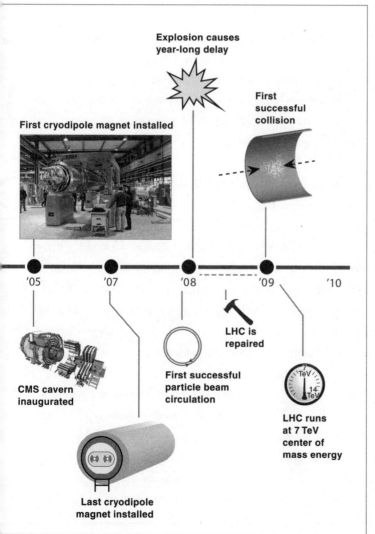

[FIGURE 28] Brief outline of the LHC's history.

lions of protons collided with the injection energy of 900 GeV. These events meant that after years of waiting, experiments could begin taking data—recording the results of the first proton collisions in the LHC ring. Scientists from ALICE, one of the smaller experiments, even submitted a preprint (a paper before publication) on November 28.

Not too long afterward, a modest acceleration was applied to create 1.18 TeV proton beams, the highest-energy circulating beams ever. Only a week after the first LHC collisions, on November 30, these higher-energy protons collided. The net center of mass energy of 2.36 TeV exceeded the highest energies ever achieved before, breaking Fermilab's eight-year-old record.

Three LHC experiments registered beam collisions and tens of thousands of such collisions occurred over the next few weeks. Those collisions won't be used to discover new physical theories, but they were incredibly useful for determining that the experiments in fact worked and could be used to study Standard Model *backgrounds*—events that don't indicate anything new, but could potentially interfere with real discoveries.

Experimenters everywhere shared the satisfaction of the LHC's having reached record energies. Remarkably, the LHC did it just in the nick of time—the machine had been scheduled to shut down from the middle of December until March of the following year, so it was either December or several more months' delay. Jeff Richman, a Santa Barbara experimenter who works on the LHC, joyfully shared this fact at the dark matter conference we were both attending, since he had made a bet with a Fermilab physicist as to whether the LHC would achieve higher energy collisions than Fermilab's Tevatron before the close of 2009. His cheerful demeanor made it clear who had won.

On December 18, 2009, the wave of excitement was temporarily suspended when the LHC shut down after this commissioning run. Lyn Evans concluded his talk discussing the plans for 2010, when he promised a sizable increase in energy. The plan was to go up to 7 TeV before the end of the year—a substantial increase in energy over anything before. He was enthusiastic and confident—as turned out to be justified when indeed the machine came back on line at this higher energy. After so many ups and downs, the LHC was finally working according to plan. (See Figure 28 for an abbreviated timeline.) The LHC should continue to run through 2012 at 7 TeV, or possibly a bit higher energy, before shutting down for at least a year to prepare for raising the energy to as close as possible to the LHC's 14 TeV target. During this and the following runs, the LHC will also try to raise the intensity of the beams to increase the number of collisions.

Given the smooth operation of the experiments and machines after turning back on in 2009, Lyn's closing words for his talk resonated with the audience: "The adventure of LHC construction is finished. Now let the adventure of discovery begin."

CHAPTER TEN

BLACK HOLES THAT WILL DEVOUR THE WORLD

For quite some time physicists had been looking forward to the LHC turning on. Data are essential to scientific progress, and particle physicists had been starved for high-energy data for years. Until the LHC provides answers, no one can know which of the many suggestions for what might underlie the Standard Model are on the right track. But before this book explores several of the more intriguing possibilities, we'll take a detour in these next few chapters to consider some important questions about risk and uncertainty that are critical both to understanding how to interpret the LHC's experimental studies and to many issues that are relevant in the modern world. We'll begin this excursion with the topic of LHC black holes, and how they just might have received a bit more attention than they deserved.

THE QUESTION

Physicists are currently considering many suggestions for what the LHC might ultimately find. In the 1990s, theorists and experimenters first got excited about a particular newly identified class of scenarios in which not just particle physics, but gravity itself, is modified, and would produce new phenomena at LHC energies. One interesting potential consequence of these theories attracted a good deal of attention, especially from people outside the physics community. This was the possibility of microscopic low-energy black holes. Such tiny extra-dimensional black holes might actually be produced if ideas about additional dimensions of space, such as those that Raman Sundrum and I had proposed, turn out to be correct. Physicists had optimistically predicted that such black holes—if created—could provide one verification of such ideas about modified gravity.

Mind you, not everyone was so enthusiastic about this possibility. Some people in the United States and elsewhere worried that the black holes that could be created might suck in everything on Earth. I was often asked about this potential scenario after my public lectures. Most questioners were satisfied when I explained why there was no danger. Unfortunately, however, not everyone had the opportunity to learn the whole story.

Walter Wagner, a high school teacher and a botanical garden manager in Hawaii, who is also a lawyer and was a nuclear safety officer, together with the Spaniard Luis Sancho, an author and self-described researcher on time theory, were among the most militant of the alarmists. These two went so far as to file a lawsuit in Hawaii against CERN, the U.S. Department of Energy, the National Science Foundation, and the American accelerator center Fermilab, in order to hinder the LHC's start. If the goal had been simply to delay the LHC, you might think sending a pigeon to drop a piece of baguette to gum up the works would have been simpler (this actually happened, though the bird was ostensibly an independent agent). But Wagner and Sancho were interested in a more permanent forestalling of the LHC's operation, so they pressed on.

Wagner and Sanchez were not the only ones who worried about a black hole crisis. A book by public interest trial attorney Harry V. Lehmann, which seemed to concisely summarize the concerns, was entitled *No Canary in the Quanta: Who Gets to Decide If the Large Hadron Collider Is Worth Gambling Our Planet?* A blog about it concentrated on fears from the September 2008 explosion and questioned whether the LHC could safely start again. The chief concern, however, centered not on the technology failure that was responsible for the September

19 mishap, but on the actual physical phenomena that the LHC might produce.

The purported threats that Lehmann and many others described about the "Doomsday machine" focused on black holes that they suggested could lead to the implosion of the planet. They worried about a lack of reliable risk assessment in light of the reliance on quantum mechanics in the LHC Safety Assessment Group's study—given claims by Richard Feynman and others that "no one understands quantum mechanics"—as well as uncertainties due to the many unknowns in string theory, which they thought to be relevant. Their questions involved whether it is permissible to risk the Earth for any reason, even when risks are supposed to be tiny, and who should take charge of deciding.

Though the instantaneous destruction of the Earth is certainly more apocalyptic a concern, in reality, the latter questions are more appropriate to other discussions—such as those concerning global warming. Hopefully this chapter and the next will convince you that your time is better spent worrying about the depletion of the contents of your 401(k) than fretting about the disappearance of the Earth by black holes. Although schedules and budgets posed a risk for the LHC, theoretical considerations, supplemented by careful scrutiny and investigations, demonstrated that black holes did not.

To be clear, this doesn't mean the questions shouldn't have been asked. Scientists, like everyone else, need to anticipate possible dangerous consequences of their actions. But for the question of black holes, physicists built on existing scientific theories and data to evaluate the risk, and thereby determined there was no worrisome threat. Before moving on to a more general discussion of risk in the chapter that follows, this chapter explores why anyone even considered the possibility of LHC black holes, and why the doomsday fears about them that some suggested were ultimately misguided. The details this chapter discusses aren't going to be important for the general discussion following, or even for the next part's outline of what the LHC will explore. But it serves as an example of how physicists think and anticipate, and sets the stage for the broader considerations of risk that follow.

BLACK HOLES AT THE LHC

Black holes are objects with such strong gravitational attraction that they trap anything that approaches too closely. Whatever comes within a radius known as the *event horizon* of the black hole gets engulfed and becomes imprisoned inside. Even light, which seems rather inconspicuous, succumbs to a black hole's enormous gravitational field. Nothing can escape a black hole. A Trekkie friend jokes that they are the "perfect Borgs." Any object that encounters a black hole gets assimilated, since the laws of gravity dictate that "resistance is futile."

Black holes form when enough matter gets concentrated inside a small enough region that the force of gravity becomes indomitable. The size of the region required to make a black hole depends on the amount of mass. Smaller mass must congregate in a proportionately smaller region, while larger mass can be distributed over a larger region. Either way, when the density is enormous and a critical mass is within the required volume, the gravitational force becomes irresistible and a black hole is formed. Classically (which means according to calculations that ignore quantum mechanics), these black holes grow as they accrete nearby matter. Also according to such classical calculations, these black holes wouldn't decay.

Before the 1990s, no one thought about creating black holes in a laboratory since the minimum mass required to make a black hole is enormous compared to a typical particle mass or the energies of current colliders. After all, black holes embody very strong gravity, whereas the gravitational force of any individual particle that we know of is negligible—far less than other forces such as electromagnetism. If gravity jibes with our expectations, then in a universe composed of three dimensions of space, particle collisions at accessible energies fall far short of the requisite energy. Black holes do, however, exist throughout the universe—in fact they seem to sit at the center of most large galaxies. But the energy required to create a black hole is at least fifteen orders of magnitude—a one followed by fifteen zeroes—bigger than anything a lab will create.

So why did anyone even mention the possibility of black hole creation at the LHC? The reason is that physicists realized that space and gravity could be very different from what we have observed so far. Gravity might spread not just in the three spatial dimensions we know, but also in as-yet-invisible additional dimensions that have so far eluded detection. Those dimensions have had no identifiable effect on any measurement made so far. But it could be that when we reach the energies of the LHC, extra-dimensional gravity—if it exists—could manifest itself in a detectable manner.

As we will explore further in Chapter 17, the extra dimensions that were briefly introduced in Chapter 7 are an exotic idea—but have reasonable theoretical underpinnings and might even explain the extraordinary feebleness of the gravitational force we know. Gravity can be strong in the higher-dimensional world but diluted and extremely weak in the three-dimensional world that we observe, or—according to the idea Raman Sundrum and I worked out—it could vary in an extra dimension so that it is strong elsewhere but weak in our location in higher-dimensional space. We don't know yet whether such ideas are correct. They are far from certain, but as Chapter 17 will explain, they are among the leading contenders for what experimenters at the LHC might discover.

Such scenarios would imply that when we explore smaller distances at which the effect of the extra dimensions can in principle appear, a very different face of gravity could emerge. Theories involving additional dimensions suggest that the physical properties of the universe should change at the larger energies and smaller distances that we will soon explore. If extra-dimensional reality is indeed responsible for observed phenomena, then gravitational effects could be much stronger at LHC energies than previously thought. In this case, LHC results would not simply depend on gravity as we know it, but also on the stronger gravity of a higher-dimensional universe.

With such strong gravity, protons could conceivably collide in a sufficiently tiny region to trap the amount of energy necessary to create higher-dimensional black holes. These black holes, if they lasted long enough, would suck in mass and energy. If they did this forever, they would indeed be dangerous. This was the catastrophic scenario that the worriers envisioned.

Fortunately, however, classical black hole calculations—those that rely solely on Einstein's theory of gravity—are not the whole story. Stephen Hawking has many accomplishments to his name, but one of his signature discoveries was that quantum mechanics provides an escape hatch for matter trapped in black holes. Quantum mechanics allows black holes to decay.

The surface of a black hole is "hot," with a temperature that depends on its mass. Black holes radiate like hot coals, sending off energy in all directions. They still absorb everything that comes too close, but quantum mechanics tells us that particles evaporate from a black hole's surface through this Hawking radiation, carrying away energy so that it slowly goes back out. The process allows even a large black hole to eventually radiate away all its energy and disappear.

Because the LHC would have at best just barely enough energy to make a black hole, the only black holes it could conceivably form would be small ones. If a black hole started off small and hot, such as one that could potentially be produced at the LHC, it would pretty much disappear immediately. The decay due to Hawking radiation would very efficiently deplete it to nothing. So even if higher-dimensional black holes did form (assuming this whole story is correct in the first place), they wouldn't stick around long enough to do any damage. Big black holes evaporate slowly, but tiny black holes are very hot and lose their energy almost right away. In this respect, black holes are rather strange. Most objects, coals for instance, cool down as they radiate. Black holes, on the other hand, heat up. The smallest ones are the hottest, and therefore radiate the most efficiently.

Now I'm a scientist—so I have to insist on rigor. Technically, a potential caveat to the above argument based on Hawking radiation and black hole decay does exist. We understand black holes only when they are sufficiently big, in which case we know precisely the equations that describe their gravitational system. The well-tested laws of gravity give a reliable mathematical description for black holes. However, we have no such credible formulation of what extremely small black holes would look like. For these very tiny black holes, quantum mechanics would come into play—not just for their evaporation, but in describing the nature of the objects themselves.

No one really knows how to solve systems in which both quantum mechanics and gravity play an essential role. String theory is physicists' best attempt, but we don't yet understand all its implications. This means that in principle there could be a loophole. Extremely tiny black holes, which we will understand only with a theory of quantum gravity, are unlikely to behave the same way as the big black holes we derive using classical gravity. Perhaps such very tiny black holes don't decay at the rates we expect.

Even this isn't a serious loophole however. Few people, if any, worried about these objects. Only black holes that can grow to be big can possibly be dangerous. Small black holes can't accrete enough matter to pose any problem. The only potential risk is that the tiny objects could grow to a dangerous size before evaporating. Yet even without knowing exactly what these objects are, we can estimate how long they should last. These estimates yield lifetimes that are so significantly less than would be required for a black hole to be dangerous that even the very unlikely events on the tails of distributions would still be extremely safe. Small black holes wouldn't behave very differently from familiar unstable heavy particles. Like these short-lived particles, small black holes would very rapidly decay.

Some did, however, still worry that Hawking's derivation, although consistent with all known laws of physics, could be wrong and that black holes might be completely stable. After all, Hawking radiation has never been tested by observations since the radiation from known black holes is too weak to see. Physicists are rightfully skeptical of these objections since they would then have to throw away not only Hawking radiation, but also many other independent and well-tested aspects of our physical theories. Furthermore, the logic underlying Hawking radiation directly predicts other phenomena that have been observed, giving us further confidence in its validity.

Nonetheless, Hawking radiation has never been seen. So to be super-safe, physicists asked the question: If Hawking radiation was somehow not correct and the black holes the LHC might create were stable and never decayed, would they be dangerous then?

Fortunately, even stronger proof exists that black holes pose no danger. The argument makes no assumptions about black hole decay and is not theoretical but is based instead solely on observations of the cosmos. In June 2008, two physicists, Steve Giddings and Michelangelo Mangano,[39] and soon afterward, the LHC Safety Assessment Group,[40] wrote explicit empirically based papers that convincingly ruled out any black hole disaster scenario. Giddings and Mangano calculated the rate at which black holes could form and what their impact would already have been in the universe if they were indeed stable and didn't decay. They observed that even though we haven't yet produced the energies required to create black holes—even higher-dimensional black holes—at accelerators here on Earth, the requisite energies are reached quite frequently in the cosmos. Cosmic rays—highly energetic particles—travel through space all the time, and they often collide with other objects. Although we have no way to study their consequences in detail as we can with experiments on Earth, these collisions frequently have energies at least as high as that which the LHC will achieve.

So if extra-dimensional theories are correct, black holes might then form in astrophysical objects—even the Earth or the Sun. Giddings and Mangano calculated that for some models (the rate depends on the number of additional dimensions), black holes simply grow too slowly to be dangerous: even over the course of billions of years, most black holes would remain extremely small. In other cases, black holes could indeed accrete enough matter to grow big—but they often carried charge. If these had indeed been dangerous, they would have been trapped in the Earth and in the Sun, and both of the objects would have disappeared long ago. Since the Earth and Sun seem to have remained intact, the charged black holes—even those that rapidly accrete matter—can't have dangerous consequences.

So the only possibly dangerous scenario that remains is that black holes don't carry charge but could grow big sufficiently quickly to be a threat. In that case, the Earth's gravitational pull—the only force that could slow them down—wouldn't be sufficiently strong to stop them. Such black holes would pass right through the Earth so we couldn't use the Earth's existence to draw any conclusions about their potential danger.

However, Giddings and Mangano ruled out even that case too, since the gravitational attraction of much denser astrophysical objects— namely, neutron stars and white dwarfs—is sufficiently strong to stop black holes before they could escape. Ultra-high-energy cosmic rays hitting dense stars with strong gravitational interactions would have already produced exactly the sorts of black holes that are potentially possible at the LHC. Neutron stars and white dwarfs are much denser than the Earth—so very dense that their gravity alone would suffice to stop black holes in their interior. If the black holes had been produced and had been dangerous, they would have already destroyed these objects that we know have lasted billions of years. The number of them in the sky tells us that even if black holes exist, they certainly are not dangerous. Even if black holes were formed, they must have disappeared almost immediately—or at worst left tiny innocuous stable remnants. They wouldn't have had sufficient time to do any damage.

On top of that, in the process of accreting matter and destroying such objects, black holes would have released large amounts of visible light, which no one has ever seen. The existence of the universe as we know it and the absence of any signal of white dwarf destruction is very convincing proof that any black holes the LHC could possibly make cannot be dangerous. Given the state of the universe, we can conclude that the Earth is in no danger from LHC black holes.

I'll now give you a moment to breathe a sigh of relief. But I'll nonetheless briefly continue with the black hole story—this time from my perspective as someone who works on related topics such as the extra dimensions of space necessary for low-energy black holes to be created.

Before the black hole controversy blew up in the news, I'd already become interested in the topic. I have a colleague and friend in France who used to work at CERN but now works on an experiment called Auger, which studies cosmic rays as they descend through our atmosphere toward Earth. He complained to me that the LHC takes away resources that can be used to study the same energy scales in his cosmic rays. Since his experiment is far less precise, the only type of events it might find would be those with dramatic signatures such as decaying black holes.

So along with a postdoctoral fellow at Harvard at the time, Patrick Meade, I set out to calculate the number of such events they might observe. With a more careful calculation, we found that the number was much less than physicists had originally optimistically predicted. I say "optimistic" since we are always excited about the idea of evidence for new physics. We weren't concerned about disasters on the Earth—or in the cosmos, which I hope you now agree were not a real threat.

After recognizing that Auger wouldn't discover tiny black holes, even if higher-dimensional explanations of particle physics phenomena were correct, our calculations made us curious about the claims other physicists had made that black holes could be produced in abundance at the LHC. We found that those rates were overestimates as well. Although the rough ballpark estimates had indicated that in these scenarios, the LHC would copiously produce black holes, our more detailed calculations demonstrated that this was not the case.

Patrick and I had not been concerned about dangerous black holes. We had wanted to know whether small, harmless, rapidly decaying higher-dimensional black holes could be produced and thereby signal the presence of higher-dimensional gravity. We calculated this could rarely happen, if at all. Of course, if possible, the production of small black holes could have been a fantastic verification of the theory Raman and I had proposed. But as a scientist, I'm obliged to pay attention to calculations. Given our results, we couldn't entertain false expectations. Patrick and I (and most other physicists) don't expect even small black holes to appear.

That's how science works. People have ideas, work them out roughly, and then they or others go back and check the details. The fact that the initial idea had to be modified after further scrutiny is not a mark of ineptitude—it's just a sign that science is difficult and progress is often incremental. Intermediate stages involve forward and backward adjustments until we settle theoretically and experimentally on the best ideas. Sadly, Patrick and I didn't finish our calculations in time to prevent the black hole controversy from permeating the newspapers and leading to a lawsuit.

We did realize, however, that whether or not black holes could ultimately be produced, other interesting signatures of strongly interacting particles at the LHC might provide important clues about the underlying nature of forces and gravity. And we would see these other signals of higher dimensions at lower energies. Until we see these other exotic signals, we know there is no chance for making black holes. But these other signals themselves might eventually illuminate some aspects of gravity.

This work exemplifies another important aspect of science. Even though paradigms might shift dramatically at different ranges of scales, we rarely suddenly encounter such abrupt shifts in the data itself. Data that was already available sometimes precipitated changes in paradigms, such as when quantum mechanics ultimately explained known spectral lines. But often small deviations from predictions in active experiments are preludes to more dramatic evidence to come. Even dangerous applications of science take time to develop. Scientists might be held accountable in some respects for the nuclear weapons era, but none of them suddenly discovered a bomb by surprise. Understanding the equivalence of mass and energy wasn't enough. Physicists had to work very hard to configure matter into its dangerous explosive form.

Black holes could even possibly be worthy of worry if they could grow to be large, which calculations and observations demonstrated won't happen. But even if they could, small ones—or at least the gravitational effects on particle interactions just discussed—would nonetheless signal the presence of a shift in gravity first.

In the end, black holes don't pose any danger. But just in case, I'll promise to take full responsibility if the LHC creates a black hole that gobbles up the planet. Meanwhile, you can do what my freshman seminar students suggested and check out http://hasthelargehadroncollider destroyedtheworldyet.com.

CHAPTER ELEVEN

RISKY BUSINESS

Nate Silver, the creator of the blog FiveThirtyEight—the most successful predictor of the results of the 2008 presidential election—came to interview me in the fall of 2009 for a book he was writing about forecasting. At that time we faced an economic crisis, an apparently unwinnable war in Afghanistan, escalating health-care costs, potentially irreversible climate change, and other looming threats. I agreed to meet—a bit in the spirit of tit for tat—since I was interested to learn Nate's views on probability and when and why predictions work.

I was nonetheless somewhat puzzled at being chosen for the interview since my expertise was predicting the results of particle collisions, which I doubt that people in Vegas, never mind the government, were betting on. I thought perhaps Nate would ask about black holes at the LHC. But despite the by then defunct lawsuit that suggested possible dangers, I really doubted Nate would be asking about that scenario, given the far more genuine threats listed above.

Nate in fact wasn't interested in this topic. He asked far more measured questions about how particle physicists make speculations and predictions for the LHC and other experiments. He is interested in forecasting, and scientists are in the business of making predictions. He wanted to learn more about how we choose our questions and the methods we use to speculate about what might happen—questions we will soon address more fully.

Nonetheless, before considering LHC experiments and speculations for what we might find, this chapter continues our discussion of risk. The strange attitudes about risks today and the confusions about when and how to anticipate them certainly merit some consideration. The news reports the myriad bad consequences of unanticipated or unmitigated problems on a daily basis. Perhaps thinking about particle physics and separation by scale can shed some light on this complicated subject. The LHC black hole lawsuit was certainly misguided, but both this and the truly pressing issues of the day can't help but alert us to the importance of addressing the subject of risk.

Making particle physics predictions is very different from evaluating risk in the world, and we can only skim the surface of the realities pertinent to risk evaluation and mitigation in a single chapter. Furthermore, the black hole example won't readily generalize since the risk is essentially nonexistent. Nonetheless, it does help guide us in identifying some of the relevant issues when considering how to evaluate and account for risks. We'll see that although black holes at the LHC were never a menace, misguided applications of forecasting often are.

RISK IN THE WORLD

When physicists considered predictions for black holes at the LHC, we extrapolated existing scientific theories to as yet unexplored energy scales. We had precise theoretical considerations and clear experimental evidence that allowed us to conclude that nothing disastrous could happen, even if we didn't yet know what would appear. After careful investigations, all scientists agreed that the risk of danger from black holes was negligible—with no chance that they could be a problem, even over the lifetime of the universe.

This is quite different from how other potential risks are addressed. I'm still a bit mystified how economists and financiers a few years back could fail to anticipate the looming financial crisis—or even after the crisis had been averted possibly set the stage for a new one. Economists and financiers did not share a uniform consensus in their prognoses of smooth sailing, yet no one intervened until the economy teetered on collapse.

In the fall of 2008, I participated in a panel at an interdisciplinary conference. Not for the first or last time, I was asked about the danger of black holes. The vice-chairman of Goldman Sachs International, who was seated to my right, joked to me that the real black hole risk everyone was facing was the economy. And the analogy was remarkably apt.

Black holes trap anything nearby and transform it through strong internal forces. Because black holes are characterized entirely by their mass, charge, and a quantity called angular momentum, they don't keep track of what went in or how it got there—the information that went in appears to be lost. Black holes release that information only slowly, through subtle correlations in the radiation that leaks out. Furthermore, large black holes decay slowly whereas small ones disappear right away. This means that whereas small black holes don't last very long, large ones are essentially too big to fail. Any of this ring a bell? Information—plus debts and derivatives—that went into banks became trapped and was transformed into indecipherable, complicated assets. And after that, information—and everything else that went in—was only slowly released.

With too many global phenomena today, we really are doing uncontrolled experiments on a grand scale. Once, on the radio show *Coast to Coast,* I was asked whether I would proceed with an experiment—no matter how potentially interesting—if it had a chance of endangering the entire world. To the chagrin of the mostly conservative radio audience, my response was that we are already doing such an experiment with carbon emissions. Why aren't more people worried about that?

As with scientific advances, rarely do abrupt changes happen without any advance indicators. We don't know that climate will change cataclysmically, but we have already seen indications of melting glaciers and changing weather patterns. The economy might have suddenly failed in 2008, but many financiers knew enough to leave the markets in advance of the collapse. New financial instruments and high carbon levels have the potential to precipitate radical changes. In such real-world situations, the question isn't whether risk exists. In these cases we need to determine how much caution to exercise if we are to properly account for possible dangers and decide on an acceptable level of caution.

CALCULATING RISK

Ideally, one of the first steps would be to calculate risks. Sometimes people simply get the probabilities wrong. When John Oliver interviewed Walter Wagner, one of the LHC litigants, about black holes on *The Daily Show*, Wagner forfeited any credibility he might have had when he said the chance of the LHC destroying the Earth was 50–50 since it either will happen or it won't. John Oliver incredulously responded that he "wasn't sure that's how probability works." Happily, John Oliver is correct, and we can make better (and less egalitarian) probability estimates.

But it's not always easy. Consider the probability of detrimental climate change—or the probability of a bad situation in the Middle East, or the fate of the economy. These are much more complex situations. It's not merely that the equations that describe the risks are difficult to solve. It's that we don't even necessarily know what the equations are. For climate change, we can do simulations and study the historical record. For the other two, we can try to find analogous historical situations, or make simplified models. But in all three cases, huge uncertainties plague any predictions.

Accurate and trustworthy predictions are difficult. Even when people do their best to model everything relevant, the inputs and assumptions that enter any particular model might significantly affect a conclusion. A prediction of low risk is meaningless if the uncertainties associated with the underlying assumptions are much greater. It's critical to be thorough and straightforward about uncertainties if a prediction is to have any value.

Before considering other examples, let me recount a small anecdote that illustrates the problem. Early in my physics career, I observed that the Standard Model allowed for a much wider range of values for a particular quantity of interest than had been previously predicted, due to a quantum mechanical contribution whose size depended on the (then) recently measured and surprisingly large value of the top quark mass. When presenting my result at a conference, I was asked to plot my new prediction as a function of top quark mass. I refused, knowing there were several different contributions and the remaining uncertainties allowed for too broad a range of possibilities to permit such a simple curve. However, an "expert" colleague underestimated the uncertainties and made such a plot (not unlike many real-world predictions made today), and—for a while—his prediction was widely referenced. Eventually, when the measured quantity didn't fall within his predicted range, the disagreement was correctly attributed to his overly optimistic uncertainty estimate. Clearly, it's better to avoid such embarrassments, both in science and in any real-world situation. We want predictions to be meaningful, and they will be only if we are careful about the uncertainties that we enter.

Real-world situations present even more intractable problems, requiring us to be still more careful about uncertainties and unknowns. We have to be cautious about the utility of quantitative predictions that cannot or do not take account of these issues.

One stumbling block is how to properly account for systemic risks, which are almost always difficult to quantify. In any big interconnected system, the large-scale elements involving the multiple failure models arising from the many interconnections of the smaller pieces are often the least supervised. Information can be lost in transitions or never attended to in the first place. And such systemic problems can amplify the consequences of any other potential risks.

I saw this kind of structural issue firsthand when I was on a committee addressing NASA safety. To accommodate the necessity of appeasing diverse congressional districts, NASA sites are spread throughout the country. Even if any individual site takes care of its piece of equipment, there is less institutional investment in the connections. This then becomes true for the larger organization as well. Information can easily get lost in reporting between different sublayers. In an email to me from the NASA and aerospace industry risk-analyst Joe Fragola, who ran the study, "My experience indicates that risk analyses performed without the joint activity between the subject matter experts, the system integration team and the risk analysis team are doomed to be inadequate. In particular, so called 'turn-key' risk analyses become so much actuarial exercise and are only of academic interest." Too often there is a trade-off between breadth and detail, but both are essential in the long term.

One dramatic consequence of such a failure (among others) was the BP incident in the Gulf of Mexico. In a talk at Harvard in February 2011, Cherry Murray, a Harvard dean and member of the National Commission on the BP Deepwater Horizon Oil Spill and Offshore Drilling, cited management failure as one major contributor to the BP incident. Richard Sears, the commission's senior science and engineering adviser and former vice president for Deepwater Services at Shell Oil Co., described how BP management addressed one problem at a time, without ever formulating the big picture in what he called "hyper-linear thinking."

Although particle physics is a specialized and difficult enterprise, its goal is to isolate its simple underlying elements and make clear predictions based on our hypotheses. The challenge is to access small distances and high energies, not to address complicated interconnections. Even though we don't necessarily know which underlying model is correct, we can predict—given a particular model—what sorts of events should occur when, for instance, protons collide with each other at the LHC. When small scales get absorbed into larger ones, effective theories appropriate to the larger scales tell us exactly how the smaller scales enter, as well as the errors we can make by ignoring small-scale details.

In most situations, however, this neat separation by scale that we introduced in Chapter 1 doesn't readily apply. Despite the sometimes shared methods, in the words of more than one New York banker, "Finance is not a branch of physics." In climate or banking, knowledge of small-scale interactions can often be essential to determining large-scale results.

This lack of scale separation can have disastrous consequences. Take as an example the collapse of Barings Bank. Before its failure in that year, Barings, founded in 1762, was Britain's oldest merchant bank. It had financed the Napoleonic wars, the Louisiana Purchase, and the Erie Canal. Yet in 1995, the bad bets made by a sole rogue trader at a small office in Singapore brought it nearly to financial ruin.

More recently, the machinations of Joseph Cassano at AIG led to its near destruction and the threat of major financial collapse for the world as a whole. Cassano headed a relatively small (400-person) unit within the company called AIG Financial Products, or AIGFP. AIG had made reasonably stable bets until Cassano started employing credit-default swaps (a complex investment vehicle promoted by various banks) to hedge the bets made on collateralized debt obligations.

In what seems in retrospect to be a pyramid scheme of hedging, his group ratcheted up $500 billion in credit-default swaps, more than $60 billion of which were tied to subprime mortgages.[41] If subunits had been absorbed into larger systems as they are in physics, the smaller piece would have yielded information or activity at a higher level in a controlled manner that a midlevel supervisor could readily handle. But in an unfortunate and unnecessarily excessive violation of separation of scales, Cassano's machinations went virtually unsupervised and infiltrated the entire operation. His activities weren't regulated as securities, they weren't regulated as gaming, and they weren't regulated as insurance. The credit-default swaps were distributed all over the globe, and no one had worked through the potential implications. So when the subprime mortgage crisis hit, AIG wasn't prepared and it imploded with losses. American taxpayers subsequently were left to bail the company out.

Regulators attended to conventional safety issues (to some extent) concerning the soundness of individual institutions, but they didn't assess the system as a whole, or the interconnected risks built into it. More complex systems with overlapping debts and obligations call for a better understanding of these interconnections and a more comprehensive way of evaluating, comparing, and deciding risks and the tradeoffs for possible benefits.[42] This challenge applies to most any large system—as does the time frame that is deemed relevant.

This brings us to a further factor that makes calculating and dealing with risk difficult: our psyches and our market and political systems apply different logic to long-term risks and short-term ones—sometimes sensibly, but often greedily, so. Most economists and some in the financial markets understood that market bubbles don't continue indefinitely. The risk wasn't that the bubble would burst—did anyone really think that housing prices would continue doubling within short time frames forever?—but that the bubble would burst in the imminent future. Riding or inflating a bubble, even one that you know is unsustainable, isn't necessarily shortsighted if you are prepared at any point to take your profits (or bonuses) and close up shop.

In the case of climate change, we don't actually know how to assign a number to the melting of the Greenland ice cap. The probabilities are even less certain if we ask for the likelihood that it will begin to melt within a definite time frame—say in the next hundred years. But not knowing the numbers is no reason to bury our head in the ice—or the proto-cold water.

We have trouble finding consensus on the risks from climate change and how and when to avert them when the possible environmental consequences arise relatively slowly. And we don't know how to estimate the cost of action or inaction. Were there to be a dramatic climate-driven event, we would be much more likely to take action immediately. Of course, no matter how fast we were, at that point it would be too late. This means that non-cataclysmic climate changes are worth attending to as well.

Even when we do know the likelihood of certain outcomes, we tend to apply different standards to low-probability events with catastrophic outcomes than to high-probability events with less dramatic results. We hear a lot more about airplane crashes and terrorist attacks than we do about car accidents, even though car accidents kill far more people every year. People talked about black holes even without understanding probabilities because the consequences of the disaster scenario seemed so dire. On the other hand, many small (and not so small) probabilities are neglected altogether when their low visibility keeps them under the radar. Even offshore drilling was considered completely safe by many until the Gulf of Mexico disaster actually occurred.[43]

A related problem is that sometimes the greatest benefits or costs arise from the tails of distributions—the events that are the least likely and that we know least well.[44] Ideally, we'd like our calculations to be objectively determined by midrange estimates or averages of preexisting related situations. But we don't have these data if nothing similar ever occurred or if we ignore the possibility altogether. If the costs or benefits are sufficiently high at these tail ends, they dominate the predictions—assuming that you know in advance what they are in the first place. In any case, traditional statistical methods don't apply when the rates are too low for averages to be meaningful.

The financial crisis happened because of events that were outside the range of what the experts had taken into account. Lots of people made money based on the predictable aspects, but supposedly unlikely events determined some of the more negative developments. When modeling the reliability of financial instruments, most applied the data for the previous few years without allowing for the possibility that the economy might turn down, or turn down at a far more dramatic rate. Assessments about whether to regulate financial instruments were based on a short time frame during which markets had only increased. Even when the possibility of a market drop was admitted, the assumed values for the drop were too low to accurately predict the true cost of lack of regulation to the economy. Virtually no one paid attention to the "unlikely" events that precipitated the crisis. Risks that might otherwise have been apparent therefore never came up for consideration. But even unlikely events need to be considered when they can have significant enough impact.[45]

Any risk assessment is plagued by the difficulty of evaluating the risk that the underlying assumptions are incorrect. Without such estimates, any estimate becomes subject to intrinsic prejudices. On top of the calculational problems and hidden prejudices buried in these underlying assumptions, many practical policy decisions involve unknown unknowns—factors that can't be or haven't been anticipated. Sometimes we simply can't foresee the precise unlikely event that will cause trouble. This can make any prediction attempts—that will inevitably fail to factor in these unknowns—completely moot.

MITIGATING RISK

Luckily for our search for understanding, we are extremely certain that the probability of producing dangerous black holes is minuscule. We don't know the precise numerical probability for a catastrophic outcome, but we don't need to because it's so negligible. Any event that won't happen even once in the lifetime of the universe can be safely ignored.

More generally, however, quantifying an acceptable level of risk is extremely difficult. We clearly want to avoid major risks altogether—anything that endangers life, the planet, or anything we hold dear. With risks we can tolerate, we want a way of evaluating who benefits and who stands to lose, and to have a system that would evaluate and anticipate risks accordingly.

The risk analyst Joe Fragola's comment to me about climate change, along with other potential dangers he is concerned with, was the following: "The real issue is not if these could happen, nor what their consequences would be, but rather what is their probability of occurrence and the associated uncertainty? And how much of our global resources should we allocate to address such risks based not only on the probability of occurrence but also on the probability that we might do something to mitigate them?"

Regulators often rely on so-called cost-benefit analysis to evaluate risk and determine how to deal with it. On the surface, the idea sounds simple enough. Calculate how much you need to pay versus the benefit and see if the proposed change is worth it. This might even be the best available procedure in many circumstances, but it might also dangerously generate a deceptive patina of mathematical rigor. In practice, cost-benefit analysis can be very hard to do. The problems involve not just measuring cost and benefit, which can be a challenge, but defining what we mean by cost and benefit in the first place. Many hypothetical situations involve too many unknowns to reliably calculate either, or to calculate risk in the first place. We can certainly try, but these uncertainties need to be accounted for—or at least recognized.

A sensible system that anticipates costs and risks in the near term and in the future would certainly be useful. But not all trade-offs can even be evaluated solely according to their cost. What if that which is at risk can't be replaced at all?[46] Had the creation of an Earth-eating black hole by the LHC been something that could happen with reasonable probability within our lifetime, or even within a million years, we certainly would have pulled the plug.

And even though we ultimately benefit quite a bit from research in basic science, the economic cost of abandoning a project is rarely calculable either, because the benefits are so difficult to quantify. The goals of the LHC include achieving fundamental knowledge, including a better understanding of masses and forces, and possibly even of the nature of space. The benefits also include an educated and motivated technically trained populace inspired by big questions and deep ideas about the universe and its composition. On a more practical front, we will follow the information advance CERN made with the World Wide Web, with the "grid" that will allow a global processing of information, as well as improvements in magnet technology that will be useful for medical devices such as MRIs. Possible further applications from fundamental science might ultimately be found, but these are almost always impossible to anticipate.

Cost-benefit analyses are difficult to apply to basic science. A lawyer jokingly applied a cost-benefit approach to the LHC, noting that along with the extremely tiny proposed enormous risk, the LHC also had a minuscule chance of stupendous benefits by solving all the problems of the world. Of course, neither outcome readily fits into a standard cost-benefit calculation, though—incredibly—lawyers have tried.[47]

At least science benefits from its goals being "eternal" truths. If you find the way the world works, it's true no matter how quickly or slowly you found it. We certainly don't want scientific progress to be slow. But the year's delay showed us the danger of too quickly turning on the LHC. In general, scientists try to proceed safely.

Cost-benefit analysis is riddled with difficulty for almost any complex situation—such as climate change policy or banking. Although in principle a cost-benefit analysis makes sense and there may be no fundamental objection, how you apply it makes an enormous difference. Defenders of cost-benefit analysis essentially make a cost-benefit argument to justify the approach when they ask how can we possibly do better—and they might even be right. I'm simply advocating that where we do apply the method, we do it more scientifically. We need to be clear about the uncertainties in any numbers we present. As with any scientific analysis, we need to take errors, assumptions, and biases into account and be open in presenting these.

One factor that matters a great deal for climate change issues is whether the costs or benefits refer to an individual, a nation, or the globe. The potential costs or benefits can also cross these categories, but we don't always take this into account. One reason that American politicians decided against the Kyoto Protocol was they concluded that the cost would have exceeded the benefit to Americans—American businesses in particular. However, such a calculation didn't really factor in the long-term costs of instabilities across the globe or the benefits of a regulated environment where new businesses might prosper. Many economic analyses of the costs of climate change mitigation fail to account for the potential additional benefits to the economy through innovation or to stability through less reliance on foreign nations. Too many unknowns about how the world will change are involved.

These examples also raise the question of how to evaluate and mitigate risk that crosses national borders. Suppose black holes really had posed a risk to the planet. Could someone in Hawaii constructively sue an experiment planned for Geneva? According to existing laws, the answer is no, but perhaps a successful suit could have interfered with American financial contributions to the experiment.

Nuclear proliferation is another issue where clearly global stability is at stake. Yet we have limited control over the dangers generated in other nations. Both climate change and nuclear proliferation are issues that are managed nationally but whose dangers are not restricted to the institutions or nations creating the menace. The political problem of what to do when risks cross national boundaries or legal jurisdictions is difficult. But it's clearly an important question.

As an institution that is truly international, CERN's success hinges on the shared common goals of many nations. One nation can try to minimize its own contribution, but aside from that, no individual interests are at stake. All involved nations work together since the science they value is the same. The host countries, France and Switzerland, might receive slightly greater economic advantages in labor and infrastructure, but on the whole, it's not a zero-sum game. No one nation benefits at the expense of another.

Another notable feature of the LHC is that CERN and the member states are responsible should any technical or practical problems occur. The 2008 helium explosion had to be repaired through CERN's budget. No one, especially those working at the LHC, benefits from mechanical failure or scientific disasters. Cost-benefit analyses, when applied to situations where costs and benefits aren't fully aligned and the benefactors don't have full responsibility for the risk they take on, are less useful. It is very different from applying this type of reasoning to the types of closed systems that science tries to address.

In any situation, we want to avoid moral hazards, where people's interest and risk are not aligned so they may have an incentive to take on greater risk than they would if no one else effectively contributed insurance. We need to have the right incentive structures.

Consider hedge funds, for example. The general partners get a percentage of profits from their fund each year when they make money, but they don't forfeit a comparable percentage if their fund faces losses or if they go bankrupt. Individuals keep their gains, while their employers— or taxpayers—share the losses. With these parameters, the most profitable strategy for the employees would encourage large fluctuations and instabilities. An efficient system and effective cost-benefit analysis should take into account such allocation of risks, rewards, and responsibilities. They have to factor in the different categories or scales of the people involved.

Banking, too, has obvious moral hazards where risks and benefits aren't necessarily aligned. A "too big to fail" policy combined with weak leveraging limits yields a situation in which the people who are accountable for losses (taxpayers) are not the same as those who stand to benefit the most (bankers or insurers). One can debate whether bailouts were essential in 2008, but preventing the situation in the first place by aligning risk with responsibility seems like a good idea.

Furthermore, at the LHC, all data about the experiments and risks are readily available. The safety report is on the web. Anyone can read it. Certainly any institution that would expect a bailout were it to fail, or even one that simply speculates in a potentially unstable fashion, should provide enough data to regulatory institutions so that the relative weight of benefits against risks can potentially be evaluated. Ready access to reliable data should help mortgage experts or regulators or others antici-pate financial or other potential disasters in the future.

Though not in itself a solution, another factor that could at least im-prove or clarify the analyses would again be to take "scale"—in terms of categories of those subject to benefits and risks, as well as time ranges—into account. The question of scale translates into the issue of who is involved in a calculation: is it an individual, an organization, a govern-ment, or the world, and are we interested in a month, a year, or a decade? A policy that is good for Goldman Sachs might not ultimately benefit the economy as a whole—or the individual whose mortgage is currently under water. That means that even if there were perfectly accurate cal-culations, they would guarantee the right result only if they were applied to the correct carefully thought through question.

When we make policy or evaluate costs versus benefits, we tend to neglect the possible benefits of global stability and helping others—not just in a moral sense, but in the long-term financial sense as well. In part, this is because these gains are difficult to quantify, and in part it is due to the challenge in making evaluations and creating robust regulations in a world that changes quickly. Still, it's clear that regulations that consider all possible benefits, not just those to an individual or an institution or a state, will be more reliable, and may even lead to a better world.

The time frame can also influence the computed cost or benefit for policy decisions as do the assumptions the deciding parties make, as we saw with the recent financial crisis. Time scales matter in other ways as well, since acting too hastily can increase risk while rapid transac-tions can enhance benefits (or profits). But even though fast trades can make pricing more efficient, lightning-fast transactions don't necessarily benefit the overall economy. An investment banker explained to me how important it was to be able to sell shares at will, but even so he couldn't explain why they needed to be able to sell them after owning them a few seconds or less—aside from the fact that he and his bank make more money. Such trades create more profits for bankers and their institutions in the short term, but they aggravate existing weaknesses in the financial sector in the long-term. Perhaps even with a short-term competitive disadvantage, a system that inspires more confidence could be more profitable in the long term and therefore prevail. Of course, the banker I mentioned made $2 billion for his institution in a single year, so his employers might not agree on the wisdom of my suggestion. But anyone who ultimately pays for this profit might.

THE ROLE OF EXPERTS

Many people take away the wrong lesson and conclude that the absence of reliable predictions implies an absence of risk. In fact, quite the opposite applies. Until we can definitively rule out particular assumptions or methods, the range of possible outcomes is within the realm of possibility. Despite the uncertainties—or perhaps because of them—with so many models predicting dangerous results, the probability of something very bad happening with climate or with the economy—or with offshore drilling—is not negligibly small. Perhaps one can argue that the chances are small within a definite time frame. However, in the long run, until we have better information, too many scenarios lead to calamitous results to ignore the dangers.

People interested only in the bottom line rally against regulation while those who are interested in safety and predictability argue for it. It is too easy to be tempted to come down on one side or the other, since figuring out where to draw the line is a daunting—if not an impossible—task. As with calculating risk, not knowing the deciding point doesn't mean there is none or that we shouldn't aim for the best approximation. Even without the insights necessary to make detailed predictions, structural problems should be addressed.

This brings us to the last important question: Who decides? What is the role of experts, and who gets to evaluate riskiness?

Given the money and bureaucracy and careful oversight involved in the LHC, we can expect that risks were adequately analyzed. Furthermore, at its energies, we aren't even really in a new regime where the basic underpinnings of particle physics should fail. Physicists are confident the LHC is safe, and we look forward to the results from particle collisions.

This isn't to say that scientists don't have a big responsibility. We always need to ensure that scientists are responsible and are attentive to risks. We'd like to be as certain with respect to all scientific enterprises as we were with the LHC. If you are creating matter or microbes or anything else that has not existed before (or drilling deeper or otherwise exploring new frontiers on the Earth for that matter), you need to be certain of not doing anything dramatically bad. The key is to do this rationally, without unfounded fearmongering that would impede progress and benefits. This is true not just for science but for any potentially risky endeavor. The only answer to imagined unknowns and even to "unknown unknowns" is to heed as many reasonable viewpoints as possible and to have the freedom to intervene if necessary. As anyone in the Gulf of Mexico will attest to, you need to be able to turn off the spigots if something goes wrong.

Early in the previous chapter, I summarized some of the objections that bloggers and skeptics made about the methods physicists used for black hole calculations, including relying on quantum mechanics. Hawking did indeed use quantum mechanics to derive black hole decay. Yet despite Feynman's statement that "no one understands quantum mechanics," physicists understand its implications, even if we don't have a deep philosophical insight into why quantum mechanics is true. We believe quantum mechanics because it explains data and solves problems that are impenetrable with classical physics.

When physicists debate quantum mechanics, they don't dispute its predictions. Its repeated success has forced generations of astonished students and researchers to accept the theory's legitimacy. Debates today about quantum mechanics concern its philosophical underpinnings. Is there some other theory with more familiar classical premises that nonetheless predicts the bizarre hypotheses of quantum mechanics? Even if people make progress on such issues, it would make no difference to quantum mechanical predictions. Philosophical advances could affect the conceptual framework we use to describe predictions—but not the predictions themselves.

For the record, I find major advances on this front unlikely. Quantum mechanics is probably a fundamental theory. It is richer than classical mechanics. All classical predictions are a limiting case of quantum mechanics, but not vice versa. So it's hard to believe that we will ultimately interpret quantum mechanics with classical Newtonian logic. Trying to interpret quantum mechanics in terms of classical underpinnings would be like me trying to write this book in Italian. Anything I can say in Italian I can say in English, but because of my limited Italian vocabulary the reverse is far from true.

Still, with or without agreement on philosophical import, all physicists agree on how to apply quantum mechanics. The wacky naysayers are just that. Quantum mechanical predictions are trustworthy and have been tested many times. Even without them, we still have alternative experimental evidence (in the form of the Earth and Sun and neutron stars and white dwarfs) that the LHC is safe.

LHC alarmists also objected to the purported use of string theory. Indeed, using quantum mechanics was just fine but relying on string theory would not have been. But the conclusions about black holes never needed string theory anyway. People do try to use string theory to understand the interior of black holes—the geometry of the apparent singularity at the center where according to general relativity energy becomes infinitely dense. And people have done string-theory-based calculations of black hole evaporation in nonphysical situations that support Hawking's result. But the computation of black hole decay relies on quantum mechanics and not on a complete theory of quantum gravity. Even without string theory, Hawking could do his calculations. The very questions some bloggers posed reflected the absence of sufficient scientific understanding to weigh the facts.

A more generous interpretation of this objection is as resistance not to the science itself but to scientists with "faith-based" beliefs in their theories. After all, string theory is beyond the experimentally verifiable regime of energies. Yet many physicists think it's right and continue to work on it. However, the variety of opinions about string theory—even within the scientific community—nicely illustrates just the opposite point. No one would base any safety assessment on string theory. Some physicists support string theory and some do not. Yet everyone knows it is not yet proven or fully fleshed out. Until everyone agreed on string theory's validity and reliability, trusting string theory for risky situations would be foolhardy. As concerns our safety, the inaccessibility of string theory's experimental consequences is not the only reason that we don't yet know if it's correct—it's also the reason it isn't required to predict most real-world phenomena we will encounter in our lifetimes.

Yet despite my confidence that it was okay to rely on experts when evaluating potential risks from the LHC, I recognize the potential limitations of this strategy and don't quite know how to address them. After all, "experts" told us that derivatives were a way of minimizing risk, not creating potential crises. "Expert" economists told us that deregulation was essential to the competitiveness of American business, not to the potential downfall of the American economy. And "experts" tell us only those in the banking sector understand their transactions sufficiently well to address its woes. How do we know when experts are thinking broadly enough?

Clearly experts can be shortsighted. And experts can have conflicts of interest. Are there any lessons from science here?

I don't think it is my bias that leads me to say that in the case of LHC black holes, we examined the full range of potential risks that we could logically envision. We thought about both the theoretical arguments and also the experimental evidence. We thought about situations in the cosmos where the same physical conditions applied, yet did not destroy any nearby structure.

It would be nice to be so sanguine that economists do similar comparisons to existing data. But the title of Carmen Reinhart and Kenneth Rogoff's book *This Time Is Different* suggests otherwise. Although economic conditions are never identical, some broad measures do indeed repeat themselves in economic bubbles.

The argument made by many today that no one could anticipate the dangers of deregulation also doesn't stand up. Brooksley Born, the former chairperson of the Commodity Futures Trading Commission, which oversees futures and commodity options markets, did point out the dangers of deregulation—actually she rather reasonably suggested that potential risks be explored—but she was shouted down. There was no solid analysis of whether caution was justified (as it clearly turned out to be) but only a partisan view that moving slowly would be bad for business (as it would have been for Wall Street in the short term).

Economists speaking out about regulation and policy might have a political as well as a financial agenda and that can interfere with doing the right thing. Ideally, scientists pay more attention to the merits of arguments, including those regarding risk, than politics. LHC physicists made serious scientific inquiries to ensure no disasters would occur.

Although perhaps only financial experts understand the details of a particular financial instrument, anyone can consider some basic structural issues. Most people can understand why an overly leveraged economy is unstable, even without predicting or even understanding the precise trigger that might cause a collapse. And most anyone can understand that giving the banks hundreds of billions of dollars with few or no constraints is probably not the best way to spend taxpayers' money. And even a faucet is built with a reliable means of turning it off—or at least a mop and plan in place to clean up any mess. It's hard to see why the same shouldn't apply to deep-sea oil rigs.

Psychological factors enter when we count on experts, as the *New York Times* economics columnist David Leonhardt explained in 2010 when attributing Mr. Greenspan's and Mr. Bernanke's errors to factors that were "more psychological than economic." He explained, "They got trapped in an echo chamber of conventional wisdom" and "fell victim to the same weakness that bedeviled the engineers of the Challenger space shuttle, the planners of the Vietnam and Iraq wars, and the airline pilots who have made tragic cockpit errors. They didn't adequately question their own assumptions. It's an entirely human mistake."[48]

The only way to address complicated issues is to listen broadly, even to the outliers. Despite their ability to predict that the economy could collapse into a black hole, self-interested bankers were content to ignore warnings so long as they could. Science is not democratic in the sense that we all get together and vote on the right answer. But if anyone has a valid scientific point, it will ultimately be heard. People will often pay attention to the discoveries and insights from more prominent scientists first. Nonetheless, an unknown who makes a good point will eventually gain an audience.

With the ear of a well-known scientist, an unknown might even be listened to right away. That is how Einstein could present a theory that shook scientific foundations almost immediately. The German physicist Max Planck understood the implications of Einstein's relativistic insights and was fortuitously in charge of the most important physics journal at the time.

Today we benefit from the rapid spread of ideas over the Internet. Any physicist can write a paper and have it sent out through the physics archive the next day. When Luboš Motl was an undergraduate in the Czech Republic, he solved a scientific problem that a prominent scientist at Rutgers was working on. Tom Banks paid attention to good ideas, even if they came from an institution he had never heard of before. Not everyone is so receptive. But so long as a few people pay attention, an idea, if good and correct, will ultimately enter scientific discourse.

LHC engineers and physicists sacrificed time and money for safety. They wanted to economize as much as possible, but not at the expense of danger or inaccuracy. Everyone's interests were aligned. No one benefits from a result that doesn't stand the test of time.

The currency in science is reputation. There are no golden parachutes.

FORECASTING

I hope we all now agree that we shouldn't be worrying about black holes—though we do have much else to worry about. In the case of the LHC, we are and should be thinking about all the good things it can do. The particles created there will help us answer deep and fundamental questions about the underlying structure of matter.

To briefly return to my conversation with Nate Silver, I realized how special our situation is. In particle physics, we can restrict ourselves to simple enough systems to exploit the methodical manner in which new results build on old ones. Our predictions sometimes originate in models we know to be correct based on existing evidence. In other cases, we make predictions based on models we have reasons to believe might exist and use experiments to winnow down the possibilities. Even then— without yet knowing if these models will prove correct—we can anticipate what the experimental evidence would be, should the idea turn out to be realized in the world.

Particle physicists exploit our ability to separate according to scale. We know small-scale interactions can be very different from those that occur on large scales, but they nonetheless feed into large-scale interactions in a well-defined way, giving consistency with what we already know.

Forecasting is very different in almost all other cases. For complex systems, we often have to simultaneously address a range of scales. That can be true not only for social organizations, such as a bank in which an irresponsible trader could destabilize AIG and the economy, but even in other sciences. Predictions in those cases can have a great deal of variability.

For example, the goals of biology include predicting biological patterns and even animal and human behavior. But we don't yet fully understand all the basic functional units or the higher-level organization by which elementary elements produce complex effects. We also don't know all the feedback loops that threaten to make separating interactions by scale impossible. Scientists can make models, but without better understanding the critical underlying elements or how they contribute to emergent behavior, modelers face a quagmire of data and competing possibilities.

A further challenge is that biological models are designed to match preexisting data, but we don't yet know the rules. We haven't identified all the simple independent systems, so it is difficult to know which—if any—model is right. When I spoke with my neuroscientist colleagues, they described the same problem. Without qualitatively new measurements, the best that the models can do is to match all existing data. Since all the surviving models must agree with the data, it is difficult to decisively determine which underlying hypothesis is correct.

It was interesting to talk to Nate about the kind of things he tries to predict. A lot of recent popular books present shaky hypotheses that give predictions that work—except when they don't. Nate is a lot more scientific. He first became famous for his accurate predictions of baseball games and elections. His analysis was based on careful statistical evaluations of similar situations in the past, where he included as many variables that he could manage to apply historical lessons to as precisely as he could.

He now has to choose wisely where to apply his methods. But he realizes that the kinds of correlations he focuses on can be tricky to interpret. You can say an engine on fire caused a plane crash, but it's not a surprise to find an engine on fire in a plane going down. What really was the initial cause? You have the same issue when you connect a mutated gene to cancer. It doesn't necessarily cause the disease even if it is correlated with it.

He is aware of other potential pifalls too. Even with large amounts of data, randomness and noise may enhance or suppress the interesting underlying signals. So Nate won't work on financial markets or earthquakes or climate. Although in all likelihood he could predict overall trends, the short-term predictions would be inherently uncertain. Nate now studies other places where his methods shed light such as how best to distribute music and movies, as well as questions such as the value of NBA superstars. But he acknowledges that only very few systems can be so accurately quantified.

Nonetheless, Nate told me that forecasters do make one other type of prediction. Many of them do metaforecasting—predicting what people will try to predict.

CHAPTER TWELVE

MEASUREMENT AND UNCERTAINTY

Familiarity and comfort with statistics and probability help when evaluating scientific measurements, not to mention many of the difficult issues of today's complex world. I was reminded of the virtue of probabilistic reasoning when, a few years back, a friend was frustrated by my "I don't know" response to his question about whether or not I planned to attend an event the following evening. Fortunately for me, he was a gambler and mathematically inclined. So instead of exasperatingly insisting on a definite reply, he asked me to tell him the odds. To my surprise, I found that question a lot simpler to deal with. Even though the probability estimate I gave him was only a rough guess, it more closely reflected my competing considerations and uncertainties than a definite yes or no reply would have done. In the end, it felt like a more honest response.

Since then I've tried this probabilistic approach out on friends and colleagues when they didn't think they could reply to a question. I've found that most people—scientists or not—have strong but not irrevocable opinions that they frequently feel more comfortable expressing probabilistically. Someone might not know if he wants to go to the baseball game on the Thursday three weeks from now. But if he knows that he likes baseball and doesn't think he has any work trips coming up, yet hesitates because it's during the week, he might agree he is 80 percent likely to, even if he can't give a definite yes. Although just an estimate, this probability—even one he makes up on the spot—more accurately reflects his true expectation.

In our conversation about science and how scientists operate, the screenwriter and director Mark Vicente observed how he was struck by the way that scientists hesitate to make definite unqualified statements of the sort most other people do. Scientists aren't necessarily always the most articulate, but they aim to state precisely what they do and don't know or understand, at least when speaking about their field of expertise. So they rarely just say yes or no, since such an answer doesn't accurately reflect the full range of possibilities. Instead, they speak in terms of probabilities or qualified statements. Ironically, this difference in language frequently leads people to misinterpret or underplay scientists' claims. Despite the improved precision that scientists aim for, nonexperts don't necessarily know how to weigh their statements—since anyone other than a scientist with as much evidence in support of their thesis wouldn't hesitate to say something more definite. But scientists' lack of 100 percent certainty doesn't reflect an absence of knowledge. It's simply a consequence of the uncertainties intrinsic to any measurement—a topic we'll now explore. Probabilistic thinking helps clarify the meaning of data and facts, and allows for better-informed decisions. In this chapter, we'll reflect on what measurements tell us and explore why probabilistic statements more accurately reflect the state of knowledge—scientific or otherwise—at any given time.

SCIENTIFIC UNCERTAINTY

Harvard recently completed a curricular review to try and determine the essential elements of a liberal education. One of the categories the faculty considered and discussed as part of a science requirement was "empirical reasoning." The teaching proposal suggested the university's purpose should be to "teach how to gather and assess empirical data, weigh evidence, understand estimates of probabilities, draw inferences from the data when available [so far, so good], and also to recognize when an issue cannot be settled on the basis of the available evidence."

The proposed wording of the teaching requirement—later clarified— was well intentioned, but it belied a fundamental misunderstanding of how measurements work. Science generally settles issues with some degree of probability. Of course we can achieve high confidence in any particular idea or observation and use science to make sound judgments. But only infrequently can anyone absolutely settle an issue—scientific or otherwise—on the basis of evidence. We can collect enough data to trust causal relationships and even to make incredibly precise predictions, but we can generally do it only probabilistically. As Chapter 1 discussed, uncertainty—however small—allows for the potential existence of interesting new phenomena that remain to be discovered. Rarely is anything 100 percent certain, and no theory or hypotheses will be guaranteed to apply under conditions where tests have not yet been performed.

Phenomena can only ever be demonstrated with a certain degree of precision in a set domain of validity where they can be tested. Measurements always have some probabilistic component. Many science measurements rely on the assumption that an underlying reality exists that we can uncover with sufficiently precise and accurate measurements. We use measurements to find this underlying reality as well as we can (or as well as necessary for our purposes). This then permits statements such as that an interval centered on a collection of measurements contains the true value with 95 percent probability. In that case, we might colloquially say we are confident with 95 percent confidence. Such probabilities tell us the reliability of any particular measurement and the full range of possibilities and implications. You can't fully understand a measurement without knowing and evaluating its associated uncertainties.

One source of uncertainty is the absence of infinitely precise measuring instruments. Such a precise measurement would require a device calibrated with an infinite number of decimal places. The measured value would have an infinite number of carefully measured numbers after the decimal place. Experimenters can't make such measurements—they can only calibrate their tools to make them as accurate as possible with available technology, just as the astronomer Tycho Brahe did so expertly more than four centuries ago. Increasingly advanced technology results in increasingly precise measuring devices. Even so, measurements will never achieve infinite accuracy, despite the many advances that have occurred over time. Some *systematic uncertainty,*[49] characteristic of the measuring device itself, will always remain.

Uncertainty doesn't mean that scientists treat all options or statements equally (though news reports frequently make this mistake). Only rarely are probabilities 50 percent. But they do mean that scientists (or anyone aiming for complete accuracy) will make statements that tell what has been measured and what it implies in a probabilistic way, even when those probabilities are very high.

When scientists and wordsmiths are extremely careful, they use the words *precision* and *accuracy* differently. An apparatus is *precise* if, when you repeat a measurement of a single quantity, the values you record won't differ from each other very much. Precision is a measure of the degree of variability. If the result of repeating a measurement doesn't vary a lot, the measurements are precise. Because more precisely measured values span a smaller range, the average value will more rapidly converge if you make repeated measurements.

Accuracy, on the other hand, tells you how close your average measurement is to the correct result. In other words, it tells whether there is *bias* in a measuring apparatus. Technically speaking, an intrinsic error in your measuring apparatus doesn't reduce its precision—you would make the same mistake every time—though it would certainly reduce your accuracy. *Systematic uncertainty* refers to the unbeatable lack of accuracy that is intrinsic to the measuring devices themselves.

Nonetheless, in many situations, even if you could construct a perfect measuring instrument, you would still need to make many measurements to get a correct result. That is because the other source of uncertainty[50] is *statistical,* which means that measurements usually need to be repeated many times before you can trust the result. Even an accurate apparatus won't necessarily give the right value for any particular measurement. But the average will converge to the right answer. Systematic uncertainties control the accuracy of a measurement while statistical uncertainty affects its precision. Good scientific studies take both into account, and measurements are done as carefully as possible on as large a sample as is feasible. Ideally, you want your measurements to be both accurate and precise so that the expected absolute error is small and you trust the values you find. This means you want them to be within as narrow a range as possible (precision) and you want them to converge to the correct number (accuracy).

One familiar (and important) example where we can consider these notions is tests of drug efficacy. Doctors often won't say or perhaps they don't know the relevant statistics. Have you ever been frustrated by being told, "Sometimes this medicine works; sometimes it doesn't"? Quite a bit of useful information is suppressed in this statement, which gives no idea of how often the drug works or how similar the population they tested it on is to you. This makes it very difficult to decide what to do. A more useful statement would tell us the fraction of times a drug or procedure has worked on a patient with similar age and fitness level. Even in the cases when the doctors themselves don't understand statistics, they can almost certainly provide some data or information.

In fairness, the *heterogeneity* of the population, with different individuals responding to drugs in different ways, makes determining how a medicine will work a complicated question. So let's first consider a simpler case in which we can test on a single individual. Let's use as an example the procedure for testing whether or not aspirin helps relieve your headache.

The way to figure this out seems pretty easy: take an aspirin and see if it works. But it's a little more complicated than that. Even if you get better, how do you know it was the aspirin that helped? To ascertain whether or not it really worked—that is, whether your headache was less painful or went away faster than without the drug—you would have to be able to compare how you feel with and without the drug. However, since you either took aspirin or you didn't, a single measurement isn't enough to tell you the answer you want.

The way to tell is to do the test many times. Each time you have a headache, flip a coin to decide whether to take an aspirin or not and record the result. After you do this enough, you can average out over all the different types of headaches you had and the varying circumstances in which you had them (maybe they go away faster when you're not so sleepy) and use your statistics to find the right result. Presumably there is no bias in your measurement since you flipped a coin to decide and the population sample you used was just yourself so your result will correctly converge with enough self-imposed tests.

It would be nice to always be able to learn whether drugs worked with such a simple procedure. However, most drugs are treating more serious illnesses than headaches—perhaps even ones that lead to death. And many drugs have long-term effects, so you couldn't do repeated short-term trials on a single individual even if you wanted to.

So usually when biologists or doctors test how well a drug works, they don't simply study a single individual, even though for scientific purposes at least they would prefer to do so. They then have to contend with the fact that people respond differently to the same drug. Any medicine produces a range of results, even when tested on a population with the same degree of severity of a disease. So the best scientists can do in most cases is to design studies for a population as similar as possible to any given individual they are deciding whether or not to give the drug to. In reality, however, most doctors don't design the studies themselves, so similarity to their patient is hard for them to guarantee.

Doctors might want instead to try to use pre-existing studies where no one did a carefully designed trial but the results were based simply on observations of existing populations, such as the members of an HMO. They would then face the challenge of making the correct interpretation. With such studies, it can be difficult to ensure that the relevant measurement establishes causality and not just association or correlation. For example, someone might mistakenly conclude that yellow fingers cause lung cancer because they noticed many lung cancer patients have yellow fingers.

That's why scientists prefer studies in which treatments or exposures are randomly assigned. For example, a study in which people take a drug based on a coin toss will be less dependent on the population sample since whether or not any patient receives treatment depends only on the random outcome of a coin flip. Similarly, a randomized study could in principle teach about the relationships among smoking, lung cancer, and yellow fingers. If you were to randomly assign members of a group to either smoke or refrain from smoking, you would determine that smoking was at least one underlying factor responsible for both yellow fingers and lung cancer in the patients you observed, whether or not one was the cause of the other. Of course, this particular study would be unethical.

Whenever possible, scientists aim to simplify their systems as much as possible so as to isolate the specific phenomena they want to study. The choice of a well-defined population sample and an appropriate control group are essential to both the precision and accuracy of the result. With something as complicated as the effect of a drug on human biology, many factors enter simultaneously. The relevant question is then how reliable do the results need to be?

THE OBJECTIVE OF MEASUREMENTS

Measurements are never perfect. With scientific research—as with any decision—we have to determine an acceptable level of uncertainty. This allows us to move forward. For example, if you are taking a drug you hope will mitigate your nagging headache, you might be satisfied to try it even if it significantly helps the general population only 75 percent of the time (as long as the side effects are minimal). On the other hand, if a change in diet will reduce your already low likelihood of heart disease by a mere two percent of your existing risk, decreasing it from five percent to 4.9 percent, for example, that might not worry you enough to convince you to forgo your favorite Boston cream pie.

For public policy, decision points can be even less clear. Public opinion usually occupies a gray zone where people don't necessarily agree on how accurately we should know something before changing laws or implementing restrictions. Many factors complicate the necessary calculations. As the previous chapter discussed, ambiguity in goals and methods make cost-benefit analyses notoriously difficult, if not impossible, to reliably perform.

As *New York Times* columnist Nicholas Kristof wrote in arguing for prudency about potentially dangerous chemicals (BPA) in foods or containers, "Studies of BPA have raised alarm bells for decades, and the evidence is still complex and open to debate. That's life: in the real world, regulatory decisions usually must be made with ambiguous and conflicting data."[51]

None of these issues mean that we shouldn't aim for quantitative evaluations of costs and benefits when assessing policy. But they do mean that we should be clear about what the assessments mean, how much they can vary according to assumptions or goals, and what the calculations have and have not taken into account. Cost-benefit analyses can be useful but they can also give a false sense of concreteness, certainty, and security that can lead to misguided applications in society.

Fortunately for physicists, the questions we ask are usually a lot simpler—at least to formulate—than they are for public policy. When we're dealing with pure knowledge without an immediate eye to applications, we make different types of inquiries. Measurements with elementary particles are a lot simpler, at least in principle. All electrons are intrinsically the same. You have to worry about statistical and systematic error, but not the heterogeneity of a population. The behavior of one electron is representative of them all. But the same notions of statistical and systematic error apply, and scientists try to minimize these whenever feasible. However, the lengths to which they will go to accomplish this depends on the questions they want to answer.

Nonetheless, even in "simple" physics systems, given that measurements won't ever be perfect, we need to decide the accuracy to aim for. At a practical level, this question is equivalent to asking how many times an experimenter should repeat a measurement and how precise he needs his measuring device to be. The answer is up to him. The acceptable level of uncertainty depends on the question he asks. Different goals require different degrees of accuracy and precision.

For example, atomic clocks measure time with stability of one in 10 trillion, but few measurements require such a precise knowledge of time. Tests of Einstein's theory of gravity are an exception—they use as much precision and accuracy as can be attained. Even though all tests so far demonstrate that the theory works, measurements continue to improve.

With higher precision, as-yet-unseen deviations representing new physical effects might appear that were impossible to see with previous less precise measurements. If so, these deviations would give us important insights into new physical phenomena. If not, we would trust that Einstein's theory was even more accurate than had been demonstrated before. We would know we can confidently apply it over a greater regime of energy and distances and with a higher degree of accuracy. If you were sending a man to the Moon, on the other hand, you would want to understand physical laws sufficiently well that you aim your rocket correctly, but you wouldn't need to include general relativity—and you certainly would not need to account for the even smaller potential effects representing possible deviations.

ACCURACY IN PARTICLE PHYSICS

In particle physics, we search for the underlying rules that govern the smallest and most fundamental components of matter we can detect. An individual experiment is not measuring a mishmash of many collisions happening at once or repeatedly interacting over time. The predictions we make apply to single collisions of known particles colliding at a definite energy. Particles enter the collision point, interact, and fly through detectors, usually depositing energy along the way. Physicists characterize particle collisions by the distinctive properties of the particles flying out—their mass, energy, and charges.

In this sense, despite the technical challenges of our experiments, particle physicists have it lucky. We study systems that are as basic as possible so that we can isolate fundamental components and laws. The idea is to make experimental systems that are as clean as existing resources permit. The challenge for physicists is reaching the required physical parameters rather than disentangling complex systems. Experiments are difficult because science has to push the frontiers of knowledge in order to be interesting. They are therefore often at the outer limit of the energies and distances accessible to technology.

In truth, particle physics experiments aren't all that simple, even when studying precise fundamental quantities. Experimenters presenting their results face one of two challenges. If they do see something exotic, they have to be able to prove it cannot be the result of mundane Standard Model events that occasionally resemble some new particle or effect. On the other hand, if they don't see anything new, they have to be certain of their level of accuracy in order to present a more stringent new limit on what can exist beyond known Standard Model effects. They have to understand the sensitivity of the measuring apparatus sufficiently well to know what they can rule out.

To be sure of their result, experimenters have to be able to distinguish those events that can signal new physics from the *background* events that arise from the known physical particles of the Standard Model. This is one reason we need many collisions to make new discoveries. The presence of lots of collisions ensures enough events representing new physics to distinguish them from "boring" Standard Model processes they might resemble.

Experiments therefore require adequate statistics. Measurements themselves have some intrinsic uncertainties necessitating their repetition. Quantum mechanics tells us that the underlying events do too. Quantum mechanics implies that no matter how cleverly we design our technology, we can compute only the probability that interactions occur. This uncertainty exists, no matter how we make a measurement. That means that the only way to accurately measure the strength of an interaction is to repeat the measurement many times. Sometimes this uncertainty is smaller than measurement uncertainty and too small to matter. But sometimes we need to take it into account.

Quantum mechanical uncertainty tells us, for example, that the mass of a particle that decays is an intrinsically uncertain quantity. The principle tells us that no energy measurement can possibly be exact when a measurement takes a finite time. The time of the measurement will necessarily be shorter than the lifetime of the decaying particle, which sets the amount of variation expected for the measured masses. So if experimenters were to find evidence of a new particle by finding the particles it decayed into, measuring its mass would require that they repeat the measurement many times. Even though no single measurement would be exact, the average of all the measurements would nonetheless converge to the correct value.

In many cases, the quantum mechanical mass uncertainty is less than the systematic uncertainties (intrinsic error) of the measuring devices. When that is true, experimenters can ignore the quantum mechanical uncertainty in mass. Even so, a large number of measurements are required to ensure the precision of a measurement due to the probabilistic nature of the interactions involved. As was the case with drug testing, large statistics help get us to the right answer.

It's important to recognize that the probabilities associated with quantum mechanics are not completely random. Probabilities can be calculated from well-defined laws. We'll see this in Chapter 14 in which we discuss the W boson mass. We know the overall shape of the curve describing the likelihood that this particle with a given mass and a given lifetime will emerge from a collision. Each energy measurement centers around the correct value, and the distribution is consistent with the lifetime and the uncertainty principle. Even though no single measurement suffices to determine the mass, many measurements do. A definite procedure tells us how to deduce the mass from the average value of these repeated measurements. Sufficiently many measurements ensure that the experimenters determine the correct mass within a certain level of precision and accuracy.

MEASUREMENTS AND THE LHC

Neither the use of probability to present scientific results nor the probabilities intrinsic to quantum mechanics imply that we don't know anything. In fact, it is often quite the opposite. We know quite a lot. For example, the *magnetic moment of the electron* is an intrinsic property of an electron that we can calculate extremely accurately using *quantum field theory*, which combines together quantum mechanics and special relativity and is the tool used to study the physical properties of elementary particles. My Harvard colleague Gerald Gabrielse has measured the

magnetic moment of the electron with 13 digits of accuracy and precision, and it agrees with the prediction at nearly this level. Uncertainty
enters only at the level of less than one in a trillion and makes the magnetic moment of the electron the constant of nature with the most accurate agreement between theoretical prediction and measurement.

No one outside of physics can make such an accurate prediction
about the world. But most people with such a precise number would say
they definitely know the theory and the phenomena it predicts. Scientists, while able to make much more accurate statements than most anyone else, nonetheless acknowledge that measurements and observations,
no matter how precise, still leave room for as-yet-unseen phenomena and
new ideas.

But they can also state a definite limit to the size of those new phenomena. New hypotheses could change predictions, but only at the level
of the present measurement uncertainty or less. Sometimes the predicted new effects are so small that we have no hope of ever encounter-
ing them in the lifetime of the universe—in which case even scientists
might make a definite statement such as "that won't ever happen."

Clearly Gabrielse's measurement shows that quantum field theory is
correct to a very high degree of precision. Even so, we can't confidently
state that quantum field theory or particle physics or the Standard Model
is all that exists. As explained in Chapter 1, new phenomena whose effects appear only at different energy scales or when we make even more
precise measurements can underlie what we see. Because we haven't yet
experimentally studied those regimes of distance and energy, we don't
yet know.

LHC experiments occur at higher energies than we have ever studied before and therefore open up new possibilities in the form of new
particles or interactions that the experiments search for directly, rather
than through only indirect effects that can be identified only with extremely precise measurements. In all likelihood, LHC measurements
won't reach sufficiently high energy to see deviations from quantum field
theory. But they could conceivably reveal other phenomena that would
predict deviations to Standard Model predictions for measurements at

the level of current precision—even the well-measured magnetic moment of the electron.

For any given model of physics beyond the Standard Model, any predicted small discrepancies—where the inner workings of an as-yet-unseen theory would make a visible difference—would be a big clue as to the underlying nature of reality. The absence of such discrepancies so far tells us the level of precision or how high an energy we need to find something new—even without knowing the precise nature of potential new phenomena.

The real lesson of effective theories, introduced in the opening chapter, is that we only fully understand what we are studying and its limitations at the point where we see them fail. Effective theories that incorporate existing constraints not only categorize our ideas at a given scale, but they also provide systematic methods for determining how big new effects can be at any specific energy.

Measurements concerning the electromagnetic and weak forces agree with Standard Model predictions at the level of 0.1 percent. Particle collision rates, masses, decay rates, and other properties agree with their predicted values at this level of precision and accuracy. The Standard Model therefore leaves room for new discoveries, and new physical theories can yield deviations, but they must be small enough to have eluded detection up to now. The effects of any new phenomena or underlying theory must have been too small to have been seen already—either because the interactions themselves are small or because the effects are associated with particles too heavy to be produced at the energies already probed. Existing measurements tell us how high an energy we require to directly find new particles or new forces, which can't cause bigger deviations to measurements than current uncertainties allow. They also tell us how rare such new events have to be. By increasing measurement precision sufficiently, or doing an experiment under different physical conditions, experimenters search for deviations from a model that has so far described all experimental particle physics results.

Current experiments are based on the understanding that new ideas build upon a successful effective theory that applies at lower energies.

Their goal is to unveil new matter or interactions, keeping in mind that physics builds knowledge scale by scale. By studying phenomena at the LHC's higher energies, we hope to find and fully understand the theory that underlies what we have seen so far. Even before we measure new phenomena, LHC data will give us valuable and stringent constraints on what phenomena or theories beyond the Standard Model can exist. And—if our theoretical considerations are correct—new phenomena should eventually emerge at the higher energies the LHC now studies. Such discoveries would force us to extend or absorb the Standard Model into a more complete formulation. The more comprehensive model would apply with greater accuracy over a larger range of scales.

We don't know which theory will be realized in nature. We also don't know when we will make new discoveries. The answers depend on what is out there, and we don't yet know that or we wouldn't have to look. But for any particular speculation about what exists, we know how to calculate how we might discover the experimental consequences and estimate when it might occur. In the next couple of chapters, we'll look into how LHC experiments work, and in Part IV that follows, we'll consider how physicists make models and predictions for what they might see.

CHAPTER THIRTEEN

THE CMS AND ATLAS EXPERIMENTS

In August 2007, the Spanish physicist and CERN theory group leader Luis Alvarez-Gaume enthusiastically encouraged me to join a tour of the ATLAS experiment that the experimental physicists Peter Jenni and Fabiola Gianotti were planning for the visiting Nobel Prize winner T. D. Lee and a few others. It was impossible to resist the infectious enthusiasm of Peter and Fabiola, who at the time were spokesperson and deputy

[FIGURE 29] Looking down from the platform above into the ATLAS pit, with the tubes that transported materials down in view.

spokesperson of the experiment, and who generously shared an expertise and familiarity with all the details of the experiment that suffused all of their words.

My fellow visitors and I donned our helmets and entered the LHC tunnel. Our first stop was a landing where we could stare down at the gaping pit beneath, as is shown in the photo in Figure 29. Witnessing the gargantuan cavern with its vertical tubes that would transport pieces of the detector from the place where we stood to the floor 100 meters below got me hooked. My fellow ATLAS tourists and I eagerly anticipated the experience we had in store.

After the first stop, we proceeded to the floor down below that housed the not-yet-completed ATLAS detector. The nice thing about the unfinished state was that you could see the detector's innards, which would eventually be closed up and shielded from view—at least until the LHC turns off for an extended period of time for maintenance and repairs. So we had the opportunity to stare directly at the elaborate construction, which was impressively colorful and big—larger even than the nave of the Cathedral of Notre Dame.

But the size was not in itself the most magnificent aspect. Those of us who grew up in New York or any other big city are not necessarily overly impressed by enormous construction projects. What makes the ATLAS experiment so imposing is that this huge detector is composed of many small detection elements—some designed to measure distances with a precision at the level of microns. The irony of the LHC detectors is that you need such big experiments to accurately measure the smallest distances. When I now show an image of the detector in public lectures, I feel compelled to emphasize that ATLAS is not only big, but it is also precise. This is what makes it so amazing.

A year later, in 2008, I returned to CERN and saw the construction progress ATLAS had made. The ends of the detector that had been open the previous year were now closed up. I also took a spectacular tour of CMS, the LHC's second general-purpose detector, along with the physicist Cinzia da Via and my collaborator, Gilad Perez, who appears in Figure 30.

[**FIGURE 30**] My colleague, Gilad Perez, in front of part of the layered CMS muon detector/magnet return yoke.

Gilad hadn't yet visited an LHC experiment, so I had the opportunity to relive my first experience through his excitement. We took advantage of the lax supervision to clamber around and even look down a beam pipe. (See Figure 31.) Gilad noted this could be the place where extra-dimensional particles get created and provide evidence for a theory I had proposed. But whether it will be evidence for this model or some other one, it was nice to be reminded that this beam pipe was where insight into new elements of reality would soon emerge.

Chapter 8 introduced the LHC machine that accelerates protons and collides them together. This chapter focuses on the two general-purpose LHC detectors—CMS and ATLAS—that will identify what comes out of the collisions. The remaining LHC experiments—ALICE, LHCb, TOTEM, ALFA, and LHCf—are designed for more specialized

[**FIGURE 31**] Cinzia da Via (*left*) walking past the location where we could stare down the beam pipe and see inside (*right*).

purposes, including better understanding the strong nuclear force and making precise measurements of bottom quarks. These other experiments will most likely study Standard Model elements in detail, but they are unlikely to discover the new high energy beyond the Standard Model physics that is the LHC's primary goal. CMS and ATLAS are the chief detectors that will make the measurements that will, we hope, reveal new phenomena and matter.

This chapter contains a good amount of technical detail. Even theorists like me don't need to know all these facts. Those of you interested only in the new physics that we might discover or the LHC concepts in general might choose to jump ahead. Still, the LHC experiments are clever and impressive. Omitting these details wouldn't do justice to the enterprise.

GENERAL PRINCIPLES

In some sense, the ATLAS and CMS detectors are the logical evolution of the transformation Galileo and others instigated several centuries ago. Since the invention of the microscope at that time, successively advanced technology has allowed physicists to indirectly study increasingly remote distances. The study of small sizes has repeatedly revealed underlying structure of matter that can only be observed with very tiny probes.

Experiments at the LHC are designed to study substructure and interactions with a range a hundred thousand trillion times smaller than a centimeter. This is about a factor of ten smaller in size than anything any experiment has ever looked at before. Although previous high-energy collider experiments, such as those running at the Tevatron at Fermilab in Batavia, Illinois, were based on similar principles to these LHC detectors, the record energy and collision rate that the new detectors faced posed many novel challenges that forced their unprecedented size and complexity.

Like telescopes in space, the detectors, once built, are essentially inaccessible. They are enclosed deep underground and subject to large amounts of radiation. No one can access the detector while the machine is running. Even when it is not, reaching any particular detector element is extremely difficult and time-consuming. For this reason, the detectors were built to last at least a decade, even with no maintenance. However, long shut-down periods are planned for every two years of LHC running, during which time physicists and engineers will have access to many of the detector components.

In one important respect, however, particle experiments are very different from telescopes. Particle detectors don't need to point in a particular direction. In some sense they look in all directions at once. Collisions happen and particles emerge. The detectors record any event that has the potential to be interesting. ATLAS and CMS are general-purpose detectors. They don't record just one type of particle or event or focus on particular processes. These experimental apparatuses are designed to absorb the data from the broadest possible range of interactions and energies. Experimenters with enormous computational power at their disposal try to unambiguously extricate information about such particles and their decay products from the "pictures" experiments record.

More than 3,000 people from 183 scientific institutes, representing 38 countries, participate in the CMS experiment—building and operating the detector and analyzing the data. The Italian physicist

Guido Tonelli—originally deputy spokesperson—now heads the collaboration.

In a break from CERN's legacy of male physicists presiding, the impressive Italian donna Fabiola Gianotti also transitioned from deputy to spokesperson, this time for ATLAS, the other general-purpose experiment. She is well deserving of the role. She has a mild-mannered, friendly, and polite demeanor—yet her physics and organizational contributions have been tremendous. What makes me really jealous, however, is that she is also an excellent chef—maybe forgivable for an Italian with enormous attention to detail.

ATLAS too involves a gigantic collaboration. More than 3,000 scientists from 174 institutes in 38 countries participated in the ATLAS experiment (December 2009). The collaboration was initially formed in 1992 when two proposed experiments—EAGLE (Experiment for Accurate Gamma, Lepton, and Energy Measurements) and ASCOT (Apparatus with Super Conducting Toroids) joined together with a design combining features of both with some aspects of proposed SSC detectors. The final proposal was presented in 1994, and it was funded two years later.

The two experiments are similar in basic outline, but different in their detailed configurations and implementations, as is illustrated in some detail in Figure 32. This complementarity gives each experiment slightly different strengths so that physicists can cross-check the two experiments' results. With the extreme challenges involved in particle physics discoveries, two experiments with common search targets will have much more credibility when they confirm the findings of each other. If they both come to the same conclusion, everyone will be much more confident.

The presence of two experiments also introduces a strong element of competition—something my experimenter colleagues frequently remind me about. The competition pushes them to get results more quickly and more thoroughly. The members of the two experiments also learn from each other. A good idea will find its way to both experiments, even if im-

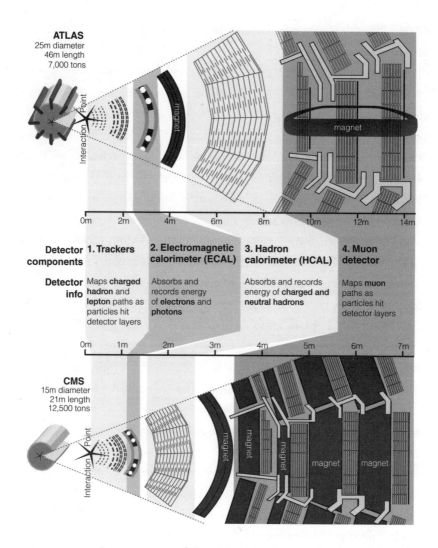

ATLAS
25m diameter
46m length
7,000 tons

Detector components	1. Trackers	2. Electromagnetic calorimeter (ECAL)	3. Hadron calorimeter (HCAL)	4. Muon detector
Detector info	Maps **charged hadron** and **lepton** paths as particles hit detector layers	Absorbs and records energy of **electrons** and **photons**	Absorbs and records energy of **charged and neutral hadrons**	Maps **muon** paths as particles hit detector layers

CMS
15m diameter
21m length
12,500 tons

[**FIGURE 32**] Cross sections of the ATLAS and CMS detectors. Note the overall sizes have been rescaled.

plemented somewhat differently in each. This competition and collaboration, coupled with the redundancy of having two independent searches relying on somewhat different configurations and technology, underlies the decision to have two experiments with common goals.

I am often asked when the LHC will run my experiments and search for the particular models that my collaborators and I have proposed. The answer is right away—but they are looking for everyone else's proposals too. Theorists help by introducing new search targets and new strategies for finding stuff. Our research aims to identify ways to find whatever new physical elements or forces are present at higher energies, so that physicists will be able to find, measure, and interpret the results and thereby gain new insights into underlying reality—whatever it might be. Only after data is recorded do the thousands of experimenters, who are split up into analysis teams, study whether the information fits or rules out my models or any others that are potentially interesting.

Theorists and experimenters then examine the data that gets recorded to see whether they conform to any particular type of hypothesis. Even though many particles last only a fraction of a second and even though we don't witness them directly, experimental physicists use the digital data that compose these "pictures" to establish which particles form the core of matter and how they interact. Given the complexity of the detectors and data, experimenters will have a lot of information to contend with. The rest of this chapter gives a sense of what, exactly, that information will be.

THE ATLAS AND CMS DETECTORS

So far we have followed LHC protons from their removal from hydrogen atoms to their acceleration to high energy in the 27 km ring. Two completely parallel beams will never intersect, and neither will the two beams of protons traveling in opposite directions within them. So at several locations along the ring, dipole magnets divert them from their path while quadrupole magnets focus them so that the protons in the two beams meet and interact within a region less than 30 microns across. The points at the center of each detector where proton-proton collisions occur are known as the interaction points.

Experiments are set up concentrically around each of these interaction points to absorb and record the many particles that are emitted by the frequent proton collisions. (See Figure 33 for a graphic of the CMS detector.) The detectors are cylindrically shaped because even though

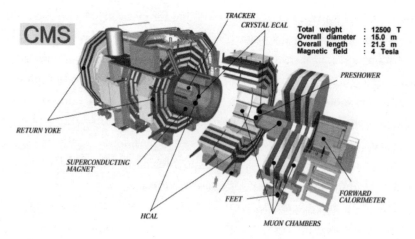

[**FIGURE 33**] Computer image of CMS broken up to reveal individual detector components. (Graphic courtesy of CERN and CMS)

the proton beams travel in opposite directions at the same speed, the collisions tend to contain a lot of forward motion in both directions. In fact, because individual protons are much smaller than the beam size, most of the protons don't collide at all but continue straight down the beam pipe with only mild deflection. Only the rare event where individual protons collide head-on are of interest.

That means that although most particles continue to travel along the beam direction, the potentially interesting events contain a spray of particles that travel significantly transversely to the beam. The cylindrical detectors are designed to detect as much of these interaction products as possible, taking into account the large spread of particles along the beam direction. The CMS detector is located around one proton collision point below ground at Cessy in France, close to the Geneva border, while the ATLAS interaction region is under the Swiss town of Meyrin, very near the main CERN complex. (See Figure 34 for a simulation of particles coming out of a collision and emanating through a cross section of the ATLAS detector.)

Standard Model particles are characterized by their mass, spin, and the forces through which they interact. No matter what is ultimately created, both experiments rely on detecting it through known Standard

Model forces and interactions. That's all that's possible. Particles with no such charges would leave the interaction region without a trace.

But when experiments measure Standard Model interactions, they can identify what passed through. So that's what the detectors are designed to do. Both CMS and ATLAS measure the energy and momentum of photons, electrons, muons, taus, and strongly interacting particles, which get subsumed into jets of closely aligned particles traveling in the same direction. Detectors emanating from the proton collision region are designed to measure energy or charge in order to identify particles, and they contain sophisticated computer hardware, software, and electronics to deal with the overwhelming abundance of data. Experimenters identify charged particles since they interact with other charged stuff that we know how to find. They also find anything that interacts via the strong force.

The detector components all ultimately rely on wires and electrons

[FIGURE 34] Simulation of an event in the ATLAS detector showing the transverse spray of particles though the detector layers. (Note that the person gives a sense of scale, but collisions don't happen when people are in the cavern.) The distinctive toroidal magnets are clearly visible. (Courtesy of CERN and ATLAS)

produced through interactions with the material in the detector to record what passed through. Sometimes charged particle showers occur because many electrons and photons are produced and sometimes material is simply ionized with charges recorded. But either way wires record the signal and send it along for it to be processed and analyzed by physicists at their computers.

Magnets are also critical to both detectors. They are essential to measuring both the sign of the charges and the momenta of charged particles. Electromagnetically charged particles bend in a magnetic field according to how fast they are moving. Particles with bigger momenta tend to go straighter, and particles with opposite charges bend in opposite directions. Because particles at the LHC have such large energies (and momenta), the experiments need very strong magnets to have a chance of measuring the small curvature of the energetic charged particle tracks.

The Compact Muon Solenoid (CMS) apparatus is the smaller in size of the two large general-purpose detectors, but it is heavier, weighing in at a whopping 12,500 metric tons. Its "compact" size is 21 meters long by 15 meters in diameter—smaller than ATLAS but still big enough to cover the area of a tennis court.

The distinguishing element in CMS is its strong magnetic field of 4 tesla, which the "solenoid" piece of the name refers to. The solenoid in the inner part of the detector consists of a cylindrical coil six meters in diameter made up of superconducting cable. The magnetic return yoke that runs through the outer part of the detector is also impressive and contributes most of the huge weight. It contains more iron than Paris's Eiffel Tower.

You might also wonder about the word "muon" in the name CMS (I did too when I first heard it). Rapidly identifying energetic electrons and muons, which are heavier counterparts of electrons that penetrate to the outer reaches of the detector, can be important for new particle detection—since these energetic particles are sometimes produced when heavy objects decay. Since they don't interact via the strong nuclear force, they are more likely to be something new—since protons won't automatically make them. These readily identifiable particles could therefore indicate the presence of an interesting decaying particle that has emerged from the collision. The magnetic field in CMS was initially designed with special attention paid to energetic muons so that it could trigger on them. This means it will record the data from any event involving them, even when it is forced to throw a lot of other data out.

ATLAS, like CMS, features its magnet in its name since a big magnetic field is also critical to its operation. As noted earlier, ATLAS is the acronym for A Toroidal LHC ApparatuS. The word "toroid" refers to the magnets, whose field is less strong than that of CMS but extends over an enormous region. The huge magnetic toroids help make ATLAS the larger of the two general-purpose detectors and in fact the largest experimental apparatus ever constructed. It is 46 meters long and 25 meters in diameter and fits rather snugly into its 55-meter-long, 40-meter-high cavern. At 7,000 metric tons, ATLAS is a little more than half the weight of CMS.

To measure all the particle properties, increasingly large cylindrical detector components emanate from the region where collisions occur. The CMS and ATLAS detectors both contain several embedded pieces designed to measure the trajectory and charges of the particles as they pass through. Particles emerging from the collision first encounter the *inner trackers* that precisely measure the paths of charged particles close to the interaction point, next the *calorimeters* that measure energy deposited by readily stopped particles, and finally the *muon detectors* that are at the outer edges and measure the energy of highly penetrating muons. Each of these detector elements has multiple layers to increase the precision for each measurement. We'll now tour the experiments from the innermost detectors to the outermost as measured radially from the beams and explain how the spray of particles leaving a collision turns into recorded identifiable information.

THE TRACKERS

The innermost portions of the apparatuses are the trackers that record the positions of charged particles as they leave the interaction region so that their paths can be reconstructed and their momenta measured. In both ATLAS and CMS, the tracker consists of several concentric components.

The layers closest to the beams and interaction points are the most finely segmented and generate the most data. Silicon *pixels*, with extremely tiny detector elements, sit in this innermost region, starting at a few centimeters from the beam pipe. They are designed for extremely precise tracking very close to the interaction point where the particle density is highest. Silicon is used in modern electronics because of the fine detail that can be etched into each tiny piece, and particle detectors use it for the same reason. Pixel elements at ATLAS and CMS are designed to detect charged particles with extremely high resolution. By connecting the dots to one another and to the interaction points from which they emerged, experimenters find the paths the particles followed in the innermost region very near to the beam.

The first three layers of the CMS detector—out to 11 centimeter radius—consist of 100 by 150 micrometer pixels, 66 million in total. ATLAS's inner pixel detector is similarly precise. The smallest unit that can be read out in the ATLAS innermost detector is a pixel of size 50 by 400 micrometers. The total number of ATLAS pixels is about 82 million, a little more than the number in CMS.

The pixel detectors, with their tens of millions of elements, require elaborate electronic readouts. The extent and speed required for the readout systems, as well as the huge radiation the inner detectors will be subjected to, were two of the major challenges for both of the detectors. (See Figure 35.)

Because there are three layers in these inner trackers, they record three *hits* for any long-lasting enough charged particle that passes through. These tracks will generally continue to an outer tracker beyond the pixel layers to create a robust signal that can be definitively associated with a particle.

My collaborator Matthew Buckley and I paid a good deal of attention to the geometry of the inner trackers. We realized that by sheer coincidence, some conjectured new charged particles that decay via the weak force into a neutral partner would leave a track that's only a few

[FIGURE 35] Cinzia da Via and an engineer, Domenico Dattola, stand-ing on scaffolding in front of one of the bulkheads of the CMS silicon tracker, to which the cables are connected.

centimeters long. That means that in these special cases, tracks might extend *only* through the inner tracker so that the information read out here would be all there is. We considered the additional challenges faced by experimenters who had only the pixels—the innermost layers of the inner detector—to rely on.

Most charged particles, however, live long enough to make it to the next tracker component, so detectors record a much greater length path. Therefore, outside the inner pixel detectors with fine resolution in two directions are silicon strips with asymmetric size in the two directions, much coarser in one of the two. The longer strips are consistent with the cylindrical shape of the experiment and make covering a larger area (remember the area gets far bigger with bigger radius) feasible.

The CMS silicon tracker consists of a total of 13 layers in the cen-tral region and 14 layers in the forward and backward regions. After the first three finely pixilated layers we just described, the next four layers, consisting of silicon strips, extend to 55 centimeters radius. The detector elements here are 10-centimeter-long, 180-micrometer-wide strips. The remaining six layers are even less precise in the coarser orientation, consisting of strips up to 20 centimeters long and varying in width between 80 and 205 micrometers, with the strips extending out to a radius of 1.1 meters. The total number of strips in the CMS inner detector is 9.6 million. These strips are essential to reconstructing the tracks of most charged particles that pass through. In total, CMS has silicon covering essentially the area of a tennis court—a significant advance over the previous largest silicon detector of only two square meters.

The ATLAS inner detector extends to a slightly smaller radius of one meter and is seven meters long longitudinally. As with CMS, outside the three inner silicon pixel layers, the Semiconductor Tracker (SCT) consists of four layers of silicon strips. In ATLAS's case, they are 12.6 centimeters by 80 micrometers in size. The total area of the SCT is also enormous, covering 61 square meters. Whereas the pixel detectors are useful for reconstructing fine measurements near the interaction points, the SCT is most critical to overall track reconstruction because of the large region it covers with high precision (albeit in one direction).

Unlike CMS, the outer detector of the ATLAS apparatus is not made of silicon. The transition radiation tracker (TRT), the outermost component of the inner detector, consists of tubes filled with gas and acts as both a tracking device and a transition radiation detector. Charged particle tracks are measured when they ionize the gas in the straws, which are 144 centimeters by 4 millimeters in size, with wires down the center to detect the ionization. Here again there is highest resolution in the transverse direction. The straws measure the tracks with a precision of 200 micrometers, which is less precise than with the innermost tracker but covers a far greater region. The detectors also discriminate among particles moving very close to the speed of light that produce so-called *transition radiation*. This discriminates among particles of different mass, since lighter particles will generally be moving faster. This helps identify electrons.

If you're finding all these details a bit overwhelming, keep in mind that this is more information than even most physicists need to know.

They give a sense of the magnitude and precision, and are of course important to anyone working on a particular detector component. But even those who have extreme familiarity with one component don't necessarily keep track of all the others, as I accidentally learned when trying to track down some detector photos and make sure some diagrams were precise. So don't feel too badly if you don't get it all the first time. Though some experts coordinate the overall operation, even many experimenters don't necessarily have every detail at their fingertips.

THE ELECTROMAGNETIC CALORIMETER (ECAL)

Once through the three types of trackers, the next section of detector a particle encounters on its outward radial journey is the electromagnetic calorimeter (ECAL), which records the energy deposited by charged and neutral particles that stop there—electrons and photons in particular—and the position where they left it. The detection mechanism looks for the spray of particles that incident electrons or photons produce when they interact with the detector material. This piece of the detector yields both precise energy and position tracking information for these particles.

The material used for the ECAL in the CMS experiment is a wonder to behold. It is made of lead tungstate crystals, chosen because they are dense but optically clear—exactly what you want for stopping and detecting electrons and photons as they arrive. You can perhaps get a sense of this from my photograph in Figure 36. The reason they are fascinating is their incredible clarity. You've never seen anything this dense and this transparent. The reason they are useful is that they measure electromagnetic energy incredibly precisely, which could turn out to be critical to finding the elusive Higgs particle as Chapter 16 will describe.

The ATLAS detector uses lead to stop electrons and photons. Interactions in this absorbing material transform the energy from the initial charged track into a shower of particles whose energy will then be detected. Liquid argon, which is a noble gas that doesn't chemically interact with other elements and is very resistant to radiation, is then

[**FIGURE 36**] Photograph of the lead tungstate crystal that is used in CMS's electromagnetic calorimeter.

used to sample the energy of the shower to deduce the incident particle energy.

Despite my theoretical inclinations, I was fascinated to see this detector element at ATLAS on my tour. Fabiola participated in the pioneering development and construction of this calorimeter's novel geometry with radial layers of accordion-shaped lead plates separated by thin layers

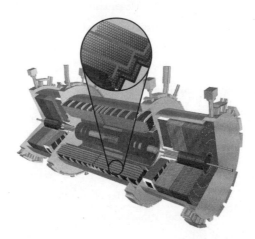

[**FIGURE 37**] The accordion-like structure of ATLAS's electromagnetic calorimeter.

of liquid argon and electrodes. She described how this geometry makes readout of the electronics much faster, since the electronics is much closer to the detector elements. (See Figure 37.)

THE HADRONIC CALORIMETER (HCAL)

Next in line along our radial outward journey from the beam pipe is the hadronic calorimeter (HCAL). The HCAL measures the energy and positions of hadronic particles—those particles that interact through the strong force—though it does so less precisely than the electron and photon energy measurements made by the ECAL. That's by necessity. The HCAL is huge. In ATLAS, for example, the HCAL is eight meters in diameter and 12 meters long. It would be prohibitively expensive to segment the HCAL with the precision of the ECAL, so the precision of the track measurement is necessarily degraded. On top of that, energy measurements are simply harder for strongly interacting particles, independent of segmentation, since the energy in hadronic showers fluctuates more.

The HCAL in CMS contains layers of dense material—brass or steel—alternating with plastic scintillator tiles that record the energy and position of the hadrons that pass through, based on the intensity of the scintillating light. The absorber material in the central region of ATLAS is iron, but the HCAL there works pretty much the same way.

MUON DETECTOR

The outermost elements in any general-purpose detector are the muon chambers. Muons, you will remember, are charged particles like electrons, but they are 200 times heavier. They don't stop in the electromagnetic or hadronic calorimeters but instead barrel straight through the thick outer region of the detector. (See Figure 38.)

Energetic muons are very useful when looking for new particles because, unlike hadrons, they are sufficiently isolated that they are relatively clean to detect and measure. Experimenters want to record all events with energetic muons in the transverse direction because muons

[FIGURE 38] CMS's magnetic return coil interlaced with its muon detector—all under construction.

are likely to be associated with the more interesting collisions. Muon detectors could also prove useful for any heavy stable charged particle that makes it to the outer reaches of the detector.

Muon chambers record the signals left by the muons that reach these outermost detectors. They are similar in some respects to the inner detector with its trackers and magnetic fields bending the muon tracks so their trajectories and momenta can be measured. However, in the muon chambers, the magnetic field is different, and the thickness of the detector is much bigger, permitting measurements of smaller curvatures and hence higher-momentum particles (high-momentum particles bend less in a magnetic field). In CMS, the muon chambers extend from about three meters to the outer radius of the detector at about 7.5 meters, while in ATLAS they extend from four meters to the outer reaches of that detector at 11 meters. These huge structures permit 50-micrometer particle track measurements.

ENDCAPS

The last detector elements to describe are the endcaps, the detectors at the forward and backward ends of the experiments. (See Figure 39 to get a sense of the overall structure.) We are no longer working our way radially outward from the beam—the muon detectors were the last step in that direction—but rather we now are proceeding along the axis of the cylindrical detectors to the two ends that cap them off. The cylindrical portions of the detectors are "capped" off there with detectors covering the end regions that ensure that as many particles as possible get recorded. Since the endcaps were the last components of the detector to be moved to their final positions, I could readily see the multiple layers that sit inside the detectors when I visited in 2009.

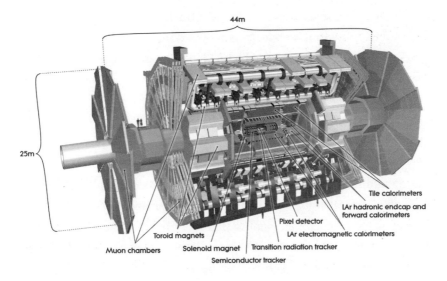

[**FIGURE 39**] Computer image of ATLAS showing its many layers and the endcaps separated. (Courtesy of CERN and ATLAS)

Detectors are placed in these end regions to ensure that LHC experiments measure all the particles' momenta. The goal is to make the experimental apparatuses *hermetic*, meaning there is coverage in all directions with no holes or missing regions. Hermetic measurements ensure that even noninteracting or very weakly interacting particles can be discovered. If "missing" transverse momentum is observed, one or more particles with no directly detectable interactions must have been produced. Such particles carry momentum, and the momentum they take away makes experimenters aware of their existence.

If you know the detector is measuring all the transverse momentum, and the momentum perpendicular to the beam doesn't appear to be conserved after a collision, then something must have disappeared undetected and carried away momentum. Detectors, as we have seen, measure momentum in the perpendicular directions very carefully. The calorimeters in the forward and backward regions ensure hermeticity by guaranteeing that very little energy or momentum perpendicular to the beam can escape unnoticed.

The CMS apparatus has steel absorbers and quartz fibers in the end regions, which separate the particle tracks better because they are denser. The brass in the endcaps is recycled material—it was originally used in Russian artillery shells. The ATLAS apparatus uses liquid-argon calorimeters in the forward region to detect not only electrons and photons but also hadrons.

MAGNETS

The remaining pieces of both detectors that remain to be described in more detail are the magnets that give both experiments their names. A magnet is not a detector element in that it doesn't record particle properties. But magnets are essential to particle detection because they help determine momentum and charge, properties that are critical to identifying and characterizing particle tracks. Particles bend in magnetic fields, so their tracks appear to be curved rather than straight. How much and in which direction they bend depends on their energies and charges.

CMS's enormous solenoidal magnet made of refrigerated superconducting niobium-titanium coils is 12.5 meters long and six meters in diameter. This magnet is the defining feature of the detector and is the largest magnet of its type ever made. The solenoid has coils of wire surrounding a metal core, generating a magnetic field when electricity is applied. The energy stored in this magnet is the same as that generated by a half-metric ton of TNT. Needless to say, precautions have been taken in case the magnet quenches and suddenly loses superconductivity. The solenoid's successful 4-tesla test was completed in September 2006, but it will be run at a slightly lower field—3.8 tesla—to ensure greater longevity.

The solenoid is sufficiently big to enclose the tracking and calorimeter layers. The muon detectors, on the other hand, are on the outer perimeter of the detector, outside the solenoid. However, the four layers of muon detector are interlaced with a huge iron structure surrounding the magnetic coils that contains and guides the field, ensuring uniformity and stability. This magnetic return yoke, 21 meters long and 14 meters in diameter, reaches to the full seven-meter radius of the detector. In effect, it also forms part of the muon system since the muons should be the only known charged particles to penetrate the 10,000 metric tons of iron and cross the muon chambers (though in reality energetic hadrons will sometimes also get in, creating some headaches for the experimenters). The magnetic field from the yoke bends the muons in the outer detector. Since the amount muons bend in the field depends on their momenta, the yoke is vital to measuring muons' momenta and energy. The structurally stable enormous magnet plays another role as well. It supports the experiment and protects it from the giant forces exerted by its own magnetic field.

The ATLAS magnet configuration is entirely different. In ATLAS, two different systems of magnets are used: a 2-tesla solenoid enclosing the tracking systems and huge toroidal magnets in the outer regions interleaved with the muon chambers. When you look at pictures of ATLAS (or the experiment itself), the most notable elements are these eight huge toroidal structures (seen in Figure 34) and the two additional toroids that cap the ends. The magnetic field they create stretches 26 meters along the beam axis and extends from the start of the muon spectrometer 11 meters in the radial direction.

Among the many interesting stories I heard when visiting the ATLAS experiment was how when the magnets were originally lowered by the construction crews, they started off in a more oval configuration (when viewed from the side). The engineers had factored in gravity before installing them so they correctly anticipated that after some time, due to their own weight, the magnets would become more round.

Another story that impressed me was about how ATLAS engineers factored in a slight rise of the cavern floor of about one millimeter per year caused by the hydrostatic pressure from the cavern excavation. They designed the experiment so that the small motion would put the machine in optimal position in 2010, when the initial plan was to have the first run at full capacity. With the LHC delays, that hasn't been the case. But by now, the ground under the experiment has settled to the point that the experiment has stopped moving, so it will remain in the correct position throughout operation. Despite Yogi Berra's admonition that it's "tough to make predictions, especially about the future,"[52] the ATLAS engineers got it right.

COMPUTATION

No description of the LHC is complete without describing its enormous computational power. In addition to the remarkable hardware that goes into the trackers, calorimeters, muon systems, and magnets we just considered, coordinated computation around the world is essential to dealing with the overwhelming amount of data the many collisions will generate.

Not only is the LHC seven times higher in energy than the Tevatron—the highest-energy collider before—but it also generates events at a rate 50 times faster. The LHC needs to handle what are essentially extremely high resolution pictures of events that are happening at a rate of up to about a billion collisions per second. The "picture" of each event contains about a megabyte of information.

This would be way too much data for any computing system to deal with. So trigger systems make decisions on the fly about which data to keep and which to throw away. By far the most frequent collisions are just ordinary proton interactions that occur via the strong force. No one cares about most of these collisions, which represent known physical processes but nothing new.

The collisions of protons are analogous in some respects to two bean-bags colliding. Because beanbags are soft, most of the time they wilt and hang and don't do anything interesting during the collision. But occasionally when beanbags bang together, individual beans hit each other with great force—maybe even so much so that individual beans collide and the bags themselves break. In that case, individual colliding beans will fly off dramatically since they are hard and collide with more local-ized energy, while the rest of the beans will fly along in the direction in which they started.

Similarly, when protons in the beam hit each other, the individual subunits collide and create the interesting event, whereas the rest of the ingredients of the proton just continue in the same direction down the beampipe.

However unlike bean collisions, in which the beans simply collide and change directions, when protons bang into each other, the ingre-dients inside—quarks, antiquarks, and gluons—collide together—and when they do the original particles can convert into energy or other types of matter. And, whereas at lower energies, collisions involve pri-marily the three quarks that carry the proton charge, at higher energies virtual effects due to quantum mechanics create significant gluon and antiquark content, as we saw earlier in Chapter 6. The interesting colli-sions are those in which any of these subcomponents of the protons hit each other.

When the protons have high energy, so do the quarks, antiquarks, and gluons inside them. Nonetheless, that energy is never the entire en-ergy of the proton. In general, it is a mere fraction of the total. So more often than not, quarks and gluons collide with too small a fraction of the proton's energy to make heavy particles. Due possibly to a smaller interaction strength or to the heavier mass expected for new particles, interesting collisions involving as-yet-unseen particles or forces occur at a much lower rate than "boring" Standard Model collisions.

As with the beanbags, most of the collisions therefore are uninterest-ing. They involve either protons just glancing off each other or protons colliding to produce Standard Model events that we already know should be there and that won't teach us much. On the other hand, predictions tell us that roughly one-billionth as often as that the LHC might produce a new exciting particle such as the Higgs boson.

The upshot is that only in a small but lucky fraction of the time does the good stuff get made. That's why we need so many collisions in the first place. Most of the events are nothing new. But a few rare events could be very special and informative.

It's up to the *triggers*—the hardware and software designed to iden-tify potentially interesting events—to ferret these out. One way to under-stand the enormity of this task (once you account for different possible channels) is as if you had a 150-megapixel (the amount of data from each bunch crossing) camera that can snap pictures at a rate of 40 million per second (the bunch crossing rate). This amounts to about a billion physics events per second, when you account for the 20 to 25 events expected to occur during each bunch crossing. The trigger would be the analog of the device responsible for keeping only the few interesting pictures. You might also think of the triggers as spam filters. Their job is to make sure that only interesting data make it to the experimenters' computers.

The triggers need to identify the potentially interesting collisions and discard the ones that won't contain anything new. The events themselves—what leaves the interaction point and gets recorded in the detectors—must be sufficiently distinguishable from usual Standard Model processes. Knowing when the events look special tells us which events to keep. This makes the rate for readily recognizable new events even smaller still. The triggers have a formidable task. They are responsi-ble for winnowing down the billion events per second to the few hundred that have a chance of being interesting.

A combination of hardware and software "gates" accomplishes this mission. Each successive trigger level rejects most of the events it re-ceives as uninteresting, leaving a far more manageable amount of data.

These data in turn get analyzed by the computer systems at 160 academic institutions around the globe.

The first-level trigger is hardware based—built into the detectors—and does a gross pass at identifying distinctive features, such as selecting events containing energetic muons or large transverse energy depositions in the calorimeters. While waiting a few microseconds for the result of the level-one trigger, the data from each bunch crossing are held in buffer. The higher-level triggers are software based. The selection algorithms run on a large computer cluster near the detector. The first-level trigger reduces the billion per second event rate to about 100,000 events per second, which the software triggers further reduced by a factor of about a thousand to a few hundred events.

Each event that passes the trigger carries a huge amount of information—the readouts of the detector elements we just discussed—of more than a megabyte. With a few hundred events per second, the experiments keep well over 100 megabytes of disk space per second, which amounts to over a petabyte, which is 10^{15} bytes, or one quadrillion bytes (how often do you get to use that word?), the equivalent of hundreds of thousands of DVDs worth of information, each year.

Tim Berners-Lee first developed the World Wide Web to deal with CERN data and let experimenters around the world share information on a computer in real time. The LHC Computing Grid is CERN's next major computational advance. The Grid was launched late in 2008—after extensive software development—to help handle the enormous amounts of data that the experimenters intend to process. The CERN Grid uses both private fiber-optic cables and high-speed portions of the public Internet. It is so named because data aren't associated with any single location but are instead distributed in computers around the world—much as the electricity in an urban area isn't associated with one particular power plant.

Once the trigger-happy events that made it through are stored, they are distributed via the Grid all over the globe. With the Grid, computer networks all over the globe have ready access to the redundantly stored data. Whereas the web shares information, the Grid shares computational power and data storage among the many participating computers.

With the Grid, tiered computing centers process the data. Tier 0 is CERN's central facility where the data get recorded and reprocessed from their raw form to one more suitable for physics analyses. High-bandwidth connections send the data to the dozen large national computing centers constituting Tier 1. Analysis groups can access these data if they choose to do so. Fiber-optic cables connect Tier 1 to the roughly 50 Tier 2 analysis centers located at universities, which have enough computing power to simulate physics processes and do some specific analyses. Finally, any university group can do Tier 3 analyses, where most of the real physics will ultimately be extracted.

At this point, experimenters anywhere can go through their data to sleuth out what the high-energy proton collisions might reveal. This can be something new and exciting. But in order to establish whether or not this is the case, the first task for the experiments—which we'll explore further in the following chapter—is deducing what was there.

CHAPTER FOURTEEN

IDENTIFYING PARTICLES

The Standard Model of particle physics, compactly categorizes our current understanding of elementary particles and their interactions (summarized in Figure 40).[53] It includes particles like the up and down quarks and the electrons that sit at the core of familiar matter, but it also accommodates a number of other heavier particles that interact through the same forces, but which are not commonly found in nature—particles that we can study carefully only at high-energy collider experiments. Most of the Standard Model's ingredients, such as the particles the LHC is currently studying, were rather thoroughly buried until the clever experimental and theoretical insights that revealed them in the latter half of the twentieth century.

At the LHC, the ATLAS and CMS experiments are designed to detect and identify Standard Model particles. The real goal, of course, is to go beyond what we already know—to find new ingredients or forces that address outstanding mysteries. But to do so, physicists need to be able to distinguish Standard Model background events and identify the Standard Model particles into which any exotic new particles might decay. Experimenters at the LHC are like detectives who analyze data to piece together clues and ascertain what was there. They will be able to deduce the existence of something new only after they have ruled out everything that is familiar.

Having toured the general-purpose experiments, we will now revisit

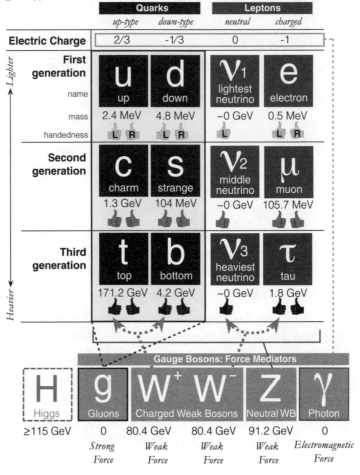

[**FIGURE 40**] The elements of the Standard Model of particle physics, with masses shown. Also shown are separate left- and right-handed particles. The weak force that changes particle type acts only on the left-handed ones.

them in this chapter to better understand how LHC physicists identify individual particles. A bit more familiarity with the particle physics status quo and how Standard Model particles are found will help when we discuss the discovery potential of the LHC in Part IV.

FINDING LEPTONS

Particle physicists divide the elementary matter particles of the Standard Model into two categories. One type is called leptons, which includes particles such as the electron that don't experience the strong nuclear force. The Standard Model also includes two heavier versions of the electron, which have the same charge but much bigger masses, and which are called the *muon* and the *tau*. It turns out that every Standard Model matter particle has three versions, all with the same charge but with each successive *generation* heavier than the next. We don't know why there should be three versions of these particles, all with the same charges. The Nobel Prize–winning physicist Isidor Isaac Rabi, on hearing of the muon's existence, notably expressed his bafflement with the exclamation, "Who ordered that?"

The lighter leptons are the easiest to find. Although both electrons and photons deposit energy in the electromagnetic calorimeter, because the electron is charged and the photon is not, the electron is readily distinguished from a photon. Only an electron leaves a a track in the inner detector before depositing energy in the ECAL.

Muons too are relatively straightforward to identify. Like all the other heavier Standard Model particles, muons decay so quickly that they aren't found in ordinary matter, so we rarely find them on Earth. However, muons live long enough to travel to the outer reaches of the detectors before they decay. They therefore leave long clearly visible tracks throughout that experimenters can match up from the inner detector to the outer muon chambers. Because muons are the only Standard Model particle to reach these outer detectors and leave a visible signal, they are easy to pick out.

Though visible, taus are not quite so simple to find. The tau is a charged lepton like the electrons and the muon, but it is even heavier. Like most heavy particles, it too is unstable, which is to say it decays— leaving only other particles in its wake. A tau rapidly decays into a lighter charged lepton and two particles called neutrinos or into a single neutrino along with a particle called a pion that experiences the strong force. Experimenters study these decay products—the particles the initial particle decayed into—to figure out whether a heavy decaying particle was responsible for their presence and if so, what its properties are. Even though the tau doesn't directly leave a track, all the information the experiments record about the decay products helps identify it and its properties.

· The electron, muon, and the even heavier tau lepton have charge −1, the opposite charge of a positively charged proton. Colliders also produce the antiparticles associated with these charged leptons—the positron, antimuon, and antitau. These antiparticles carry charge +1, and leave similar-looking tracks in the detectors. However, because of their opposite charges, they curve in the opposite direction in the presence of a magnetic field.

In addition to the three types of charged leptons just described, the Standard Model also includes neutrinos, which are leptons that don't carry electric charge at all. Whereas the three charged leptons experience both the force of electromagnetism and the weak nuclear force, neutrinos have zero charge and are therefore impervious to the electric force. Until the 1990s, experimental results indicated that neutrinos had zero mass. One of the most interesting discoveries in that decade was the extremely tiny but nonvanishing masses of neutrinos, which provided important information about the structure of the Standard Model.

Although neutrinos are very light and therefore well within the energy reach of colliders, they are impossible to directly detect at the LHC because they have no electric charge and therefore interact only weakly—so weakly that although more than 50 trillion neutrinos from the Sun pass through you every second, you really have no idea until someone tells you.

In spite of their invisibility, the physicist Wolfgang Pauli conjectured neutrinos existed as a "desperate way out" to explain where the energy went when neutrons decay. Without the neutrino carrying off some of the energy, it appeared that energy conservation was violated by this process, since the proton and the electron that were detected after the decay didn't add up to the same energy as the neutron that went in. Even well-established physicists such as Niels Bohr were willing at the time to sacrifice their principles and accept that energy could be lost. Pauli was more faithful to known physical premises and conjectured instead that energy is indeed conserved, but experimenters just couldn't see the charge-neutral particle that carried the remaining energy off. He turned out to be right.

Pauli named his then-hypothetical particle the neutron, but the name has since been used for other purposes—namely, the neutral partner of the proton that sits inside a nucleus. So Enrico Fermi, the Italian physicist who developed the theory of the weak interactions but is perhaps best known for helping develop the first nuclear reactor, gave it the cutish name neutrino, which in Italian means "little neutron." It's of course not a little neutron, but—like a neutron—it carries no electric charge. And a neutrino is indeed much lighter than a neutron.

As with all the other types of Standard Model particles, three types of neutrinos exist. Each charged lepton—the electron, muon, and tau— has an associated neutrino that it interacts with via the weak nuclear force.[54]

We have already seen how to find electrons, muons, and taus. So the remaining experimental question about leptons is how experimenters find neutrinos. Because neutrinos have no electric charge and interact so weakly, they escape the detector without leaving any trace at all. How does anyone at the LHC tell they were there?

Momentum (which is velocity times mass when particles move slowly but is more like energy moving in a particular direction when the particle travels near the speed of light) is conserved in all directions. As with energy, we have never found any evidence that momentum can be lost. So if the momentum of the particles measured in the detector is less than the momentum that went in, some other particle (or particles) must have escaped, carrying away the missing momentum in the process. This type of logic led Pauli to deduce the existence of neutrinos in the first place (in his case in nuclear beta decay), and to this day it's how we learn of the existence of weakly interacting particles that seem to be invisible.

At hadron colliders, experimenters measure all the momentum transverse to the beam and calculate if something is missing. They focus on momentum transverse to the beam since a lot of momentum is carried away by particles that head down the beam pipe and is therefore too difficult to keep track of. The momentum perpendicular to the initial protons is much simpler to measure and account for.

Since the initial collision has essentially zero total momentum transverse to the beam, so too should the final state. So if measurements don't agree with expectations, experimenters can "detect" that something is missing. The only remaining question is how to distinguish which of the many potential noninteracting particles it was. For Standard Model processes, we know neutrinos will be among the undetected elements. Based on the neutrino's known weak force interactions that we will get to shortly, physicists calculate and predict the rate at which neutrinos should be produced. In addition, physicists already know what the decay of a W boson should look like—for example, an isolated electron or muon whose transverse momentum carries energy comparable to half the W mass is fairly unique. So using momentum conservation and theoretical input, neutrinos can be "found." Clearly, there are fewer identifying tags on these particles than ones we see directly. Only a combination of theoretical considerations and missing energy measurements can tell us what was there.

It's important to keep such ideas in mind when we consider new discoveries. Similar considerations apply for other novel particles without any charges, or with charges so weak that they can't be directly detected. Only a combination of missing energy and theoretical input can be used in those cases to deduce what was there. That's why hermeticity—detecting as much momentum as possible—is so important.

FINDING HADRONS

We've now considered leptons (electrons, muons, taus, and their associated neutrinos). The remaining category of particles in the Standard Model have the name *hadrons*—particles that interact through the strong nuclear force. This category includes all particles made from quarks and gluons, such as protons and neutrons and other particles called *pions*.

Hadrons have internal structure—they are bound states of quarks and gluons held together by the strong nuclear force.

However, the Standard Model doesn't list the many possible bound states. It lists the more fundamental particles that get bound together into hadronic states—namely, the quarks and gluons. In addition to the up and down quarks that sit inside protons and neutrons, heavier quarks called *charm* and *strange* and *top* and *bottom* exist as well. As with the charged and neutral leptons, the heavier quarks have charges identical to their lighter counterparts—the up and down quarks. The heavier quarks are also not readily found in nature. Colliders are needed to study them too.

Hadrons (which interact via the strong force) look very different from leptons (which don't) in particle collisions. That is primarily because quarks and gluons have such strong interactions that they never appear in isolation. They are always in the middle of a jet that might contain the original particle, but will always include a bunch of others that also experience the strong force. Jets don't contain single particles, but a spray of strongly interacting particles "protecting" the initial one, as can be seen in Figure 41. Even if not present in the initial event, the strong interactions will create many new quarks and gluons from the quark or gluon that initiated the jet in the first place. Proton colliders produce a lot of jets since protons themselves are made of strongly interacting particles. Such particles produce sprays of many additional strongly interacting particles that travel alongside them. They also sometimes create quarks and gluons that go off in different directions and form their own independent jets.

The quote I used in *Warped Passages* from the "Jet Song" in *West Side Story* actually describes hadronic jets quite well:

> *You're never alone,*
> *You're never disconnected!*
> *You're home with your own:*
> *When company's expected,*
> *You're well protected.*

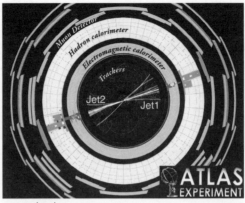

cross section view

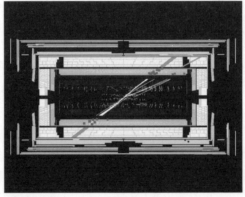

side view

[**FIGURE 41**] Jets are sprays of strongly interacting particles that develop around quarks and gluons. The picture shows their detection in trackers and the hadronic calorimeter. (Modified version of photo courtesy of CERN)

Quarks—and most gang members—won't be found on their own, but in the midst of related strongly interacting companions.

Jets generally leave visible tracks, since some of the particles in jets are charged. And when a jet reaches the calorimeters, it deposits its energy. Careful experimental studies, as well as analytic and computer calculations, help experimenters deduce the properties of the hadrons that created the jets in the first place. Even so, strong interactions and jets make quarks and gluons more subtle. You don't measure the quark or gluon itself, but the jet in which it resides. That makes most quark and gluon jets indistinguishable from each other. They all deposit lots of energy and leave many tracks. (See Figure 42 for a schematic of how detectors identify key Standard Model particles.)

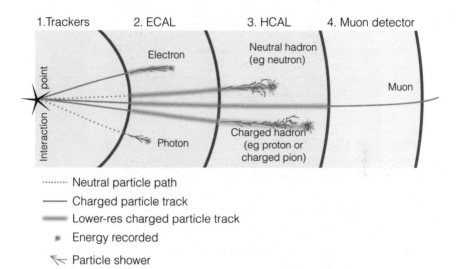

1.Trackers   2. ECAL   3. HCAL   4. Muon detector

Electron

Neutral hadron
(eg neutron)

Muon

Interaction point

Photon

Charged hadron
(eg proton or
charged pion)

········ Neutral particle path
——— Charged particle track
━━━ Lower-res charged particle track
  ✷ Energy recorded
  ✄ Particle shower

[FIGURE 42] A summary of how Standard Model particles are distinguished in the detectors. Neutral particles don't register in the trackers. Both charged and neutral hadrons can leave small deposits in the ECAL but deposit most of their energy in the HCAL. Muons go through to the outer detector.

Even after measuring a jet's properties, telling which of the different quarks or gluons initiated the jets is challenging if not impossible. The bottom quark—which is the heaviest quark with the same charge as the *down* (as well as the heavier *strange*) quark—is an exception to the rule. The reason the bottom quark is special is that it decays more slowly than the other quarks. Other unstable quarks decay essentially immediately after they are produced, so their decay products appear to start their tracks at the interaction point where the protons collided. Bottom quarks, on the other hand, last long enough (about one and a half picoseconds, or enough time to travel about a half millimeter at the light speed at which they travel) to leave a track a noticeably large distance

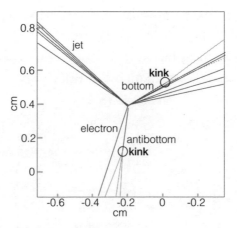

[**FIGURE 43**] Hadrons made of bottom quarks live long enough to leave a visible track in the detector before decaying into other charged particles. This can leave a kink in the silicon vertex detectors, which can be used to identify bottom quarks. The ones here came from top quark decays.

from the interaction point. The inner silicon detectors detect this *displaced vertex,* as illustrated in Figure 43.

When experimenters reconstruct a track from a bottom quark decay, it doesn't extend back to the initial interaction point in the center of the event. Instead the track seems to originate from the place in the inner tracker where the bottom quark decayed, leaving a *kink* in tracks that is the juncture between the bottom quark that came in and the decay product that came out.[55] With the fine segmentation of the silicon detectors, experimenters can view detailed tracks in the region close to the beam, and successfully identify bottom quarks a significant fraction of the time.

The other type of quark that is distinctive from an experimental vantage point is the *top* quark, which is special because it is so heavy. The top quark is the heaviest of the three quarks that have the same charge as the up quark (the other one is called *charm*). Its mass is about 40 times heavier than the differently charged bottom quark and more than 30,000 times the mass of the up quark, which has the same charge as the top.

Top quarks are sufficiently heavy that their decay products leave distinct tracks. When lighter quarks decay, the decay products, like the initial particle, travel so close to the speed of light that they are rushed along together into what appears to be a single jet—even if the jet had its origin in two or more distinct decay products. Unless they are extremely energetic, top quarks, on the other hand, visibly decay into bottom quarks and W bosons (the charged weak gauge bosons) and can be identified by finding both of them. Because the top quark's heavy mass implies that it interacts most closely with the Higgs particle and other particles involved in weak scale physics that we are hoping to soon understand, the properties of top quarks and their interactions might provide valuable clues to physical theories underlying the Standard Model.

FINDING THE WEAK FORCE CARRIERS

Before we finish discussing how to identify Standard Model particles, the final particles to consider are the weak gauge bosons, the two Ws and the Z, that communicate the weak nuclear force. The weak gauge bosons have the peculiar property that, unlike the photon or gluons, they have nonvanishing mass. The masses associated with the weak gauge bosons that communicate the weak force pose some major fundamental mysteries. The origin of this mass—as with the masses of the other elementary particles this chapter has discussed—is rooted in the Higgs mechanism that we will get to shortly.

Because the Ws and Z are heavy, these gauge bosons decay. This means that the W and Z bosons, as with the top quark and other unstable heavy particles, can be identified only by finding the particles into which they decay. Because heavy new particles are also likely to be unstable, we'll use the weak gauge boson decays to exemplify one other interesting property of decaying particles.

A W boson interacts with all particles that are sensitive to the weak force (namely, all the particles we have discussed). That gives the W plenty of decay options. It can decay into any charged lepton (the electron, the muon, or the tau) and their associated neutrino. It can also decay into an up and down quark or into a charm and strange quark pair, as illustrated in Figure 44.

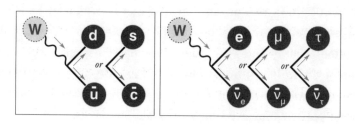

[**FIGURE 44**] The W boson can decay into a charged lepton and its associated neutrino, or into an up and down quark, or a charm and strange quark. In reality, the physical particles are superpositions of different types of quarks or neutrinos. This allows the W to some-times decay into particles from different generations simultaneously.

Particle masses are also critical in determining allowed decays. A particle can decay only into other particles whose masses add up to a smaller mass than the initial particle. Although the W also interacts with the top and bottom quarks, the top quark is heavier than the W, so this decay isn't allowed.[56]

Let's consider the W decaying into two quarks, since in that case the experimenters measure both decay products (not true for lepton and neutrino since the neutrino is "missing"). Because energy and momentum are conserved, measuring the total energy and momentum of both *final state* quarks tells us the energy and momentum of the particle that decayed into them, namely, the W.

At this point both Einstein's special theory of relativity combined with quantum mechanics make the story a bit more interesting. Einstein's special theory of relativity tells us how mass is related to energy and momentum. Most people know the shorthand $E = mc^2$. This formula holds for particles at rest if m is interpreted as m_0, the intrinsic mass of a particle when it's stationary. Once particles move, they have momentum and the more complete formula $E^2 - p^2c^2 = m_0^2c^4$ comes into play.[57] With this formula, the energy and momentum let experimenters deduce the particle's mass, even when the initial particle has long since disappeared via its decay. Experimenters add up all the momentum and energy and apply this equation. The initial mass is then determined.

The reason quantum mechanics comes into play is more subtle. A particle won't always seem to have exactly its real and true mass. Because the particle can decay, the quantum mechanical uncertainty relation, which says that it takes infinitely long to precisely measure energy, tells us that the energy for any particle that doesn't live forever can't be precisely known. The energy can be off by an amount that will be bigger when the decay is faster and the lifetime shorter. This means that in any given measurement, the mass can be close to—but not precisely—the true average value. Only with many measurements can experimenters deduce both the mass—the value that is most probable and to which the average will converge—and the lifetime, since it is the length of time a particle exists before decaying that determines the spread in measured masses. (See Figure 45.) This is true for the W boson, and also for any other decaying particle.

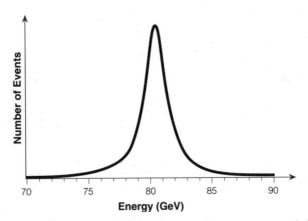

[FIGURE 45] Measurements of decaying particles center around their true masses, but allow for a spread of mass values according to their lifetime. The figure shows this for the W gauge boson.

When experimenters piece together what they measure, using the methods this chapter has described, they might find a Standard Model particle. (See Figure 46 for a summary of Standard Model particles and their properties.)[58] But they might also end up identifying something entirely new. The hope is that the LHC will create new exotic particles that will yield insights into the underlying nature of matter—or even space itself. The next part of the book explores some of the more interesting possibilities.

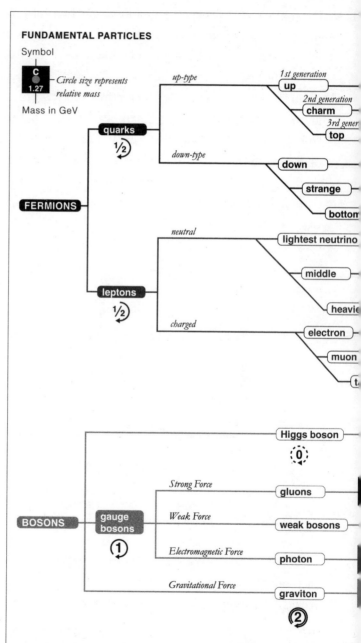

[FIGURE 46]
A summary of Standard Model particles, organized according to type and mass. The gray circles (sometimes inside the squares) give particle masses. We see the mysterious variety of the elements of the Standard Model.

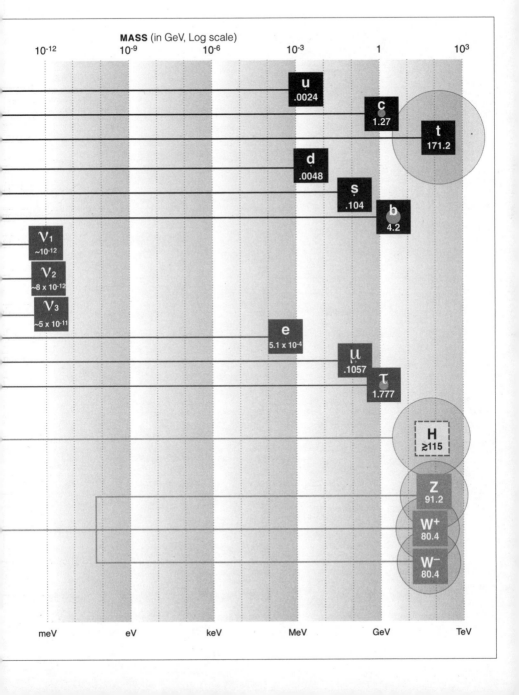

Part IV:

MODELING, PREDICTING, AND ANTICIPATING RESULTS

CHAPTER FIFTEEN

TRUTH, BEAUTY, AND OTHER SCIENTIFIC MISCONCEPTIONS

In February 2007, the Nobel Prize–winning theoretical physicist Murray Gell-Mann spoke at the elite TED conference in California, where innovators working in science, technology, literature, entertainment, and other forefront arenas gather once a year to present new developments and insights about a wide variety of subjects. Murray's crowd-pleasing talk, which was rewarded with a standing ovation, was on the topic of truth and beauty in science. The basic premise of the talk can best be summarized with his words, which echo those of John Keats: "Truth is beauty and beauty is truth."

Gell-Mann had good reasons to believe his grand statement. He had made some of his most significant Nobel Prize–winning discoveries about quarks by searching for an underlying principle that could elegantly organize the seemingly random set of data that experiments had discovered in the 1960s. In Murray's experience, the search for beauty—or at least simplicity—had also led to truth.

No one in the audience disputed his claim. After all, most people love the idea that beauty and truth go together and that the search for one will more often than not reveal the other. But I confess that I have always found this assumption a little slippery. Although everyone would love to believe that beauty is at the heart of great scientific theories, and that the truth will always be aesthetically satisfying, beauty is at least in part a subjective criterion that will never be a reliable arbiter of truth.

The basic problem with the identification of truth and beauty is that it does not always hold—it holds only when it does. If truth and beauty were equivalent, the words "ugly truth" would never have entered our vocabulary. Even though those words weren't specifically directed toward science, observations about the world are not always beautiful. Darwin's colleague Thomas Huxley nicely summarized the sentiment when he said "science is organized common sense where many a beautiful theory was killed by an ugly fact."[59]

To make matters more difficult, physicists have to allow for the disconcerting observation that the universe and its elements are not entirely beautiful. We observe a plethora of messy phenomena and a zoo of particles that we'd like to understand. Ideally, physicists would love to find a simple theory capable of explaining all such observations that uses only a spare set of rules and the fewest possible fundamental ingredients. But even when searching for a simple, elegant, unifying theory—one that can be used to predict the result of any particle physics experiment—we know that even if we find it, we would need many further steps to connect it to our world.

The universe is complex. New ingredients and principles are generally needed before we can connect a simple, spare formulation to the more complicated surrounding world. Those additional ingredients might destroy the beauty present in the initial proposed formulation, much as earmarks all too often interfere with a congressional bill's initial idealistic legislation.

Given the potential pitfalls, how do we go about trying to go beyond what we know? How do we try to interpret as-yet-unexplained phenomena? This chapter is about the idea of beauty and the role of aesthetic criteria in science, and the advantages and disadvantages of beauty as a guide. It also introduces *model building*, which uses a bottom-up approach to science, while paying attention to aesthetic criteria in attempts to guess what comes next.

BEAUTY

I recently spoke with an artist who humorously remarked how one of the great ironies of modern science is that today's researchers seem more likely than contemporary artists to present beauty as their goal. Of course, artists haven't abandoned aesthetic criteria, but they are at least as likely to talk about discovery and invention when discussing their work. Scientists cherish those other attributes too, but they simultaneously strive to find the elegant theories they often find most compelling.

Yet despite the value many scientists place on elegance, they can have divergent notions about what is simple and beautiful. Just as you and your neighbor might violently disagree over the artistic merits of a contemporary artist such as Damien Hirst, different scientists find distinct aspects of science satisfying.

Together with like-minded researchers, I prefer to search for underlying principles that illuminate connections among superficially disparate observed phenomena. Most of my string theory colleagues study specific solvable theories in which they use difficult mathematical formulations to tackle toy problems (problems not necessarily relevant to any real physical setup) that might only later find applications to observable physical phenomena. Other physicists prefer to focus only on theories with a concise elegant formalism that generate many experimental predictions which they can systematically calculate. And others simply like computing.

Interesting principles, advanced mathematics, and complicated numerical simulations are all part of physics. Most scientists value all of them, but we choose our priorities according to what we find most pleasing or most likely to lead to scientific advances. In reality, we often also choose our approach according to which method best suits our unique inclinations and talents.

Not only do current views of beauty vary. As is also true with art, attitudes evolve over time. Murray Gell-Mann's own specialty, quantum chromodynamics, presents an excellent case in point.

Gell-Mann's conjecture about the strong nuclear force was based on a brilliant insight about how the many particles that were constantly being discovered in the 1960s could be organized into sensible patterns that would explain their abundance and types. He hypothesized the existence of more basic elementary particles known as quarks, which he suggested carry a new type of charge. The strong nuclear force would then influence any object that carried the conjectured charge, and cause quarks to bind together to form neutral objects—much as the electric force binds electrons with charged nuclei to form neutral atoms. If true, all the particles being discovered could be interpreted as bound states of these quarks—aggregate objects that have no net charge.

Gell-Mann realized that if there were three different types of quarks, each of which carried a distinct color charge, many such combinations of neutral bound states would form. And these many combinations could (and did) correspond to the plethora of particles that were being found. Gell-Mann thereby had found a beautiful explanation for what seemed like an inexplicable mess of particles.

However, when Murray—as well as the physicist (and later neurobiologist) George Zweig—first proposed this idea, people didn't even believe it was a proper scientific theory. The reason is somewhat technical but interesting. Particle physics calculations rely on particles not interacting when they are far apart, so that we can compute the finite effects of the interactions that occur when they are close together. With this assumption, any interaction can be entirely captured by the local forces that apply when the interacting particles are in close proximity.

The force that Gell-Mann had conjectured, on the other hand, was stronger when particles were farther apart. That meant that quarks would always interact, even when very distant. According to the then-reigning criteria, Gell-Mann's guess didn't even correspond to a true theory that could be used for reliable calculations. Because quarks always interact, even their so-called *asymptotic states*—the states involving quarks that are far away from everything else—are very complicated. In an apparent concession to ugliness, the asymptotic states they postulated weren't the simple particles you'd like to see in a calculable theory.

Initially, no one knew how to organize calculations among these complicated strongly bound states. However, today's physicists think

quite oppositely about the strong force. We now understand it much better than we did when the idea was first proposed. David Gross, David
Politzer, and Frank Wilczek won the Nobel Prize for what they called
"asymptotic freedom." According to their calculations, the force is strong
only at low energies. At high energies, the strong force is not much more
powerful than other forces, and calculations work just as they should.
In fact, some physicists today think theories such as the strong force,
which become much weaker at high energies, are the *only* well-defined
theories, since the interaction strength won't grow to infinite strength at
high energy as it might otherwise do.

Gell-Mann's theory of the strong force is an interesting example of
the interplay between aesthetic and scientific criteria. Simplicity was his
initial guide. But hard scientific calculations and theoretical insights were
necessary before everyone could agree on the beauty of his suggestion.

This, of course, isn't the only example. Many of our most trusted theories have aspects so superficially ugly and uncompelling that even respected and well-established scientists rejected them initially. Quantum
field theory, which combines quantum mechanics and special relativity,
underlies all of particle physics. Yet the Nobel Prize–winning Italian scientist Enrico Fermi (among others) rejected it at first. For him, the problem was that although quantum field theory formalizes and systematizes
all calculations and makes many correct predictions, it involves calculational techniques that even some of today's physicists view as baroque.
Various aspects of the theory are quite beautiful and lead to remarkable
insights. Other features we just have to put up with, even though we
aren't so enamored with all their intricacies.

This story has repeated itself many times since. Beauty is often
agreed on only a posteriori. Weak interactions violate parity symmetry. This means that particles spinning to the left interact differently
from those spinning to the right. The breaking of such a fundamental
symmetry as left-right equivalence seems innately disturbing and unat-
tractive. Yet this very asymmetry is what is responsible for the range
of masses we see in the world, which is in turn necessary for structure
and life. It was considered ugly at first, yet now we know it is essen-

tial. Although perhaps ugly in itself, parity symmetry breaking leads to beautiful explanations of more complicated phenomena essential to all the matter we see.

Beauty is not absolute. An idea might appeal to its creator but be cumbersome or messy from someone else's perspective. Sometimes I'll be quite taken with the beauty of a conjecture I've come up with largely because I know of all the other ideas people had thought of before that hadn't worked. But being better than what came before doesn't guarantee beauty. Having made my share of models that satisfied this criterion, but were nonetheless met with skepticism and confusion from colleagues who were less familiar with the topic my model addressed, I now think a better criterion for a good idea might be that even someone who never studied the problem can recognize its appeal.

The reverse is sometimes true as well—good ideas are rejected because their inventors consider them too ugly. Max Planck didn't believe in photons, which he thought to be an unpleasant concept, even though he initiated the train of logic that led to their conjecture. Einstein thought the expanding universe that followed from his equations of general relativity couldn't be true, in part because it contradicted his aesthetic and philosophical predispositions. Neither of these ideas might have seemed the most beautiful at the time, but the laws of physics and the universe in which they applied didn't really care.

LOOKING GOOD

Given the evolving and uncertain nature of beauty, it's worth considering some of the features that might make an idea or an image objectively beautiful in a way that has some universal appeal. Perhaps the most basic question about aesthetic criteria is whether humans even have any universal criteria for what is beautiful—in any context—be it art or science.

No one yet knows the answer. Beauty, after all, involves taste, and taste can be a subjective criterion. Nonetheless, I find it hard to believe that humans don't share some common aesthetic criteria. I often notice a striking uniformity in people's opinions about which piece of art in a given exhibit is the best or even which exhibits people choose to go see. Of course this doesn't prove anything since we all share a time and place. Beliefs about beauty are difficult to isolate from the specific cultural context or time period in which they originate so it's difficult to isolate innate from learned values or judgments. In some extreme cases, people might all agree that something looks nice or appears unpleasant. And in some rare instances, everyone might agree on the beauty of an idea. But even in those few cases, people don't necessarily agree about all the details.

Even so, some aesthetic criteria do appear to be universal. Any beginning art class will teach about balance. Michelangelo's *David* in the Accademia Gallery in Florence exemplifies this principle. *David* stands gracefully. He's never going to tip over or fall apart. People search for balance and harmony where they can find it. Art, religion, and science all promise people the opportunity to access these qualities. But of course balance might also be simply an organizing principle. Art is also fascinating when it defies our notions of balance, as we see in early Richard Serra sculptures. (See Figure 47.)

[FIGURE 47] These early Richard Serra sculptures illustrate that sometimes art is more interesting when it appears to be slightly off balance. (Copyright © 2011 by Richard Serra/Artists Rights Society [ARS], New York.)

Symmetry is also often considered essential to beauty, and art and architecture frequently exhibit the order that it generates. Something has symmetry if you can change it—for example, by rotating it, reflecting it in a mirror, or interchanging its parts—so that the transformed system is indistinguishable from the initial one. Symmetry's harmoniousness is probably one reason that religious symbols often have it on display. The Christian cross, the Jewish star, the dharma wheel of Buddhism, and the crescent of Islam are all examples and are illustrated in Figure 48.

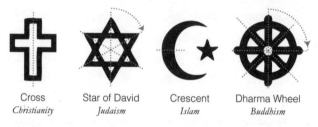

Cross	Star of David	Crescent	Dharma Wheel
Christianity	*Judaism*	*Islam*	*Buddhism*

[**FIGURE 48**] Religious symbols frequently embody symmetries.

More expansively, Islamic art, which forbids representation and relies on geometric forms, is notable for its use of symmetry. The Taj Mahal in India is a magnificent example. I haven't spoken to anyone who's visited the Taj Mahal and wasn't taken with its masterful orderliness, shape, and

[**FIGURE 49**] The architecture of the Chartres Cathedral and the ceiling of the Sistine Chapel both embody symmetry.

symmetry. The Alhambra in southern Spain, which also incorporates Moorish art and its interesting symmetry patterns, may be one of the most beautiful buildings still standing today.

Recent art, such as the work of Ellsworth Kelly or Bridget Riley, exhibits symmetry explicitly and geometrically. Gothic or Renaissance art and architecture—see the Chartres Cathedral and the roof of the Sistine Chapel, for example—exquisitely exploited symmetry as well. (See Figure 49.)

However, art is often most beautiful when it is not completely symmetrical. Japanese art is notable for its elegance, but also for the well-defined breaking of symmetry. Japanese paintings and silk screens have a clear orientation that draws one's eye across the pictures as one can see in Figure 50.

[FIGURE 50] Japanese art is interesting in part because of its asymmetry.

Simplicity is another and sometimes related criterion that might help when evaluating beauty. Some simplicity arises from symmetries, but underlying order can be present, even in the absence of manifest symmetry. Jackson Pollock pieces have an underlying simplicity in the density of paint, though the impression might first seem chaotic. Although the individual splashes of paint seem completely random, his most famous and successful pieces have a fairly uniform density of each color that enters the work.

Simplicity in art can frequently be deceptive. I once tried to sketch a few Matisse cutouts, his simplest works, which he created when he was old and frail. Yet when I tried to reproduce them, I realized that they weren't so simple—at least not for my unskilled hand. Simple elements can embody more structure than we superficially observe.

In any case, beauty isn't found only in simple basic forms. Some admired works of art, such as those of Raphael or Titian, involve rich complex canvases with many internal elements. After all, complete simplicity can be mind-numbing. When we look at art, we prefer something interesting that guides our eye. We want something simple enough to follow, but not so simple as to be boring. This seems to be how the world is constructed as well.

BEAUTY IN SCIENCE

Aesthetic criteria are difficult to pin down. In science—as in art—there are unifying themes but no absolutes. Yet even though aesthetic criteria for science might be poorly defined, they are nonetheless useful and omnipresent. They help guide our research, even if they provide no guarantee of success or truth.

Aesthetic criteria that we apply to science resemble those that were just outlined for art. Symmetries certainly play an important role. They help us organize our calculations and often relate disparate phenomena. Interestingly, as with art, symmetries are usually only approximate. The best scientific descriptions frequently respect the elegance of symmetric theories while incorporating the symmetry breaking necessary to make predictions about our world. The symmetry breaking enriches the ideas it encompasses, which thereby yield more explanatory power. And, as is often true for art, the theories that incorporate broken symmetries can be even more beautiful and interesting than those that are perfectly symmetrical.

The Higgs mechanism, which is responsible for elementary particle masses, is an excellent example. As will be explained in the following chapter, the Higgs mechanism very eloquently explains how the symmetries associated with the weak force can be slightly broken. We haven't yet discovered the Higgs boson—the particle that would provide incontrovertible evidence that the idea is correct. But the theory is so beautiful and so uniquely satisfies criteria required by both experiments and theory that most physicists believe it is realized in nature.

Simplicity is another important subjective criterion for theoretical physicists. We have a deep-rooted belief that simple elements underlie the complicated phenomena we see. Such a search for simple basic elements of which all reality is composed or resembles began long ago. In ancient Greece, Plato imagined perfect forms—geometric shapes and ideal beings that objects on Earth only approximate. Aristotle, too, believed in ideal forms, but in his case, he thought that the ideals that physical objects approximate would be revealed only through observations. Religions also often postulate a more perfect or more unified state that is removed from, but somehow connected to, reality. Even the story of the fall from the Garden of Eden presupposes an idealized prior world. Although the questions and methods of modern physics are very different from those of our ancestors, many physicists, too, are seeking a simpler universe—not in philosophy or religion, but in the fundamental ingredients that constitute our world.

The search for underlying scientific truth often involves looking for simple elements from which we can construct the complex and rich phenomena we observe. Such research often involves trying to identify meaningful patterns or organizing principles. Only with a concise realization of simple and elegant ideas do most scientists expect a proposal to have the potential to be right. A starting point involving the fewest inputs has the further benefit that it promises the most predictive power. When particle physicists consider suggestions for what might lie at the heart of the Standard Model, we usually become skeptical when the realization of an idea becomes too cumbersome.

Again, as with art, physical theories can be simple in themselves, or they may be complex compositions made up of simple and predictable elements. The end point of course isn't necessarily simple, even when the initial components—and perhaps even the rules themselves—are.

The most extreme version of such pursuits is the search for a unifying theory consisting of only a few simple elements obeying a small set of rules. This quest is an ambitious—some might say an audacious—

task. Clearly an obvious impediment prevents us from readily finding an elegant theory that completely accounts for all observations: the world around us manifests only a fraction of the simplicity that such a theory should embody. A unified theory, while being simple and elegant, must somehow accommodate enough structure to match observations. We would like to believe in a single simple, elegant, and predictable theory that underlies all of physics. Yet the universe is not as pure, simple, and ordered as the theories. Even with an underlying unified description, a lot of research will be necessary to connect it to the fascinating and complex phenomena we see in our world.

Of course, we can go too far in these characterizations of beauty or simplicity. A standard joke among students in our science or math classes involves professors who repeatedly refer to well-understood phenomena as "trivial," no matter how complex they might be. The professor already knows the answer and the underlying elements and logic very well, but this is not so true for the students sitting in class. In retrospect, after they have reduced the problem to simple pieces, it can become trivial to them, too. But they first have to discover how to do that.

MODEL BUILDING

In the end, just as in life, science doesn't have just a single criterion for beauty. We merely have some intuitions—along with experimental constraints—that we use to guide our search for knowledge. Beauty—both in art and science—might have some objective aspects, but almost any application involves taste and subjectivity.

For scientists, however, there is one big difference. Ultimately experiments will decide which, if any, of our ideas are correct. Scientific advances might exploit aesthetic criteria, but true scientific progress also requires understanding, predicting, and analyzing data. No matter how beautiful a theory appears, it can still be wrong, in which case it must be thrown away. Even the most intellectually satisfying theory has to be abandoned if it doesn't work in the real world.

Nonetheless, before we reach the higher energies or distant parameters needed to determine the correct physical descriptions, physicists have no choice but to employ aesthetic and theoretical considerations to guess what lies beyond the Standard Model. In this interim, with only limited data, we rely on existing puzzles coupled with taste and organizational criteria to point the way forward.

Ideally, we'd like to be able to work through the consequences of a variety of possibilities. *Model building* is the name of the approach we use to do this. My colleagues and I explore various particle physics models, which are guesses for physical theories that might underlie the Standard Model. Our goals are simple principles that organize the complicated phenomena that appear on more readily visible scales so that we can resolve current puzzles in our understanding.

Physics model builders take the effective theory viewpoint and the desire to understand smaller and smaller distance scales very much to heart. We follow a "bottom-up" approach that starts with what we know—both the phenomena we can explain and those we find puzzling—and attempt to deduce the underlying model that explains the connections among observed elementary particle properties and their interactions.

The term "model" might evoke a physical structure such as a small-scale version of a building used to display and explore its architecture. Or you might think of numerical simulations on a computer that calculate the consequences of known physical principles—such as climate modeling or models for the spread of contagious diseases.

Modeling in particle physics is very different from either of these definitions. Particle models do, however, share some of the flair of models in magazines or fashion shows. Models, both on runways and in physics, illustrate imaginative new ideas. And people initially flock toward the beautiful ones—or at least those that are more striking or surprising. But in the end, they are drawn toward the ones that show true promise.

Needless to say, the similarities end there.

Particle physics models are guesses for what might underlie the theories whose predictions have been already tested and that we understand. Aesthetic criteria are important in deciding which ideas are worth pursuing. But so are consistency and testability of the ideas. Models characterize different underlying physical ingredients and principles that apply at distances and sizes that are smaller than those which have yet been experimentally tested. With models, we can determine the essence and consequences of different theoretical assumptions.

Models are a means of extrapolating from what is known to create proposals for more comprehensive theories with greater explanatory power. They are sample proposals that may or may not prove correct once experiments allow us to delve into smaller distances or higher energies and test their underlying hypotheses and predictions.

Bear in mind that a "theory" is different from a "model." By the word theory, I don't mean rough speculations, as in more colloquial usage. The known particles and the known physical laws they obey are components of a theory—a definite set of elements and principles with rules and equations for predicting how the elements interact.

But even when we fully understand a theory and its implications, that same theory can be implemented in many different ways, and these will have different physical consequences in the real world. Models are a way of sampling these possibilities. We combine known physical principles and elements into candidate descriptions of reality.

If you think of a theory as a PowerPoint template, a model would be your particular presentation. The theory allows animations, but the model includes only those you need to make your point. The theory would say to have a title and some bullet points, but the model would contain exactly what you want to convey and will hopefully apply well to the task at hand.

The nature of model building in physics has changed according to the questions physicists have tried to answer. Physics always involves trying to predict the largest number of physical quantities from the smallest number of assumptions, but that doesn't mean we manage to identify the most fundamental theories right away. Advances in physics are often made even before everything is understood at the most fundamental level.

In the nineteenth century, physicists understood the notions of temperature and pressure and employed them in thermodynamics and engine design long before anyone could explain these ideas at a more fundamental microscopic level as the result of the random motion of large numbers of atoms and molecules. In the early twentieth century, physicists tried to make models to explain mass in terms of electromagnetic energy. Though these models were based on strongly shared beliefs on how those systems worked, those models proved wrong. A little later, Niels Bohr made a model of the atom to explain the emission spectra that people had observed. His model was soon superseded by the more comprehensive theory of quantum mechanics, which absorbed but also improved on Bohr's core idea.

Model builders today try to determine what lies beyond the Standard Model of particle physics. Although currently referred to as the Standard Model because it has been well tested and is well understood, it was something of a guess as to how known observations might fit together at the time it was developed. Nonetheless, because the Standard Model implied predictions for how to test its premises, experiments could ultimately show it to be correct.

The Standard Model correctly accounts for all observations to date, but physicists are fairly confident that it is not complete. In particular, it leaves open the question of what are the precise particles and interactions—the elements of the Higgs sector—that are responsible for the masses of elementary particles and why it is that the particles in that sector have the particular masses that they do. Models that go beyond the Standard Model illuminate deeper potential interconnections and relationships that might address these questions. They involve specific choices of fundamental assumptions and physical concepts, as well as the distance or energy scales at which they might apply.

Much of my current research involves thinking about new models, as well as novel or more detailed search strategies that would otherwise miss new phenomena. I think about the models I originated but the full range of other possibilities as well. Particle physicists know the types of elements and rules that could be involved, such as particles, forces, and allowed interactions. But we don't know precisely which of these ingredients enters the recipe for reality. By applying known theoretical ingredients, we attempt to identify the potentially simple underlying ideas that enter into what is an ultimately complex theory.

As important, models provide targets for experimental exploration, and suggestions for how particles will behave at smaller distances than physicists have experimentally studied so far. Measurements provide clues to help us distinguish among competing candidates. We don't yet know what the new underlying theory is, but we can nonetheless characterize the possible deviations from the Standard Model. By thinking about candidate models for underlying reality and their consequences, we can predict what the LHC should reveal if the models turn out to be right. Our use of models admits the speculative nature of our ideas and recognizes the plethora of possibilities that might agree with existing data and explain as-yet-puzzling phenomena. Only some models will prove correct, but creating and understanding them is the best way to delineate the options and build up a reservoir of compelling ingredients.

Exploring models and their detailed consequences helps us establish what experimenters should search for—whatever might be out there. Models tell experimenters the interesting features that characterize new physical theories so that experimenters can test whether model builders have correctly identified the elements or the physical principles that guide the system's relationships and interactions. Any model with new physical laws that apply at measurable energies should predict new particles and new relationships among them. Observing which particles emerge from collisions and the properties they have should help determine the type of particles that exist, their masses, and their interactions. Finding new particles or measuring different interactions will confirm or rule out models that have been proposed, and pave the way for better ones.

With enough data, experiments will determine which underlying model is the right one—at least at the level of precision, distance, and energy that we can study. The hope is that at the smallest distance scales that we can probe at LHC energies, the rules for the underlying theory will be simple enough to allow us to deduce and calculate the influence of the associated physical laws.

Physicists have lively discussions about which are the best models to study and what is the most useful way to account for them in experimental searches. I'll frequently sit down with experimental colleagues and discuss with them how best to use models to guide their searches. Are benchmark points with specific parameters in particular models too specific? Is there a better way to cover all the possibilities?

LHC experiments are so challenging that without definite search targets, the results will be overwhelmed by Standard Model background. Experiments were designed and optimized with existing models in mind, but they are searching for more general possibilities as well. It is critical that experimenters are aware of a big range of models that span the possible new signatures that might emerge, since no one wants specific models to overly prejudice the searches.

Theorists and experimenters are working hard to make sure we don't miss anything. We won't know which, if any, of the different suggestions is correct until it is experimentally verified. Proposed models might be the correct description of reality, but even if they are not, they suggest interesting search strategies that tell us the distinguishing features of new as-yet-undiscovered matter. Hopes are the LHC will tell us the answers—no matter what they turn out to be—and we want to be prepared.

CHAPTER SIXTEEN

THE HIGGS BOSON

On the morning of March 30, 2010, I awoke to a flurry of e-mails about the successful 7 TeV collisions that had taken place at CERN the night before. This triumph launched the beginning of the true physics program at the LHC. The acceleration and collisions that had taken place toward the close of the previous year had been critical technical milestones. Those events were important for LHC experimenters who could finally calibrate and better understand their detectors using data from genuine LHC collisions, and not just cosmic rays that had happened to pass through their apparatus. But for the next year and a half, detectors at CERN would be recording real data that physicists could use to constrain or verify models. Finally, after its many ups and downs, the physics program at the LHC had at long last begun.

The launch proceeded almost exactly according to plan—a good thing according to my experimental colleagues, who the day before had expressed concerns that the presence of reporters might compromise the day's technical goals. The reporters (and everyone else present) did witness a couple of false starts—in part because of the zealous protection mechanisms that had been installed, which were designed to trigger if anything went even slightly awry. But within a few hours, beams circulated and collided and newspapers and websites had lots of pretty pictures to display.

The 7 TeV collisions occurred with only half the intended LHC energy. The real target energy of 14 TeV wouldn't be reached for several years. And the intended luminosity for the 7 TeV run—the number of protons that would collide each second—was much lower than designers had originally planned. Still, with these collisions, everything at the LHC was at long last on track. We could finally believe that our understanding of the inner nature of matter would soon improve. And if all went okay, in a couple of years the machine would shut down, gear up, and come back online at full capacity and provide the real answers we were waiting for.

One of the most important goals will be learning how fundamental particles acquire their mass. Why isn't everything whizzing around at the speed of light, which is what matter would do if it had zero mass? The answer to this question hinges on the set of particles that are known collectively as the *Higgs sector,* including the Higgs boson. This chapter explains why a successful search for this particle will tell us whether our ideas about how elementary particle masses arise are correct. Searches that will take place once the LHC comes back online with higher intensity and greater energy should ultimately tell us about the particles and interactions that underlie this critical and rather remarkable phenomenon.

THE HIGGS MECHANISM

No physicist questions that the Standard Model works at the energies we have studied so far. Experiments have tested its many predictions, which agree with expectations to better than one percent precision.

However, the Standard Model relies on an ingredient that no one has yet observed. The Higgs mechanism, named after the British physicist Peter Higgs, is the only way we know to consistently give elementary particles their mass. According to the basic premises of the naive version of the Standard Model, neither the gauge bosons that communicate forces nor the elementary particles, such as quarks and leptons that are essential to the Standard Model should have nonzero masses. Yet measurements of physical phenomena clearly demonstrate that they do. Elementary particle masses are critical to understanding atomic and particle physics phenomena, such as the radius of an electron's orbit in an atom or the extremely tiny range of the weak force, not to mention the formation of structure in the universe. Masses also determine how much energy is needed to create elementary particles—in accordance with the equation $E=mc^2$. Yet in the Standard Model without a Higgs mechanism, elementary particles' masses would be a mystery. They would not be allowed.

The notion that particles don't have an inalienable right to their masses might sound needlessly autocratic. You could quite reasonably expect that particles always have the option of possessing a nonvanishing mass. Yet the subtle structure of the Standard Model and any theory of forces is just that tyrannical. It constrains the types of masses that are allowed. The explanations will seem a little different for gauge bosons than for fermions, but the underlying logic for both relates to the symmetries at the heart of any theory of forces.

The Standard Model of particle physics includes the electromagnetic, weak, and strong nuclear forces, and each force is associated with a symmetry. Without such symmetries, too many oscillation modes of the gauge bosons—the particles that communicate those forces—would be predicted to be present by the theory that quantum mechanics and special relativity tells us describes them. In the theory without symmetries, theoretical calculations would generate nonsensical predictions, such as probabilities for high-energy interactions greater than one for the spurious oscillation modes. In any accurate description of nature, such unphysical particles—particles that don't actually exist because they oscillate in the wrong direction—clearly need to be eliminated.

In this context, symmetries act like spam filters, or quality control constraints. Quality requirements might specify keeping only those cars that are symmetrically balanced, for instance, so that the cars that make it out of the factory all behave as expected. Symmetries in any theory of forces also screen out the badly behaved elements. That's because interactions among the undesirable, unphysical particles don't respect the symmetries, whereas those particles that interact in a way that preserves the necessary symmetries oscillate as they should. Symmetries thereby guarantee that theoretical predictions involve only the physical particles and therefore make sense and agree with experiments.

Symmetries therefore permit an elegant formulation of a theory of forces. Rather than eliminate unphysical modes in each calculation one by one, symmetries eliminate all the unphysical particles with one fell swoop. Any theory with symmetric interactions involves only the physical oscillation modes whose behavior we want to describe.

This works perfectly in any theory of forces involving zero mass force carriers, such as electromagnetism or the strong nuclear force. In symmetric theories, predictions for their high-energy interactions all make sense and only physical modes—modes that exist in nature—get included. For massless gauge bosons, the problem with high-energy interactions is relatively straightforward to solve, since appropriate symmetry constraints remove any unphysical, badly behaved modes from the theory.

Symmetries thereby solve two problems: unphysical modes are eliminated, and the bad high-energy predictions that would accompany them are as well. However, a gauge boson with nonzero mass has an additional physical—existent in nature—mode of oscillation. The gauge bosons that communicate the weak nuclear force fall into this category. Symmetries would eliminate too many of their oscillation modes. Without some new ingredient, weak boson masses cannot respect the Standard Model symmetries. For gauge bosons with nonzero mass, we have no choice but to keep a badly behaved mode—and that means the solution to the bad high-energy behavior is not so simple. Nonetheless something is still required for the theory to generate sensible high-energy interactions.

Moreover, none of the elementary particles in the Standard Model without a Higgs can have a nonzero mass that respects the symmetries of the most naive theory of forces. With the symmetries associated with forces present, quarks and leptons in the Higgsless Standard Model would not have nonzero masses either. The reason appears to be unrelated to the logic about gauge bosons, but it can also be traced to symmetries.

In Chapter 14, we presented a table that included both left- and right-handed fermions—particles that get paired in the presence of nonzero masses. When quark or lepton masses are nonzero, they introduce interactions that convert left-handed fermions to right-handed fermions. But for left-handed and right-handed fermions to be interconvertible, they would both have to experience the same forces. Yet experiments demonstrate that the weak force acts differently on left-handed fermions than on the right-handed fermions that massive quarks or leptons could turn into. This violation of parity symmetry, which if preserved would treat left and right as equivalent for the laws of physics, is startling to everyone when they first learn about it. After all, the other known laws of nature don't distinguish left and right. But this remarkable property—that the weak force does not treat left and right the same—has been demonstrated experimentally and is an essential feature of the Standard Model.

The different interactions of left- and right-handed quarks and leptons tells us that without some new ingredient, nonzero masses for quarks and leptons would be inconsistent with known physical laws. Such nonzero masses would connect particles that carry weak charge with particles that do not.

In other words, since only left-handed particles carry this charge, weak charge could be lost. Charges would apparently disappear into the *vacuum*—the state of the universe that doesn't contain any particles. Generally that should not happen. Charges should be conserved. If charge could appear and disappear, the symmetries associated with the corresponding force would be broken, and the bizarre probabilistic predictions about high-energy gauge boson interactions that they are supposed to eliminate would reemerge. Charges should never magically disappear in this manner if the vacuum is truly empty and contains no particles or fields.

But charges can appear and disappear if the vacuum is not really empty—but instead contains a *Higgs field* that supplies weak charge to the vacuum. A Higgs field, even one that gives charge to the vacuum, isn't composed of actual particles. It is essentially a distribution of weak charge throughout the universe that happens only when the field itself takes a nonzero value. When the Higgs field is nonvanishing, it is as if the universe has an infinite supply of weak charges. Imagine that you had an infinite supply of money. You could lend or take money at will and you would always still have an infinite amount at your disposal. In a similar spirit, the Higgs field puts infinite weak charge into the vacuum. In doing so, it breaks the symmetries associated with forces and lets charges flow into and out of the vacuum so that particle masses arise without causing any problems.

One way to think about the Higgs mechanism and the origin of masses is that it lets the vacuum behave like a viscous fluid—a Higgs field that permeates the vacuum—that carries weak charge. Particles that carry this charge, such as the weak gauge bosons and Standard Model quarks and leptons, can interact with this fluid, and these interactions slow them down. This slowing down then corresponds to the particles acquiring mass, since particles without mass will travel through the vacuum at the speed of light.

This subtle process by which elementary particles acquire their masses is known as the Higgs mechanism. It tells us not only how elementary particles acquire their masses, but also quite a bit about those masses' properties. The mechanism explains, for instance, why some particles are heavy while others are light. It is simply that particles that interact more with the Higgs field have larger masses and those that interact less have smaller ones. The top quark, which is the heaviest, has the biggest such interaction. An electron or an up quark, which have relatively small masses, have much more feeble ones.

The Higgs mechanism also provides a deep insight into the nature of electromagnetism and the photon that communicates that force. The Higgs mechanism tells us that only those force carriers that interact with the weak charge distributed throughout the vacuum acquire mass. Because the W gauge bosons and the Z boson interact with these charges, they have nonvanishing masses. However, the Higgs field that suffuses the vacuum carries weak charge but is electrically neutral. The photon doesn't interact with the weak charge, so its mass remains zero. The photon is thereby singled out. Without the Higgs mechanism, there would be three zero mass weak gauge bosons and one other force carrier—also with zero mass—known as the hypercharge gauge boson. No one would ever mention a photon at all. But in the presence of the Higgs field, only a unique combination of the hypercharge gauge boson and one of the three weak gauge bosons will not interact with the charge in the vacuum—and that combination is precisely the photon that communicates electromagnetism. The photon's masslessness is critical to the important phenomena that follow from electromagnetism. It explains why radio waves can extend over enormous distances, while the weak force is screened over extremely tiny ones. The Higgs field carries weak charge—but no electric charge. So the photon has zero mass and travels at the speed of light—by definition—while the weak force carriers are heavy.

Don't be confused. Photons are elementary particles. But in a sense, the original gauge bosons were misidentified since they didn't correspond to the physical particles that have definite masses (which might be zero) and travel through the vacuum unperturbed. Until we know the weak charges that are distributed throughout the vacuum via the Higgs mechanism, we have no way to pick out which particles have nonzero mass and which of them don't. According to the charges assigned to the vacuum by the Higgs mechanism, the hypercharge gauge boson and the weak gauge boson would flip back and forth into each other as they travel through the vacuum and we couldn't assign either one a definite mass. Given the vacuum's weak charge, only the photon and the Z boson travel without changing identity as they travel through the vacuum, with the Z boson acquiring mass, whereas the photon does not. The Higgs mechanism thereby singles out the particular particle called the photon and the charge that we know as the electric charge which it communicates.

So the Higgs mechanism explains why it is the photon and not the other force carriers that has zero mass. It also explains one other property of masses. This next lesson is even a bit more subtle, but gives us deep insights into why the Higgs mechanism allows masses that are consistent with sensible high-energy predictions. If we think of the Higgs field as a fluid, we can imagine that its density is also relevant to particle masses. And if we think of this density as arising from charges with a fixed spacing, then these particles—which travel such small distances that they never hit a weak charge—will travel as if they had zero mass, whereas particles that travel over larger distances would inevitably bounce off weak charges and slow down.

This corresponds to the fact that the Higgs mechanism is associated with *spontaneous breaking* of the symmetry associated with the weak force—and that symmetry breaking is associated with a definite scale.

Spontaneous breaking of a symmetry occurs when the symmetry itself is present in the laws of nature—as with any theory of forces—but is broken by the actual state of a system. As we've argued, symmetries must exist for reasons connected to the high-energy behavior of particles in the theory. The only solution then is that the symmetries exist—but they are spontaneously broken so that the weak gauge bosons can have mass, but not exhibit bad high-energy behavior.

The idea behind the Higgs mechanism is that the symmetry is indeed part of the theory. The laws of physics act symmetrically. But the actual state of the world doesn't respect the symmetry. Think of a pencil that originally stood on end and then falls down and chooses one particular direction. All of the directions around the pencil were the same when it was upright, but the symmetry is broken once the pencil falls. The horizontal pencil thereby spontaneously breaks the rotational symmetry that the upright pencil preserved.

The Higgs mechanism similarly spontaneously breaks weak force symmetry. This means that the laws of physics preserve the symmetry, but it is broken by the state of the vacuum that is suffused with weak force charge. The Higgs field, which permeates the universe in a way that is not symmetric, allows elementary particles to acquire mass, since it breaks the weak force symmetry that would be present without it. The theory of forces preserves a symmetry associated with the weak force, but that symmetry is broken by the Higgs field that suffuses the vacuum.

By putting charge into the vacuum, the Higgs mechanism breaks the symmetry associated with the weak force. And it does so at a particular scale. The scale is set by the distribution of charges in the vacuum. At high energies, or equivalently—via quantum mechanics—small distances, particles won't encounter any weak charge and therefore behave as if they have no mass. At small distances, or equivalently high energies, the symmetry therefore appears to be valid. At large distances, however, the weak charge acts in some respects like a frictional force that would slow the particles down. Only at low energies, or equivalently large distances, does the Higgs field seem to give particles mass.

And this is exactly as we need it to be. The dangerous interactions that wouldn't make sense for massive particles apply only at high energies. At low energies particles can—and must, according to experiments—have mass. The Higgs mechanism, which spontaneously breaks the weak force symmetry, is the only way we know to accomplish this task.

Although we have not yet observed the particles responsible for the Higgs mechanism that is responsible for elementary particle masses, we do have experimental evidence that the Higgs mechanism applies in nature. It has already been seen many times in a completely different context—namely, in *superconducting* materials. Superconductivity occurs when electrons pair up and these pairs permeate a material. The so-called *condensate* in a superconductor consists of electron pairs that play the same role that the Higgs field does in our example above.

But rather than carry weak charge, the condensate in a superconductor carries electric charge. The condensate therefore gives mass to the photon that communicates electromagnetism inside the superconducting material. The mass *screens* the charge, which means that inside a superconductor, electric and magnetic fields do not reach very far. The force falls off very quickly over a short distance. Quantum mechanics and special relativity tell us that this *screening distance* inside a superconductor is the direct result of a photon mass that exists only inside the superconducting substrate. In these materials, electric fields can't penetrate farther than the screening distance because in bouncing off the electron pairs that permeate the superconductor, the photon acquires a mass.

The Higgs mechanism works in a similar fashion. But rather than electron pairs (carrying electric charge) permeating the substance, we predict there is a Higgs field (that carries weak charge) that permeates the vacuum. And instead of a photon acquiring mass that screens electric charge, we find the weak gauge bosons acquire mass that screens weak charge. Because weak gauge bosons have nonzero mass, the weak force is effective only over very short distances of subnuclear size.

Since this is the only consistent way to give gauge bosons masses, physicists are fairly confident that the Higgs mechanism applies in nature. And we expect that it is responsible not just for the gauge boson masses, but for the masses of all elementary particles. We know of no other consistent theory that permits the Standard Model weakly charged particles to have mass.

This was a difficult section with several abstract concepts. The notions of a Higgs mechanism and a Higgs field are intrinsically linked to quantum field theory and particle physics and are remote from phenomena we can readily visualize. So let me briefly summarize some of the salient points. Without the Higgs mechanism, we would have to forfeit sensible high-energy predictions or particle masses. Yet both of these are essential to the correct theory. The solution is that symmetry exists in the laws of nature, but can be spontaneously broken by the nonzero value of a Higgs field. The broken symmetry of the vacuum allows Standard Model particles to have nonzero masses. However, because spontaneous symmetry breaking is associated with an energy (and length) scale, its effects are relevant only at low energies—the energy scale of elementary particle masses and smaller (and the weak length scale and bigger). For these energies and masses, the influence of gravity is negligible and the Standard Model (with masses taken into account) correctly describes particle physics measurements. Yet because symmetry is still present in the laws of nature, it allows for sensible high-energy predictions. Plus, as a bonus, the Higgs mechanism explains the photon's zero mass as a result of its not interacting with the Higgs field spread throughout the universe.

However, successful as they are theoretically, we have yet to find experimental evidence that confirms these ideas. Even Peter Higgs has acknowledged the importance of such tests. In 2007, he said that he finds the mathematical structure very satisfying but "if it's not verified experimentally, well, it's just a game. It has to be put to the test."[60] Since we expect that Peter Higgs' proposal is indeed correct, we anticipate an exciting discovery within the next few years. The evidence should appear at the LHC in the form of a particle or particles, and, in the simplest implementation of the idea, the evidence would be the particle known as the *Higgs boson*.

THE SEARCH FOR EXPERIMENTAL EVIDENCE

"Higgs" refers to a person and to a mechanism, but to a putative particle as well. The Higgs boson is the key missing ingredient of the Standard Model.[58] It is the anticipated vestige of the Higgs mechanism that we expect that LHC experiments will find. Its discovery would confirm theoretical considerations and tell us that a Higgs field indeed permeates the vacuum. We have good reasons to believe the Higgs mechanism is at work in the universe, since no one knows how to construct a sensible theory with fundamental particle masses without it. We also believe that some evidence for it should soon appear at the energy scales the LHC is about to probe, and that evidence is likely to be the Higgs boson.

The relationship between the Higgs field, which is part of the Higgs mechanism, and the Higgs boson, which is an actual particle, is subtle— but is very similar to the relationship between an electromagnetic field and a photon. You can feel the effects of a classical magnetic field when you hold a magnet close to your refrigerator, even though no actual physical photons are being produced. A classical Higgs field—a field that exists even in the absence of quantum effects—spreads throughout space and can take a nonzero value that influences particle masses. But that nonzero value for the field can also exist even when space contains no actual particles.

However, if something were to "tickle" the field—that is, add a little energy—that energy could create fluctuations in the field that lead to particle production. In the case of an electromagnetic field, the particle that would be produced is the photon. In the case of the Higgs field, the particle is the Higgs boson. The Higgs field permeates space and is responsible for electroweak symmetry breaking. The Higgs particle, on the other hand, is created from a Higgs field where there is energy—such as at the LHC. The evidence that the Higgs field exists is simply that elementary particles have mass. The discovery of a Higgs boson at the LHC (or anywhere else it could be produced) would confirm our conviction that the Higgs mechanism is the origin of those masses.

Sometimes the press calls the Higgs boson the "God particle," as do many others who seem to find the name intriguing. Reporters like the term because people pay attention, which is why the physicist Leon Lederman was encouraged to use it in the first place. But the term is just a name. The Higgs boson would be a remarkable discovery, but not one whose moniker should be taken in vain.

Although it might sound overly theoretical, the logic for the existence of a new particle playing the role of the Higgs boson is very sound. In addition to the theoretical justification mentioned above, consistency of the theory with massive Standard Model particles requires it. Suppose only particles with mass were part of the underlying theory, but there was no Higgs mechanism to explain the mass. In that case, as the earlier part of the chapter explained, predictions for the interactions of high-energy particles would be nonsensical—and even suggest probabilities that are greater than one. Of course we don't believe that prediction. The Standard Model with no additional structures has to be incomplete. The introduction of additional particles and interactions is the only way out.

A theory with a Higgs boson elegantly avoids high-energy problems. Interactions with the Higgs boson not only change the prediction for high-energy interactions, they exactly cancel the bad high-energy behavior. It's not a coincidence, of course. It's precisely what the Higgs mechanism guarantees. We don't yet know for sure that we have correctly predicted the true implementation of the Higgs mechanism in nature, but physicists are fairly confident that some new particle or particles should appear at the weak scale.

Based on these considerations, we know that whatever saves the theory, be it new particles or interactions, cannot be overly heavy or happen at too high an energy. In the absence of additional particles, flawed predictions would already emerge at energies of about 1 TeV. So not only should the Higgs boson (or something that plays the same role) exist, but it should be light enough for the LHC to find. More precisely, it turns out that unless the Higgs boson is less than about 800 GeV, the Standard Model would make impossible predictions for high-energy interactions.

In reality, we expect the Higgs boson to be a good deal lighter than that. Current theories favor a relatively light Higgs boson—most theoretical clues point to a mass just barely in excess of the current mass bound from the LEP experiments of the 1990s, which is 114 GeV. That was the highest-mass Higgs boson LEP could possibly produce and detect, and many people thought they were on the verge of finding it. Most physicists today expect the Higgs boson mass to be very close to that value, and probably no heavier than about 140 GeV.

The strongest argument for this expectation of a light Higgs boson is based on experimental data—not simply searches for the Higgs boson itself, but measurements of other Standard Model quantities. Standard Model predictions accord with measurements spectacularly well, and even small deviations could affect this agreement. The Higgs boson contributes to Standard Model predictions through quantum effects. If it's overly heavy, those effects would be too large to get agreement between theoretical predictions and data.

Recall that quantum mechanics tells us that virtual particles contribute to any interaction. They briefly appear and disappear from whatever state you started with and contribute to the net interaction. So even though many Standard Model processes don't involve the Higgs boson at all, Higgs particle exchange influences all the Standard Model predictions, such as the rate of decay of a Z gauge boson to quarks and leptons and the ratio between the W and Z masses. The size of the Higgs's virtual effects on these *precision electroweak* tests depends on its mass. And it turns out the predictions work well only if the Higgs mass is not too big.

The second (and more speculative) reason to favor a light Higgs boson has to do with a theory called supersymmetry that we'll turn to shortly. Many physicists believe that supersymmetry exists in nature, and according to supersymmetry, the Higgs boson mass should be close to that of the measured Z gauge boson's and hence relatively light.

So given the expectation that the Higgs boson is not very heavy, you can reasonably ask why we have seen all the Standard Model particles but we have not yet seen the Higgs boson. The answer lies in the Higgs boson's properties. Even if a particle is light, we won't see it unless colliders can make it and detect it. The ability to do so depends on its properties. After all, a particle that didn't interact at all would never be seen, no matter how light it was.

We know a lot about what the Higgs boson's interactions should be because the Higgs boson and Higgs field, though different entities, interact similarly with other elementary particles. So we know about the Higgs field's interactions with elementary particles from the size of their masses. Because the Higgs mechanism is responsible for elementary particle masses, we know the Higgs field interacts most strongly with the heaviest particles. Because the Higgs boson is created from the Higgs field, we know its interactions too. The Higgs boson—like the Higgs field—interacts more strongly with the Standard Model particles that have the biggest mass.

This greater interaction between a Higgs boson and heavier particles implies that the Higgs boson would be more readily produced if you could start off with heavy particles and collide them to produce a Higgs boson. Unfortunately for Higgs boson production, we don't start off with heavy particles at colliders. Think about how the LHC might make Higgs bosons—or any particles for that matter. LHC collisions involve light particles. Their small mass tells us that the interaction with the Higgs particle is so minuscule that if there were no other particles involved in Higgs production, the rate would be far too low to detect anything for any collider we have built so far.

Fortunately, quantum mechanics provides alternatives. Higgs production proceeds in a subtle manner at particle colliders that involves virtual heavy particles. When light quarks collide together, they can make heavy particles that subsequently emit a Higgs boson. For example, light quarks can collide to produce a virtual W, the first picture in gauge boson. This virtual particle can then emit a Higgs boson. (See the first picture in Figure 51 for this production mode.) Because the W boson is so much heavier than either the up or down quarks inside the proton, its interaction with the Higgs boson is significantly greater. With enough proton collisions, the Higgs boson should be produced in this manner.

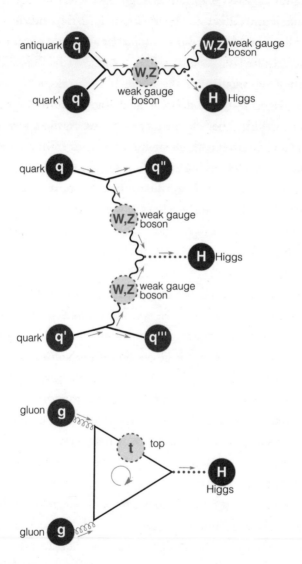

[**FIGURE 51**] Three modes of Higgs production: in order (top to bottom), Higgs-strahlung, *W Z* fusion, and *gg* fusion.

Another mode for Higgs production occurs when quarks emit two virtual weak gauge bosons, which then collide to produce a single Higgs, as seen in the second picture of Figure 51. In this case, the Higgs is produced along with two jets associated with the quarks that scatter off when the gauge bosons are emitted. Both this and the previous production mechanism produce a Higgs but also other particles. In the first case, the Higgs is produced in conjunction with a gauge boson. In the second case, which will be more important at the LHC, the Higgs boson is produced along with jets.

But Higgs bosons can also be made all by themselves. This happens when gluons collide together to make a top quark and an antitop quark that annihilate to produce a Higgs boson, as seen in the third picture. Really, the top quark and antiquark are virtual quarks that don't last a long time, but quantum mechanics tells us this process occurs reasonably often since the top quark interacts so strongly with the Higgs. This production mechanism, unlike the two we just discussed, leaves no trace aside from the Higgs particle, which then decays.

So even though the Higgs itself is not necessarily very heavy—again, it is likely to have mass comparable to the weak gauge bosons and less than that of the top quark—heavy particles such as gauge bosons or top quarks are likely to be involved in its production. Higher-energy collisions, such as those at the LHC, therefore help facilitate Higgs boson production, as does the enormous rate of particle collisions.

But even with a big production rate, another challenge to observing the Higgs boson persists—the manner in which it decays. The Higgs boson, like many other heavier particles, is not stable. Note that it is a Higgs particle, and not the Higgs field, that decays. The Higgs field spreads throughout the vacuum to give mass to elementary particles and doesn't disappear. The Higgs boson is an actual particle. It is the detectable experimental consequence of the Higgs mechanism. Like other particles, it can be produced in colliders. And like other unstable particles, it doesn't last forever. Because the decay happens essentially immediately, the only way to find a Higgs boson is to find its decay products. The Higgs boson decays into the particles with which it interacts—namely, all the particles that acquire mass through the Higgs mechanism and that are sufficiently light to be produced. When a particle and its antiparticle emerge from Higgs boson decay, those particles must each weigh less than half its mass in order to conserve energy. The Higgs particle will decay primarily into the heaviest particles it can produce, given this requirement. The problem is that this means that relatively light Higgs boson only rarely decays into the particles that are easiest to identify and observe.

If the Higgs boson defies expectations and is not light, but turns out to be heavier than twice the W boson mass (but less than twice the top quark mass), the Higgs search will be relatively simple. The Higgs with a big enough mass would decay to the W bosons or Z bosons practically all the time. (See Figure 52 for decay into Ws.) Experimenters know how to identify the Ws and Zs that would remain, so Higgs discovery wouldn't be very hard.

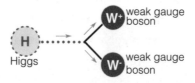

[**FIGURE 52**] A heavy Higgs boson can decay to W gauge bosons.

The next most likely decay mode in this relatively heavy Higgs scenario would involve a bottom quark and its antiparticle. However, the rate for the decay into a bottom quark and its antiparticles would be much smaller because the bottom quark has much smaller mass—and hence much smaller interaction with the Higgs boson—than the W gauge boson. A Higgs heavy enough to decay into Ws will turn into bottom quarks less than one percent of the time. Decays to lighter particles would happen less frequently still. So if the Higgs boson is relatively heavy—heavier than we expect—it will decay to weak gauge bosons. And those decays would be relatively easy to see.

However, as suggested earlier, theory coupled with experimental data about the Standard Model tell us the Higgs boson is likely to be so light that it won't decay into weak gauge bosons. The most frequent decay in this case would be into a bottom quark in conjunction with its antiparticle—the bottom antiquark (see Figure 53)—and this decay is challenging to observe. One problem is that when protons collide, lots of strongly interacting quarks and gluons are produced. And these can easily be confused with the small number of bottom quarks that will emerge from a hypothetical Higgs boson decay. On top of that, so many top quarks will be produced at the LHC that their decays to bottom quarks will also mask the Higgs signal. Theorists and experimenters are hard at work trying to see if there is any way to harness the bottom-antibottom final state of Higgs decay. Even so, despite the bigger rate, this mode probably isn't the most promising way to discover the Higgs at the LHC—though theorists and experimenters are likely to find ways to capitalize on it.

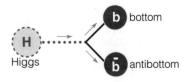

[FIGURE 53] A light Higgs boson will decay primarily to bottom quarks.

So experimenters have to investigate alternative final states from Higgs decays, even though they will occur less frequently. The most promising candidates are tau-antitau or a pair of photons. Recall that taus are the heaviest of the three types of charged leptons and are the heaviest particles aside from bottom quarks that a Higgs boson can decay into. The rate to photons is much smaller—Higgs bosons decay into photons only through quantum virtual effects—but photons are relatively easy to detect. Although the mode is challenging, experiments will be able to measure photon properties so well once enough Higgs bosons decay that they will indeed be able to identify the Higgs boson that decays into them.

In fact, because of the criticality of Higgs discovery, CMS and ATLAS put elaborate and careful search strategies in place to find photons and taus, and the detectors in both experiments were constructed with a view to detecting the Higgs boson in mind. The electromagnetic calorimeters described in Chapter 13 were designed to carefully measure photons while the muon detectors help register decays of the even heavier taus. Together these modes are expected to establish the Higgs boson's existence, and once enough Higgs bosons are detected, we'll learn about its properties.

Both production and decay pose challenges for Higgs boson discovery. But theorists and experimenters and the LHC itself should all be up to the challenge. Physicists hope that within a few years, we will be able to celebrate the discovery of the Higgs boson and learn more about its properties.

HIGGS SECTORS

So we expect to soon find the Higgs boson. In principle, it could be produced in the initial LHC run at half the intended energy, since that is more than sufficient to create the particle. However, we have seen that the Higgs boson will be produced from proton collisions only a small fraction of the time. This means that Higgs particles will be created only when there are many proton collisions—which means high luminosity. The original number of collisions that were scheduled before the LHC would shut down for a year and a half to prepare for its target energy was most likely too small to make enough Higgs bosons to see, but the plan for the LHC to run through 2012 before a year-long shutdown might permit access to the elusive Higgs boson. Certainly, when the LHC runs at full capacity, the luminosity will be high enough and the Higgs boson search will be one of its principal goals.

The search might seem superfluous if we are so confident that the Higgs boson exists (and if the pursuit is so difficult). But it's worth the effort for several reasons. Perhaps most significant, theoretical predictions take us only so far. Most people rightfully trust and believe only in scientific results that have been verified through observations. The Higgs boson is a very different particle from anything anyone has ever discovered. It would be the only fundamental *scalar* ever observed. Unlike particles such as quarks and gauge bosons, scalars—which are particles with zero spin—remain the same when you rotate or boost your system. The only spin-0 particles that have been observed so far are bound states of particles such as quarks that do have nonzero spin. We won't know for certain that a Higgs scalar exists until it emerges and leaves visible evidence in a detector.

Second, even if and when we find the Higgs boson and know for certain of its existence, we will want to know its properties. The mass is the most significant unknown. But learning about its decays is also important. We know what we expect, but we need to measure whether data agree with predictions. This will tell us whether our simple theory of a Higgs field is correct or whether it is part of a more complicated theory. By measuring the Higgs boson's properties, we will gain insights into what else might lie beyond the Standard Model.

For example, if there were two Higgs fields responsible for electroweak symmetry breaking rather than one, it could significantly alter the Higgs boson interactions that would be observed. In alternative models, the rate for Higgs boson production could be different than anticipated. And if other particles charged under Standard Model forces exist, they could influence the relative decay rates of the Higgs boson into the possible final states.

This brings us to the third reason to study the Higgs boson—we don't yet know what really implements the Higgs mechanism. The simplest model—the one this chapter has focused on so far—tells us that the experimental signal will be a single Higgs boson. However, even though we believe the Higgs mechanism is responsible for elementary particle masses, we aren't yet confident about the precise set of particles involved in implementing it. Most people still think we are likely to find a light Higgs boson. If we do, it will be an important confirmation of an important idea.

But alternative models involve more complicated Higgs sectors with an even richer set of predictions. For example, supersymmetric models—to be further considered in the following chapter—predict more particles in the Higgs sector. We would still expect to find the Higgs boson, but its interactions would differ from a model with only a single Higgs particle. On top of that, the other particles in the Higgs sector could give interesting signatures of their own if they are light enough to be produced.

Some models even suggest that a fundamental Higgs scalar does not exist but that the Higgs mechanism is implemented by a more complicated particle that is not fundamental but is rather a bound state of more elementary particles—akin to the paired electrons that give mass to the photon in a superconducting material. If this is the case, the bound state Higgs particle should be surprisingly heavy and have other interaction properties that distinguish it from a fundamental Higgs boson. These models are currently disfavored, since they are hard to match to all experimental observations. Nonetheless, LHC experimenters will search to make sure.

THE HIERARCHY PROBLEM OF PARTICLE PHYSICS

And the Higgs boson is only the tip of the iceberg for what the LHC might find. As interesting as a Higgs boson discovery will be, it is not the only target of LHC experimental searches. Perhaps the chief reason to study the weak scale is that no one thinks the Higgs boson is all that remains to be found. Physicists anticipate that the Higgs boson is but one element of a much richer model that could teach us more about the nature of matter and perhaps even space itself.

This is because the Higgs boson and nothing else leads to another enormous enigma known as the *hierarchy problem*. The hierarchy problem concerns the question of why particle masses—and the Higgs mass in particular—take the values that they do. The weak mass scale that determines elementary particle masses is ten thousand trillion times smaller than another mass scale—the Planck mass that determines the strength of gravitational interactions. (See Figure 54.)

The enormity of the Planck mass relative to the weak mass corresponds to the feebleness of gravity. Gravitational interactions depend on the *inverse* of the Planck mass. If it is as big as we know to be the case, gravity must be extremely weak.

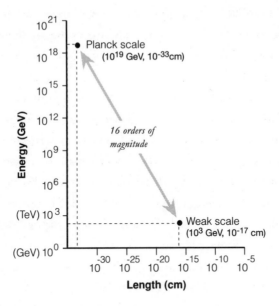

[**FIGURE 54**] The hierarchy problem of particle physics: The weak energy scale is 16 orders of magnitude smaller than the Planck scale associated with gravity. The Planck length scale is correspondingly shorter than the distances probed by the LHC.

The fact is that fundamentally, gravity is by far the weakest known force. Gravity might not seem feeble, but that's because the entire mass of the Earth is pulling on you. If you were instead to consider the gravitational force between two electrons, you would find the force of electromagnetism is 43 orders of magnitude larger. That is, electromagnetism

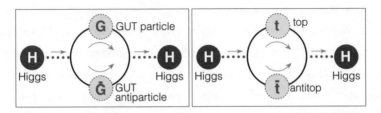

[**FIGURE 55**] Quantum contribution to the Higgs boson mass from a heavy particle—for example with GUT-scale masses—and its antiparticle (*left*) and from a virtual top quark and its antiparticle (*right*).

wins out by 10 million trillion trillion trillion. Gravity acting on elementary particles is completely negligible. The hierarchy problem in this way of thinking is: Why is gravity so much more feeble than the other elementary forces we know?

Particle physicists don't like unexplained large numbers, such as the size of the Planck mass relative to the weak mass. But the problem is even worse than an aesthetic objection to mysterious large numbers. According to quantum field theory, which incorporates quantum mechanics and special relativity, there should be barely any discrepancy at all. The urgency of the hierarchy problem, at least for theorists, is best understood in these terms. Quantum field theory indicates that the weak mass and the Planck mass constant should be about the same.

In quantum field theory, the Planck mass is significant not only because it is the scale at which gravity is strong. It is also the mass at which both gravity and quantum mechanics are essential and physics rules as we know them must break down. However, at lower energies, we do know how to do particle physics calculations using quantum field theory, which underlies many successful predictions that convince physicists that it is correct. In fact, the best measured numbers in all of science agree with predictions based on quantum field theory. Such agreement is no accident.

But the result when we apply similar principles to incorporate quantum mechanical contributions to the Higgs mass due to virtual particles is extraordinarily perplexing. The virtual contributions from just about any particle in the theory seem to give a Higgs particle a mass almost as big as the Planck mass. The intermediate particles could be heavy objects, such as particles with enormous GUT-scale masses (see left-hand-side of Figure 55) or the particles could be ordinary Standard Model particles, such as top quarks (see right-hand-side). Either way, the virtual corrections would make the Higgs mass much too large. The problem is that the allowed energies for the virtual particles being exchanged can be as big as the Planck energy. When this is true, the Higgs mass contribution too can be almost this large. In that case, the mass scale at which the symmetry associated with the weak interactions is spontaneously broken would also be the Planck energy, and that is 16 orders of magnitude—ten thousand trillion times—too high.

The hierarchy problem is a critically important issue for the Standard Model with only a Higgs boson. Technically, a loophole does exist. The Higgs mass, in the absence of virtual contributions, could be enormous and have exactly the value that would cancel the virtual contributions to just the level of precision we need. The problem is that—although possible in principle—this would mean 16 decimal places would have to be canceled. That would be quite a coincidence.

No physicist believes this fudge—or fine-tuning as we call it. We all think the hierarchy problem, as this discrepancy between masses is known, is an indication of something bigger and better in the underlying theory. No simple model seems to address the problem completely. The only promising answers we have involve extensions of the Standard Model with some remarkable features. Along with whatever implements the Higgs mechanism, the solution to the hierarchy problem is the chief search target for the LHC—and the subject of the following chapter.

CHAPTER SEVENTEEN

THE WORLD'S NEXT TOP MODEL

In January 2010, colleagues gathered at a conference in Southern California to discuss particle physics and dark matter searches in the LHC era. The organizer, Maria Spiropulu, a CMS experimenter and member of the Caltech physics department, asked me to give the first talk and outline the LHC's major issues and physics goals for the near future.

Maria wanted a dynamic conference, so she suggested we start with a "duel" among the three opening speakers. As if the term "duel" applied to three people wasn't confusing enough, the audience of invited guests posed an even greater challenge since it ranged from experts in the field to interested observers from the California technology world. Maria asked me to dig deep and look into subtle and overlooked features of current theories and experiments, while one of the attendees, Danny Hillis—a brilliant nonphysicist from the company Applied Minds—suggested I make everything as basic as possible so the nonexperts could follow.

I did what any rational person would do in the face of such contradictory and impossible-to-satisfy advice: procrastinate. The result of my web surfing was my first slide (see Figure 56), which ended up in Dennis Overbye's *New York Times* article on the subject—typo and all.

The topics referred to the subject matter that the subsequent speakers and I were scheduled to cover. But the humor in the sound effects I inserted to accompany the entrance of each of the dueling cats (which I can't reproduce here) was meant to reflect both the enthusiasm and

[**FIGURE 56**] Candidate models, as I presented on a slide at a conference.

the uncertainty associated with each of these models. Everyone at the conference, no matter how strongly convinced of an idea he or she had worked on, knew that data were coming soon. And data would be the final arbiter of who had the last laugh (or a Nobel Prize).

The LHC presents us with a unique opportunity to create new understanding and new knowledge. Particle physicists hope to soon know the answers to the deep questions we have been thinking about: Why do particles have the masses they do? What is dark matter composed of? Do extra dimensions solve the hierarchy problem? Are extra symmetries of spacetime involved? Or is there something completely unforeseen at work?

Proposed answers include models with names like supersymmetry, technicolor, and extra dimensions. The answers could turn out to be different from anything anticipated, but models give us concrete targets of what to look for. This chapter presents a few of the candidate models that address the hierarchy problem and gives a flavor of the type of explorations that the LHC will perform. Searches for these and other models happen concurrently and will provide valuable insights no matter what turns out to be the true theory of nature.

SUPERSYMMETRY

We'll begin with the bizarre symmetry called supersymmetry and the models that incorporate it. If you did a survey among theoretical particle physicists, a good fraction of them would likely say that supersymmetry solves the hierarchy problem. And if you asked experimenters what they wanted to look for, a large fraction of them would suggest supersymmetry as well.

Since the 1970s, many physicists have considered the existence of supersymmetric theories so beautiful and surprising that they believe it has to exist in nature. They have furthermore calculated that forces should have the same strength at high energy in a supersymmetric model—improving on the near-convergence that happens in the Standard Model, allowing the possibility of unification. Many theorists also find supersymmetry to be the most compelling solution to the hierarchy problem, despite the difficulty in making all the details agree with what we know.

Supersymmetric models posit that every fundamental particle of the Standard Model—electrons, quarks, and so on—has a partner in the form of a particle with similar interactions but different quantum mechanical properties. If the world is supersymmetric, then there exist many unknown particles that could soon be found—a supersymmetric partner for every known particle. (See Figure 57.)

Supersymmetric models could help solve the hierarchy problem and,

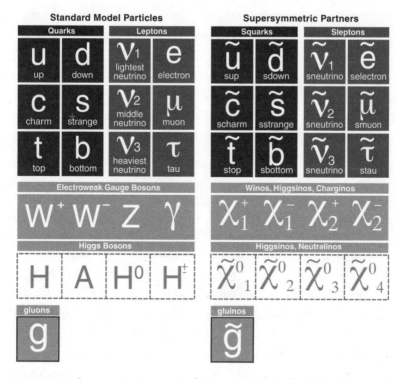

[**FIGURE 57**] In a supersymmetric theory, every Standard Model particle would have a supersymmetric partner. The Higgs sector is also enhanced beyond that of the Standard Model.

if so, would do it in a remarkable fashion. In an exactly supersymmetric model, the virtual contributions from particles and their superpartners cancel exactly. That is, if you add together all the quantum mechanical contributions from every particle in the supersymmetric model and tally their effect on the Higgs boson mass, you would find they all add up to zero. In a supersymmetric model, the Higgs boson would be massless or light, even in the presence of quantum mechanical virtual corrections. In a true supersymmetric theory, the sum of the contributions of both types of particles exactly cancel. (See Figure 58.)

This sounds miraculous perhaps but is guaranteed because supersymmetry is a very special type of symmetry. It's a symmetry of space and time—like the symmetries you are familiar with such as rotations and translations—but it extends them into the quantum regime.

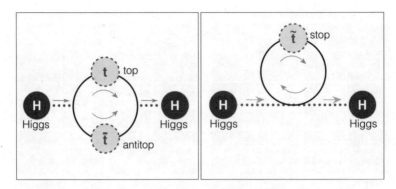

[**FIGURE 58**] In a supersymmetric model, contributions from virtual supersymmetric particles exactly cancel the Standard Model particles' contributions to the Higgs boson mass. For example, the sum of the contributions from the two diagrams above is zero.

Quantum mechanics divides matter into two very different categories—bosons and fermions. Fermions are particles that have half-integer *spin*, where spin is a quantum number that essentially tells us something like how much the particle acts as if it is spinning. Half-integer means values like 1/2, 3/2, 5/2, and so on. The quarks and leptons of the Standard Model are examples of fermions and have spin -1/2. Bosons are particles such as the force-carrying gauge bosons or perhaps the yet-to-be-discovered Higgs boson that have integral spin, indicated by whole numbers such as 0, 1, 2, and so on.

Fermions and bosons are distinguished not only by their spins. They behave very differently when there are two or more of them of the same type. For example, identical fermions with the same properties can never be found in the same place. This is what the *Pauli exclusion principle,* named after the Austrian physicist Wolfgang Pauli, tells us. This fact about fermions accounts for the structure of the periodic table that tells us that electrons, unless distinguished by some quantum number, have to orbit around the nucleus differently from each other. It is also the reason why my chair isn't falling to the center of the Earth, since the fermions in my chair can't be in the same place as the material of the Earth.

Bosons, on the other hand, behave in exactly the opposite manner. They are actually more likely to be found in the same place. Bosons can pile on top of each other—kind of like crocodiles, which is why phenomena like Bose condensates that require many particles to pile up in the same quantum mechanical state can exist. Lasers, too, rely on bosonic photons' affinity for each other. The intense beam is created by the many identical photons that shoot off together.

Remarkably, in a supersymmetric model, particles that we take to be very different—bosons and fermions—can be exchanged in such a way that the result in the end is the same as the theory you started with. Each particle has a partner particle of the opposite quantum mechanical type, but with exactly the same mass and charges. The nomenclature for the new particles is a bit funny—it never fails to elicit giggles when I speak on this topic in public. For example, the fermionic electron is paired with a bosonic *selectron*. A bosonic photon is paired with a fermionic *photino*, and a *W* is paired with a *Wino*. The new particles have related interactions to the Standard Model particles with which they are paired. But they have opposite quantum mechanical properties.

In a supersymmetric theory, the properties of each boson are related to the properties of its superpartnered fermion and vice versa. Since each particle has a partner and the interactions are carefully aligned, the theory permits this bizarre symmetry that interchanges fermions and bosons.

One way to understand the apparently miraculous cancellation of virtual contributions to the Higgs mass is that supersymmetry relates any boson to a partner fermion. In particular, supersymmetry partners the Higgs boson with a Higgs fermion, the Higgsino. Even though quantum mechanical contributions radically influence the mass of a boson, the mass of a fermion will never be much bigger than the *classical mass*, which is the mass before you account for quantum contributions you started out with—even when quantum mechanical corrections are included.

The logic is subtle, but large corrections don't occur because fermion masses involve both left-handed and right-handed particles. Mass terms allow them to convert back and forth into each other. If there were no classical mass term and they couldn't convert into each other before quantum mechanical virtual effects were included, they couldn't do so even with quantum mechanical effects taken into account. If a fermion has no mass to begin with (no classical mass), it will still have zero mass after quantum mechanical contributions are included.

No such argument applies to bosons. The Higgs boson, for example, has zero spin. So there is no sense in which we can talk about a Higgs boson spinning to the left or to the right. But supersymmetry tells us that boson masses are the same as fermion masses. So if the Higgsino mass is zero (or small), so too must be the mass of the partnered Higgs boson in a supersymmetric theory—even when quantum mechanical corrections are taken into account.

We don't yet know if this rather elegant explanation for the stability of the hierarchy and cancellation of large corrections to the Higgs mass is correct. But if supersymmetry does address the hierarchy problem, then we know a lot about what we would expect to find at the LHC. That's because we know what new particles should exist, since every known particle should have a partner. On top of that, we can estimate what the masses of the new supersymmetric particles should be.

Of course, if supersymmetry were exactly preserved in nature, we would know precisely the masses for all the superpartners. They would be identical to the mass of the particle they were paired with. However, none of the superpartners have been observed. That tells us that even if supersymmetry applies in nature, it cannot be exact. If it were, we would have already discovered the selectron and the squark and all the other supersymmetric particles a supersymmetric theory would predict.

So supersymmetry has to be *broken*, meaning the relationships that are predicted in a supersymmetric theory—though possibly approximate—cannot be exact. In a broken supersymmetric theory, every particle would still have a superpartner, but those superpartners would have different masses than their partner Standard Model particles.

However, if supersymmetry were too badly broken, it wouldn't help with the hierarchy problem, since the world would then look as if supersymmetry didn't apply to nature at all. Supersymmetry has to be broken in just such a way that we wouldn't have yet discovered evidence of supersymmetry, while the Higgs mass is nonetheless protected from large quantum mechanical contributions that would give it too big a mass.

This tells us that supersymmetric particles should have weak scale masses. Any lighter and they would have been seen, and any heavier and we would expect the Higgs mass to be heavier as well. We don't know precisely the masses since we only know the Higgs mass approximately. But we do know that if the masses were overly heavy, the hierarchy problem would persist.

So we conclude that if supersymmetry exists in nature and addresses the hierarchy problem, lots of new particles with masses in the range of a few hundred GeV to a few TeV should exist. This is precisely the range of masses the LHC is positioned to search for. The LHC, with 14 TeV of energy, should be able to produce these particles even if only a fraction of the protons' energy goes into quarks and gluons colliding together and making new particles.

The easiest particles to produce at the LHC would be the supersymmetric particles that are charged under the strong nuclear force. These particles could be made in abundance when protons collide (or more specifically the quarks and gluons within them). When these collisions happen, new supersymmetric particles that interact via the strong force can be produced. If so, they will leave very distinctive and characteristic evidence in the detectors.

These *signatures*—the experimental pieces of evidence they leave— depend on what happens to the particle after its creation. Most supersymmetric particles will decay. That's because, in general, lighter particles (such as those in the Standard Model) exist for which the total charge is the same as the heavy supersymmetric particle. If that's the case, the heavy supersymmetric particle will decay into lighter Standard Model particles in a way that conserves the initial charge. Experiments will then detect the Standard Model particles.

That's probably not sufficient to identify supersymmetry. But in almost all supersymmetric models, a supersymmetric particle won't decay solely into Standard Model particles. Another (lighter) supersymmetric particle remains at the end of the decay. That's because supersymmetric particles appear (or disappear) only in pairs. Therefore, a supersymmetric particle has to remain at the end after a supersymmetric particle has decayed—one supersymmetric particle cannot turn into none. Consequently, the lightest such particle must be stable. This lightest particle, which has nothing to decay into, is known to physicists as the lightest supersymmetric particle, the LSP.

Supersymmetric particle decays are distinctive from an experimental vantage point in that the lightest of the neutral supersymmetric particles will remain, even after the decay is complete. Cosmological constraints tell us that the LSP carries no charges, so it won't interact with any elements of the detector. This means that whenever a supersymmetric particle is produced and decays, momentum and energy will appear to be lost. The LSP will disappear from the detector and carry away momentum and energy to where it can't be recorded, leaving as its signature missing energy. Missing energy isn't specific to supersymmetry alone, but since we already know a good deal about the supersymmetric spectrum, we know both what we should and shouldn't see.

For example, suppose a squark, the supersymmetric partner of a quark, is produced. Which particles it can decay into will depend on which of the particles are lighter. One possible mode of decay will always be a squark turning into a quark and the lightest supersymmetric particle. (See Figure 59.) Recall that because decays can occur essentially immediately, the detector records only the decay products. If such a squark decay occurred, detectors would record the passage of the quark in the tracker and in the hadronic calorimeter that measures energy deposited by a strongly interacting particle. But the experiment will also measure that energy and momentum are missing. Experimenters should be able to tell that momentum is missing in the same way they can when neutrinos are produced. They would measure momentum perpendicular to the beam and find that it doesn't add to zero. One of the biggest challenges the experimenters face will be to unambiguously identify this missing momentum. After all, anything that is not detected appears to be missing. If something is wrong or mismeasured and even small amounts of energy go undetected, the missing momentum could add up to mimic an escaping supersymmetric particle's signal, even though nothing exotic was produced.

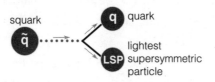

[**FIGURE 59**] A squark can decay into a quark and the lightest supersymmetric particle.

In fact, because the squark is never created on its own, but only in
conjunction with another strongly interacting object (such as another
squark or an antisquark), the experimenters will measure at least two
jets (see Figure 60 for an example). If two squarks are created by a proton
collision, they would give rise to two quarks that detectors would record.
The net missing energy and momentum would escape undetected, but
their absence would be noted and provide evidence for new particles.

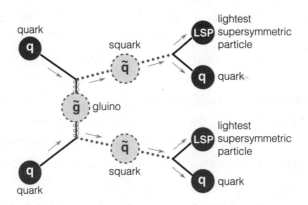

[**FIGURE 60**] The LHC might produce two squarks together, both of
which decay into quarks and LSPs, leaving a missing energy signature.

One major advantage of all the delays in the LHC schedule was that
experimenters had time to fully understand their detectors. They calibrated them so that measurements were very precise from the day the
machine went on line, so missing energy measurements should be robust. Theorists, on the other hand, had time to think about alternative

search strategies for supersymmetric and other models. For example, together with a theorist from Williams College, Dave Tucker-Smith, I found a different—but related—way to search for the squark decay just described. Our method relies on measuring only the momentum and energy of the quarks emerging from the event, with no need to explicitly measure missing momentum, which can be tricky. The great thing about the recent LHC excitement was that a number of CMS experimenters immediately ran with the idea and not only showed that it worked, but generalized and improved it within a few months. It's now part of the standard supersymmetry search strategy and the first supersymmetry search from CMS used the technique we had so recently suggested.[62]

Down the road, even if supersymmetry is discovered, experimenters won't stop there. They will try their best to determine the entire supersymmetric spectrum, and theorists will work to interpret what the results could mean. A lot of interesting theory underlies supersymmetry and the particles that could spontaneously break it. We know which supersymmetric particles should exist if supersymmetry is relevant to the hierarchy problem, but we don't yet know the precise masses they should have or how those masses arise.

Different mass spectra will make an enormous difference to what the LHC should see. Particles can only decay into other particles that are lighter. The decay chain, the sequence of possible decays of supersymmetric particles, depends on the masses—what is heavier and what is lighter. The rates of various processes also depend on particle masses. Heavier particles in general decay more quickly. And they are usually more difficult to produce since only collisions with a good deal of energy can create them. Combining all the results together could give us important insights into what underlies the Standard Model and what awaits at the next energy scales. This will be true of any analysis of new physical theories that we might find.

Nonetheless, one should keep in mind that despite supersymmetry's popularity among physicists, there are several reasons for concern about whether it truly applies to the hierarchy problem and the real world.

The first, and perhaps the most worrisome, is that we have not yet seen any experimental evidence. If supersymmetry exists, the only explanation for why we haven't yet seen evidence is that the superpartners are heavy. But a natural solution to the hierarchy problem would require that superpartners be reasonably light. The heavier the superpartners are, the more inadequate supersymmetry appears as a solution to the hierarchy problem. The fudge required is determined by the ratio of the mass of the Higgs boson to the supersymmetry breaking scale. The bigger this is, the more "fine-tuned" the theory.

Not yet having seen the Higgs boson either compounds the problem. It turns out that in a supersymmetric model, the only way to make the Higgs heavy enough to have eluded detection is to have big quantum mechanical contributions that can come only from heavy superpartners. But again, those masses need to be so heavy that the hierarchy becomes a little unnatural, even with supersymmetry.

The other problem with supersymmetry is the challenge of finding a fully consistent model that includes supersymmetry breaking and agrees with all experimental data to date. Supersymmetry is a very specific symmetry that relates many interactions and prohibits interactions that quantum mechanics would otherwise permit. Once supersymmetry is broken, the "anarchic principle" takes over. Anything that can happen will. Most models would predict decays that have either never been seen in nature or are seen only much too infrequently to agree with predictions. Because of quantum mechanics, a whole can of worms is opened once supersymmetry is broken.

Physicists might simply be missing the right answers. We certainly cannot say definitively that no good models exist or that a little fine-tuning doesn't happen. Certainly, if supersymmetry is the correct resolution of the hierarchy problem, we should find evidence for it soon at the LHC. So it is certainly worth pursuing. A discovery of supersymmetry would mean that this exotic new spacetime symmetry applies not just in a theoretical formulation on a piece of paper, but also in the real world. However, in the absence of discovery, it is also worth considering alternatives. The first we'll consider is known as *technicolor*.

TECHNICOLOR

Back in the 1970s, physicists also first considered an alternative potential solution to the hierarchy problem known as *technicolor*. Models under this rubric involve particles that interact strongly via a new force, playfully named the *technicolor force*. The proposal was that technicolor acts similarly to the strong nuclear force (which is also known as the color force among physicists), but binds particles together at the weak energy scale—not the proton mass scale.

If technicolor is indeed the answer to the hierarchy problem, the LHC wouldn't produce a single fundamental Higgs boson. Instead it would produce a bound state, something like a hadron, that would play the role of the Higgs particle. The experimental evidence in support of technicolor would be lots of bound state particles and many strong interactions—very much like the hadrons we are familiar with, but that appear only at much higher energy—at or above the weak scale.

Not yet having seen any evidence poses a significant constraint on technicolor models. If technicolor is truly the solution to the hierarchy problem, we would expect to have already found evidence—though of course we could be missing something subtle.

On top of that, model building with technicolor is even more challenging than with supersymmetry. Finding models that agree with everything we have observed in nature has posed significant challenges, and no entirely suitable model has been found.

Experimenters will nonetheless keep an open mind and search for technicolor and any other evidence of new strong forces. But hopes are not overly high. If, however, technicolor turns out to be the underlying theory of the world, maybe Microsoft Word will stop automatically spell-correcting and inserting a capital "T" whenever I write about it.

EXTRA DIMENSIONS

Neither supersymmetry nor technicolor are obviously perfect solutions to the hierarchy problem. Supersymmetric theories don't readily accommodate experimentally consistent supersymmetry breaking and deriving technicolor theories that predict the correct quark and lepton masses is even more difficult. So physicists decided to look further afield and considered ideas that are superficially even more speculative alternatives. Remember, even if an idea seems ugly or not obvious at first, only after we fully understand all the implications can we decide which idea is most beautiful—and, more importantly, correct.

The better understanding of string theory and its components that physicists gained in the 1990s led to new suggestions for addressing the hierarchy problem. These ideas were motivated by elements of string theory—though not necessarily directly derived from its very constrained structure—and involved extra dimensions of space. If extra dimensions exist—and we have reason to think they might—they could hold the key to solving the hierarchy problem. If that is indeed the case, they would give rise to experimental evidence of their existence at the LHC.

Additional spatial dimensions is an exotic concept. If the universe has such dimensions, space would be very different from what we observe in our everyday lives. In addition to the three directions—left-right, up-down, forward-backward, or alternatively longitude, latitude, and altitude—space would extend in directions no one has ever observed.

Clearly, since we don't see them, these new dimensions of space must be hidden. That could be because they are too small to directly influence

Man on tightrope   **Ant on tightrope**

[FIGURE 61] A person and a tiny ant experience a tightrope very differently. For the person, it appears to have one dimension, whereas the ant experiences two.

anything we could possibly see, as physicist Oskar Klein suggested back in 1926. The idea is that as much that owing to our limited resolution, the dimensions might be too small to discern. We might not notice a curled-up dimension that we cannot travel through—much as a tight-rope walker would view his path as one-dimensional, whereas a tiny ant on the wire might experience two, as illustrated in Figure 61.[63]

Another possibility is that dimensions can be hidden because space-time is curved or warped, as Einstein taught us will happen in the presence of energy. If the curving is sufficiently dramatic, the effects of the additional dimensions are obscured, as Raman Sundrum and I determined in 1999. This meant that warped geometry might also provide a way in which a dimension might hide.[64]

But why would we even think extra dimensions could be out there if we have never seen them? The history of physics holds many examples of finding things no one could see. No one could "see" atoms and no one could "see" quarks. Yet we now have strong experimental evidence of the existence of both.

No law of physics tells us that only three dimensions of space can exist. Einstein's theory of general relativity works for any number of dimensions. In fact, soon after Einstein completed his theory of gravity, Theodor Kaluza extended Einstein's ideas to suggest the existence of a fourth spatial dimension, and, five years later, Oskar Klein suggested how it might be curled up and differ from the familiar three.

String theory, a leading proposal for a theory combining quantum mechanics and gravity, is another reason physicists currently entertain the notion of extra dimensions. String theory does not obviously lead to the theory of gravity we are familiar with. String theory necessarily involves additional dimensions of space.

People often ask me the number of dimensions that exist in the universe. We don't know. String theory suggests six or seven extra ones. But model builders keep an open mind. It's conceivable that different versions of string theory will lead to other possibilities. In any case, dimensions model builders care about in the following discussions are only the ones that are sufficiently warped or so large that they can affect physical predictions. Other dimensions even smaller than the ones relevant to particle physics phenomena might exist, but we will ignore anything so super-tiny. We again take the effective theory approach and ignore anything too small or invisible to ever make any measurable differences.

String theory also introduces other elements—notably branes—that make for richer possibilities for the geometry of the universe, if indeed it contains extra dimensions. In the 1990s, the string theorist Joe Polchinski established that string theory was not just a theory of one-dimensional objects called strings. He, along with many others, demonstrated that higher-dimensional objects known as branes were also essential to the theory.

The word "brane" derives from "membrane." Like membranes, which

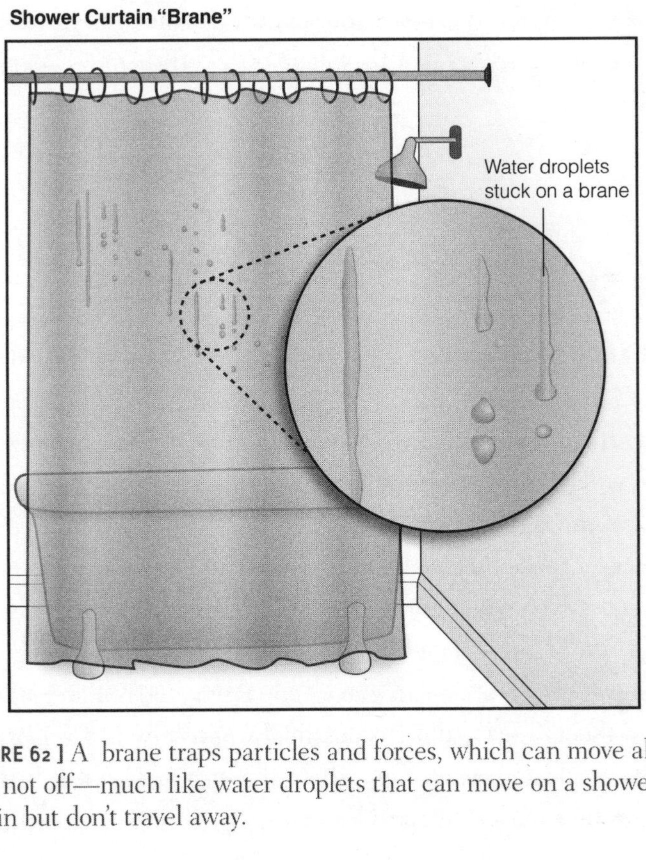

Shower Curtain "Brane"

Water droplets
stuck on a brane

[**FIGURE 62**] A brane traps particles and forces, which can move along it but not off—much like water droplets that can move on a shower curtain but don't travel away.

are two-dimensional surfaces in three-dimensional space, branes are lower-dimensional surfaces in higher-dimensional space. These branes can trap particles and forces so that they don't travel through the full higher-dimensional space. Branes in higher-dimensional space are like a shower curtain in your bathroom, which is a two-dimensional surface in a three-dimensional room. (See Figure 62.) Water droplets might travel only over the two-dimensional surface of the curtain, much as particles and forces might be stuck on the lower-dimensional "surface" of a brane.

Broadly speaking, two types of strings exist: *open strings* that have ends and *closed strings* that form loops like rubber bands. (See Figure 63.) String theorists in the 1990s realized that the ends of open strings can't be just anywhere—they have to end on branes. When particles arise from the oscillations of the open strings that are anchored to a brane, they too are confined there. Particles, the oscillations of those strings, are then stuck. As with water drops on a shower curtain, they can travel along the dimensions of the branes, but they can't travel off them.

[**FIGURE 63**] An open string with two ends, and closed string with none.

String theory suggests the existence of many types of branes, but the ones that will be of most interest for models addressing the hierarchy problem involve those that extend over three dimensions—the three physical dimensions of space that we know. Particles and forces can be trapped on these branes, even when gravity and space extend through more dimensions. (Figure 64 presents a schematic of a braneworld showing a person and a magnet on a brane, with gravity spreading both on and off it.)

String theory's extra dimensions might have physical import for the observable world and so too might three-dimensional branes. Perhaps the most important reason to consider extra dimensions is that they might

[FIGURE 64] Standard Model particles and forces can be stuck on a braneworld that lives in higher-dimensional space. In that case, my cousin Matt, the matter and stars we know, forces such as electromagnetism, and our galaxy and universe all live in its three spatial dimensions. Gravity, on the other hand, can always spread throughout all of space. (Photo courtesy of Marty Rosenberg)

affect visible phenomena, and, in particular, address outstanding puzzles such as the hierarchy problem of particle physics. Extra dimensions and branes could be the key to resolving this question—addressing the issue of why gravity is so weak.

Which brings us to what is perhaps the best reason right now to think about extra dimensions of space. They can have consequences for phenomena we are now trying to understand, and if so, we might see evidence in the imminent future.

Recall that we can phrase the hierarchy problem in two different ways. We can say it is the question of why the Higgs mass—and hence the weak scale—is so much smaller than the Planck mass. This is the question we considered when thinking about supersymmetry and technicolor. But we can also ask an equivalent question: Why is gravity so weak compared to the other known fundamental forces? The strength of gravity depends on the Planck mass scale, the enormous mass ten thousand trillion times greater than the weak scale. The bigger the Planck mass, the weaker the force of gravity. Only when masses are at or near the Planck scale is gravity strong. As long as particles are a good deal lighter than the scale set by the Planck mass, as they are in our world, the force of gravity is extremely weak.

The puzzle of why gravity is so weak is in fact equivalent to the hierarchy problem—the solution of one solves the other. But even though the problems are equivalent, phrasing the hierarchy problem in terms of gravity helps guide our thinking toward extra-dimensional solutions. We'll now delve into a couple of the leading suggestions.

LARGE EXTRA DIMENSIONS AND THE HIERARCHY

Ever since people first started thinking about the hierarchy problem, physicists thought the resolution must involve modified particle interactions at the weak energy scale of about a TeV. With only Standard Model particles, the quantum contributions to the Higgs particle mass are simply too enormous. Something has to kick in to tame the large quantum mechanical contributions to the Higgs particle mass.

Supersymmetry and technicolor are two examples in which new heavy particles might participate in high-energy interactions and cancel the contributions or prevent them from arising in the first place. Until the 1990s, all proposed solutions to the hierarchy problem could be categorized similarly, with new particles and forces and even new symmetries emerging at the weak energy scale.

In 1998, Nima Arkani-Hamed, Savas Dimopoulos, and Gia Dvali[65] proposed an alternative way of addressing the problem. They pointed out that since the problem involves not just the weak energy scale alone, but its ratio to the Planck energy scale associated with gravity, perhaps the problem lay in an incorrect understanding of the basic nature of gravity itself.

They suggested that there is in fact no hierarchy in masses at all—at least with respect to the fundamental scale of gravity compared to the weak scale. Maybe gravity is instead much stronger in the extra-dimensional universe, but is only measured to be so feeble in our three-plus-one-dimensional world because it is diluted throughout all the dimensions that we don't see. Their hypothesis was that the mass scale at which gravity becomes strong in the extra-dimensional universe is in fact the weak mass scale. In that case, we measure gravity to be minuscule in strength not because it is fundamentally weak but rather because it spreads throughout large unseen dimensions.

One way to understand this is to imagine an analogous situation with a water sprinkler. Think about the water that emerges from this sprinkler. If the water spread only in our dimensions, its impact would depend on the amount of water emerging from the hose and how far it had to travel. But if there were additional dimensions to space, the water would spread throughout those dimensions as well after emerging from the end of the hose. We would experience much less water than we would otherwise at a given distance from the source because water would also spread throughout the dimensions we don't observe. (This is illustrated schematically in Figure 65.)

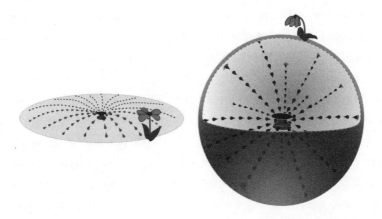

[**FIGURE 65**] The strengths of forces weaken more quickly with distance in a higher-dimensional space than in a lower-dimensional one. This is analogous to a higher-dimensional water sprinkler for which the water dilutes much more quickly with distance. The water spreads more in three dimensions than it spreads in two—in the picture, only the flower receiving water from the lower-dimensional sprinkler is adequately maintained.

If the extra dimensions were of finite size, the water would reach the boundaries of the extra dimensions and no longer spread out. But the amount of water anything would receive at any given place in the extra-dimensional space would be far less than if it had never spread out in those dimensions in the first place.

Similarly, gravity could spread into other dimensions. Even though the force wouldn't spread out forever if the dimensions have finite size, large dimensions would dilute the gravitational force we would experience in our three-dimensional world. If the dimensions were sufficiently large, we would experience very weak gravity, even though the fundamental strength of higher-dimensional gravity could be quite big. Keep in mind, however, that for this idea to work, the extra dimensions have to be enormous compared to what theoretical considerations lead us to expect, since gravity indeed appears so weakly in a three-dimensional world.

Nonetheless, the LHC will subject this idea to experimental tests. Even though the idea now seems improbable, reality and not our ease in finding models is the final arbiter of what is right. If realized in the world, these models would lead to a distinctive characteristic signature. Because higher-dimensional gravity is strong at energies of about the weak scale—the energies that the LHC will generate—particles would collide together and produce a higher-dimensional graviton—the particle that communicates the force of higher-dimensional gravity. But this graviton travels into the extra dimensions. The gravity we are familiar with is extremely weak—far too weak to produce a graviton if there are only three dimensions of space. But in this new scenario, higher-dimensional gravity would be sufficiently strong to produce a graviton at the energies reached by the LHC.

The consequence would be the production of particles known as Kaluza-Klein (KK) modes, which are the manifestation of the higher-dimensional gravitation in three-dimensional space. They are named after Theodor Kaluza and Oskar Klein, who first thought about extra dimensions in our universe. KK particles have interactions similar to those of the particles we know, but with heavier masses. These heavier masses

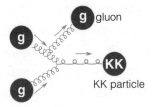

[**FIGURE 66**] In the large extra-dimensional scenario, a Kaluza-Klein partner of the graviton with momentum in the extra dimensions can be produced. If so, it will disappear from the detector, leaving as evidence missing energy and momentum.

are the result of their additional momentum in the direction of the extra dimension. If the KK mode is associated with the graviton—as the large extra dimensional scenario predicts—once produced, it would disappear from the detector. The evidence of its ephemeral visit would be the energy that would go missing. (See Figure 66, in which a KK particle is produced and takes away unseen energy and momentum.)

Of course, missing energy is also characteristic of supersymmetric models. The signals could even appear so similar that even if a discovery is made, people from both extra-dimensional and supersymmetry camps are likely to interpret the data as supporting their expectations—at least initially. But with detailed understanding of the consequences and predictions of both types of models, we will be able to determine which idea—if either—is correct. One of our goals in building models is to match experimental signatures and details to their true implications. Once we have characterized different possibilities, we know the rate and features of the signatures that follow, and we can use subtle features to distinguish among them.

In any case, at this point, along with most of my colleagues, I doubt that the large-extra-dimensional scenario is truly the solution to the hierarchy problem, though we will soon see a very different extra-dimensional example that seems much more promising. For one thing, we don't expect extra dimensions to be so large. It turns out that the extra dimensions would have to be enormous relative to the other scales posed in the problem. Even though the hierarchy between the weak scale and the gravity scale is in principle eliminated, a new hierarchy involving the new dimensions' size gets introduced in this scenario.

Even more worrisome is that in this scenario, we would expect the evolution of the universe to be very different from what has been observed. The problem is that these very large dimensions would expand along with the rest of the universe until the temperatures are very low. For a model to be a potential candidate for reality, the evolution of the universe it predicts would have to mimic that which has been observed that is consistent with only three dimensions of space. That poses a difficult challenge for scenarios with such large additional dimensions.

These challenges are not enough to definitively rule out the idea. Clever enough model builders can find solutions to most problems. But the models tend to become overly complicated and convoluted in order to agree with all observations. Most physicists are skeptical about such ideas on aesthetic grounds. Many have therefore turned to more promising extra-dimensional ideas such as the ones described in the following section. Even so, only experiments will tell us for certain whether models with large extra dimensions apply to the real world or not.

A WARPED EXTRA DIMENSION

Large extra dimensions are not the only potential solution to the hierarchy problem, even in the context of an extra-dimensional universe. Once the door was opened to extra-dimensional ideas, Raman Sundrum and I identified what seems to be a better solution[66]—one that most physicists would agree is much more likely to exist in nature. Mind you, that doesn't mean that most physicists think it is likely to be true. Many suspect that anyone would be lucky to correctly predict what the LHC will reveal or to get a model completely correct without further experimental clues. But it's an idea that probably stands as good a chance as any of being right, and—like most good models—presents clear search strategies so that theorists and experimenters can more fully exploit all the LHC's capabilities—and maybe even discover evidence that the proposal is true.

The solution that Raman and I proposed involves only a single extra dimension, and that dimension need not be large. No new hierarchy involving the dimension's size is necessary. And—as opposed to large-extra-dimensional scenarios—the universe's evolution automatically agrees with late time cosmological observations.

Although our focus is the single new dimension, additional dimensions of space might exist as well—but in this scenario they won't play any discernible role in explaining particle properties. Therefore, we can

[**FIGURE 67**] The Randall-Sundrum setup contains two branes that bound a fourth dimension of space (a fifth dimension of spacetime). In this space, the graviton wavefunction (which tells the probability of finding the graviton at any point in space) decreases exponentially from the Gravitybrane to the Weakbrane.

justifiably ignore them when investigating the hierarchy solution—in accordance with the effective theory approach—and concentrate on the consequences of the single extra dimension.

If the idea that Raman and I had is right, the LHC will soon teach us fascinating properties about the nature of space. It turns out that the universe we suggested is dramatically curved, in accordance with what Einstein taught us about spacetime in the presence of matter and energy. In technical terminology, the geometry we derived from Einstein's equations is "warped" (that really was the pre-existing technical term). What that means is that space and time vary along the single additional dimension of interest. It does so in such a way that space and time, as well as masses and energy, are all rescaled as you move from one place in extra-dimensional space to another, as we will soon get to and is illustrated in Figure 68.

One important consequence of this warped spacetime geometry is that whereas the Higgs particle would have been heavy in some other location in extra-dimensional space, it will have weak scale mass—exactly as should be the case—in the location where we reside. This might sound somewhat arbitrary, but it is not. According to our scenario, there is a brane on which we live—the Weakbrane—and a second brane where gravity is concentrated, known as the Gravitybrane—or among physicists, the Planck brane. This brane would contain another universe that is separated from us in an extra dimension. (See Figure 67.) In this scenario, the second brane would in fact be right next door—separated by an infinitesimal distance, a million trillion trillion times smaller than a centimeter.

The remarkable property that follows from the warped geometry (illustrated in the Figure 67), is that the *graviton*, the particle that communicates the force of gravity, is far more heavily weighted on the other brane than on ours. That would make gravity strong elsewhere in the other dimension, but very weak where we live. In fact, Raman and I found that gravity should be exponentially weaker in our vicinity than on the other brane, thereby giving a natural explanation for the weakness of gravity.

An alternative way of interpreting the consequences of this setup is through the geometry of spacetime, schematically illustrated in Figure

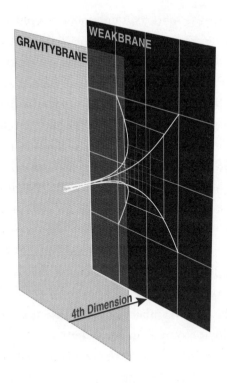

[**FIGURE 68**] Another way to understand why warped geometry solves the hierarchy problem is in terms of the geometry itself. Space, time, energy, and mass all are rescaled exponentially as you go from one brane to the next. In this scenario, it would be very natural to find that the Higgs mass is exponentially smaller than the Planck mass.

68. The scale of spacetime depends on location in the fourth spatial dimension. Masses get exponentially rescaled too—and they do so in a way that the Higgs boson mass is what it needs to be. Although one can debate the assumptions our model relies on—namely, two large flat branes bounding an extra-dimensional universe—the geometry itself follows directly from Einstein's theory of gravity once you postulate the energy carried by the branes and by the extra-dimensional space known as the bulk. Raman and I solved the equations of general relativity. And when we did, we found the geometry I just described—namely, the curved warped space in which masses get rescaled in the way required to solve the hierarchy problem.

Unlike the large extra-dimensional models, the models based on the warped geometry don't replace the old enigma of the hierarchy problem by a new one (why are the extra dimensions so large?). In the warped geometry, the extra dimension is not large. The large numbers arise from an exponential rescaling of space and time. The exponential rescaling makes the ratio of sizes—and masses—of objects enormous, even when those objects are separated only modestly in extra-dimensional space.

The exponential function isn't made up. It arises from the unique solution to Einstein's equations in the scenario we proposed. Raman and I calculated that in the warped geometry, the ratio of the strength of gravity and the weak force is the exponential of the distance between the two branes. If the separation between the two branes has a reasonable value—a few dozens or so in terms of the scale set by gravity—the right hierarchy between masses and the strength of forces naturally emerges.

In the warped geometry, the gravity we experience is weak—not because it is diluted throughout large extra dimensions—but instead because it is concentrated somewhere else: on the other brane. Our gravity arises only as the tail end of what in other regions of the extra-dimensional world feels like a very intense force.

We don't see the other universe on the other brane because the lone shared force is gravity, and gravity is too weak in our vicinity to communicate readily observable signals. In fact, this scenario can be thought of as one example of a *multiverse*, in which the stuff and elements of our world interact very weakly, or in some cases not at all, with the stuff in another world. Most such speculations cannot be tested and will be left to the realm of imagination. After all, if matter is so far distant that light couldn't reach us in the lifetime of the universe, we can't detect it. The "multiverse" scenario that Raman and I proposed is unusual in that the shared gravitational force leads to experimentally testable consequences. We don't directly access the other universe. But particles that travel in the higher-dimensional bulk can come to us.

The most obvious effect of the extra-dimensional world—in the absence of detailed searches such as those at the LHC—would be the explanation for the hierarchy of mass scales that particle physics theories need in order to successfully explain observed phenomena. This of course is not sufficient for us to know if the explanation is the one operational in the world, since it doesn't distinguish among proposed solutions.

However, the higher energy that will be achieved at the LHC should help us discover whether an extra dimension of space is just an outlandish idea or an actual fact about the universe. If our theory is correct, we would expect the LHC to produce Kaluza-Klein modes. Because of the connection to the hierarchy problem, the right energy scale to look for KK modes in this scenario is the one that will be probed at the LHC. They should have mass of about a TeV—the weak mass scale. Once the energy achieved is high enough, these heavy particles might be produced. The discovery of these KK particles would provide the key confirmation that gives us insight into a greatly expanded world.

In fact, the KK modes of the warped geometry have an important and distinctive feature. Whereas the graviton itself has extraordinarily feeble interaction strength—after all, it communicates the extremely weak gravitational force—the KK modes of the graviton interact far more strongly, almost as strongly as the force called the weak force, which is in actuality trillions of times stronger than gravity.

The reason for the KK gravitons' surprisingly strong interaction strength is the warped geometry they travel in. Owing to spacetime's dramatic curvature, the interactions of KK gravitons have far greater strength than those of the graviton that communicates the gravitational force we experience. In the warped geometry, not only do masses get rescaled, but gravitational interactions do as well. Calculations demonstrate that in the warped geometry, KK gravitons have interactions comparable to that of weak scale particles.

This means that unlike supersymmetric models, and unlike large extra-dimensional ones, the experimental evidence for this scenario will not be missing energy where the interesting particle escapes unseen. Instead, it will be a much cleaner, and easier to identify, signature, consisting of the particle decaying inside the detector into Standard Model particles that leave visible tracks. (See Figure 69, in which a KK particle is produced and decays into an electron and positron for example.)

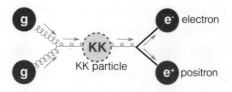

[**FIGURE 69**] In Randall-Sundrum models, a KK graviton can be produced and decay inside the detector into visible particles, such as an
electron and a positron.

This is in fact how experimenters have discovered all new heavy particles so far. They don't see the particles directly. But they observe the
particles that they decay into. That's a lot more information in principle
than would be provided by missing energy. By studying the properties of
these decay products, experimenters can figure out the properties of the
particle that was initially present.

If the warped geometry scenario is correct, we will soon see particle
pairs originating from the decay of KK graviton modes. By measuring
the energies and charges and other properties of the final state particles,
experimenters will be able to deduce the mass and other properties of
the KK particles. These identifying features, along with the relative frequency with which the particle decays into various final states, should
help experimenters determine whether they have discovered a KK graviton or some other new and exotic entity. The model tells us the nature of
the particle that should be found so that physicists can make predictions
to distinguish among the possibilities.

A friend of mine (a screenwriter who both extols and satirizes the
excesses of human nature) doesn't understand how, given the potential
implications of the discoveries that might happen, I'm not sitting on the
edge of my seat waiting for results. Whenever I see him, he insistently
asks me, "Won't the results be life-changing? Might they not confirm
your theories?" He also wants to know, "Why aren't you over there (in
Geneva) talking to people all the time?"

Of course, in some sense his instincts are right. But experimenters
already know what to look for, so much of the job of theorists is already
done. When we have new ideas about what to look for, we communicate

them. We don't necessarily have to be at CERN or even in the same room to do that. Experimenters can be found all over the United States and almost anywhere on the globe for that matter. And remote communication works pretty well—in part due to the initial Internet insight that Tim Berners-Lee had many years ago at CERN.

I also know enough to know what a challenge these searches might be, even once the LHC is fully operational. So I know we might have a bit of a wait. Fortunately for us, the KK modes just described are one of the most straightforward things experimenters can look for. The KK gravitons decay into all particles—after all, every particle experiences gravity—so experimenters can focus on the final states that they find easiest to identify.

However, there are two cautionary notes—two reasons that the searches might be more challenging than initially anticipated and why we might have to wait awhile for discovery, even if the underlying idea is correct.

One is that other candidate models of warped geometry could lead to messier experimental signatures that will be more difficult to find. Models describe the underlying framework—which here involve an extra dimension and branes. They also suggest specific implementations of the general principles the framework embodies. Our original scenario suggested that only gravity was spread throughout the higher-dimensional space known as the *bulk*. But some of us later worked on alternative implementations. In these alternative scenarios, not all particles are on branes. This would mean more KK particles since each bulk particle would have its own KK modes. But it also turns out that these KK particles would be considerably harder to find. This challenge has prompted a great deal of research into how to discover these more elusive scenarios. The investigations that followed will prove useful not only in the search for KK particles, but also for any energetic massive particles that might be present in any new model.

The other reason that searches might prove to be difficult is that KK particles could be heavier than we hope. We know the range of masses we might anticipate for KK particles, but we don't yet know the precise values. If KK particles are nice and light, the LHC will readily produce them in abundance and discovery will be easy. But if the particles are heavier, the LHC might create only a few of them. And if they turn out to be heavier still, the LHC might not produce any at all. In other words, the new particles and new interactions might only be produced or occur at higher energies than the LHC will achieve. This was always a concern for the LHC with its fixed tunnel size and constrained energy reach.

As a theorist, I can only do so much about that. The LHC energy is what it is. But we can try to find subtle clues about the existence of extra dimensions, even if the KK modes turn out to be too heavy. When Patrick Meade and I did our calculations about the production rate of possible higher-dimensional black holes, we focused not only on the negative result—the much lower black hole production rate than had originally been claimed—but also thought about what would happen if higher-dimensional gravity was strong, even if no black holes were produced. We asked whether the LHC might produce any interesting signals of higher-dimensional gravity at all. We found that even without discovering new particles or exotic objects like black holes, experimenters should be able to observe deviations from Standard Model predictions. Discovery is not guaranteed, but experimenters will do everything they can with the existing machine and detectors. In other more advanced research, colleagues have thought about improved methods for finding KK modes, even if Standard Model particles reside in the bulk.

There is also a chance that we could be lucky and that the scales for new particle masses and interactions might turn out to be lower than we anticipate. If that turns out to be the case, we would not only find KK modes sooner than expected, but we would also see other new phenomena. If string theory is the underlying theory of nature and the scale of new physics is low, the LHC could even produce—in addition to KK particles and new interactions—additional particles associated with oscillating underlying strings. These particles would be much too heavy to create under more conventional assumptions. But with warping, there is hope that some string modes will be much lighter than anticipated and could thus appear at the weak energy scale.

Clearly there are several interesting possibilities for warped geometry and we eagerly await experimental results. If the consequences of this geometry are discovered, they will change our view of the nature of the universe. But we will only know which—if any—of these possibilities is realized in nature after the LHC has done its search.

REDUX

Experiments at the LHC are currently testing all the ideas in this chapter. We hope that if any of these models are right, hints will soon appear. There might be solid evidence like KK modes, or there might be subtle changes to Standard Model processes. Either way, both theorists and experimenters are alert and waiting. Every time the LHC does or does not see something, it constrains the possibilities further. And if we're lucky, one of the ideas that have been discussed might prove right. As we learn more about what the LHC will produce and how detectors work, we will hopefully also learn more about how to extend the LHC's reach to test as large a range of possibilities as possible. And as data become available, theorists will incorporate that data into their proposals.

We don't know how long it will be before we start getting answers since we don't know what is there or what the masses and interactions might be. Some discoveries may happen within a year or two. Others could take more than a decade. Some might even require higher energies than the LHC will ever achieve. The wait is a little anxiety provoking, but the results will be mind-blowing. That should make the nail-biting worth it. They could change our view of the underlying nature of reality, or at least the matter of which we are composed. When the results are in, whole new worlds could emerge. Within our lifetimes, we just might see the universe very differently.

CHAPTER EIGHTEEN

BOTTOM-UP VERSUS TOP-DOWN

Nothing substitutes for solid experimental results. But we physicists haven't just been sitting on our thumbs for the last quarter century waiting for the LHC to turn on and produce meaningful data. We've thought long and hard about what it is that experiments should look for and what the implications of the data are likely to be. We have also studied results from experiments that have been operational during this time frame, and these have taught us details about known particles and interactions and helped orient our thinking.

This interim period has also been a tremendous opportunity to think more deeply about ideas that at least for the time being are more removed from data. Some of the more interesting and speculative models and theoretical insights of the last twenty-five years resulted from these more mathematical pursuits. I doubt that I, for one, would have thought about extra dimensions or more mathematical aspects of supersymmetry had data been more plentiful. Even if measurements that would ultimately support these ideas had been made, the implications would have taken a while to unravel without the luxury of previous mathematical pursuits.

Experiments and mathematics both lead to scientific advances. But the road to progress is rarely clear, and physicists have been divided as to the best strategy. Model builders use the "bottom-up" approach introduced in Chapter 15 to start with what is known from experiments and then address residual puzzling unexplained features—often employing more theoretical mathematical developments. The last chapter presented some specific examples of models and how they influence the searches experimenters at the LHC will perform.

Others, most notably string theorists, apply a "top-down" way of thinking, in which they start with the theory that they believe is true—namely string theory—and try to use its underlying concepts to formulate a consistent quantum theory of gravity. Top-down theories are defined at high energies and small distances. The label refers to the theoretical notion that everything can be derived from fundamental premises defined at high-energy scales. Although the name can be confusing since high energies correspond to short distances, recall that the ingredients at small distances are the fundamental building blocks of matter. In this way of thinking, everything can be derived from basic principles and fundamental ingredients, which are defined at small distances and high energies—hence the label "top-down."

This chapter is about the top-down and bottom-up approaches and the ways in which they contrast with each other. We'll explore the differences, but also reflect on how they occasionally converge to yield remarkable insights.

STRING THEORY

Unlike model builders, more mathematically inclined physicists try to work from pure theory. The hope is to start with a single elegant theory and derive the consequences, and only then apply the ideas to data. Most any attempt at a unified theory embodies such a top-down approach. String theory is perhaps the most prominent such example. It is a conjecture for the ultimate underlying framework from which all other known physics phenomena would in principle follow.

String theorists take a major leap in the physics scales they try to conquer—jumping from the weak scale to the Planck scale at which gravity becomes strong. Experiments probably won't directly test these ideas anytime soon (although the extra-dimensional models of the last chapter might be an exception). But even though string theory itself is difficult to test, elements of string theory do provide ideas and concepts that potentially observable models have incorporated.

The question physicists ask when deciding on model building versus string theory is whether to follow the Platonic approach, which tries to gain insights from some more fundamental truth, or the Aristotelian one, rooted in empirical observations. Do you take the "top-down" or the "bottom-up" approach? The choice could also be phrased as "Old Einstein versus Young Einstein." Einstein originally did thought experiments that were grounded in physical situations. Nonetheless, he also valued beauty and elegance. Even when an experimental result contradicted his ideas about special relativity, he confidently (and ultimately correctly) decided that the experiment had to be wrong since its implications would have been too ugly to believe.

Einstein became more mathematically inclined after mathematics helped him finally complete his theory of general relativity. Since mathematical advances were crucial to completing his theory, he had more faith in theoretical methods later in his career. Looking to Einstein won't resolve the issue, however. Despite his successful application of mathematics to general relativity, his later mathematical search for a unified theory never reached fruition.

The Grand Unified Theory proposed by Howard Georgi and Sheldon Glashow was also a top-down idea. GUTs, as they were known, were rooted in data—the inspiration for their conjecture was the particular set of particles and forces that exist in the Standard Model and the strength with which they interact—but the theory extrapolated from what we know to what might be happening at very distant energy scales.

Interestingly, even though the unification would happen at an energy much higher than a particle accelerator could achieve, the initial model for a GUT made a prediction that was potentially observable. The Georgi-Glashow GUT model predicted that the proton would decay. The decay would take a long time, but experimenters set up giant vats of material with the hopes that at least one of the protons inside would decay and leave a visible signal. When that didn't happen, the original GUT model was ruled out.

Since that time neither Georgi nor Glashow has chosen to work on any top-down theory that makes such a dramatic leap in energies from those we can directly access in accelerators to those so far removed that they might have only subtle experimental consequences—or likelier still, not any. They decided it would just be too ridiculously unlikely to make a correct guess about a theory so many orders of magnitude away in distance and energy from anything we currently understand.

Despite their reservations, many other physicists decided that a top-down approach was the only way to attack certain difficult theoretical issues. String theorists chose to work in a netherworld that isn't clearly traditional science but has led to a rich, if controversial, set of ideas. They understand some aspects of their theory, but they are still piecing it together—looking for the key underlying principles as they go along and develop their radical ideas.

The motivation for string theory as a theory of gravity didn't come from data, but from theoretical puzzles. String theory provides a natural candidate for the graviton, the particle quantum mechanics tells us should exist and communicate the force of gravity. It is currently the leading candidate for a fully consistent theory of quantum gravity—a theory that includes both quantum mechanics and Einstein's theory of general relativity, and that works at all conceivable energy scales.

Physicists can use known theories to reliably make predictions at small distances, such as the inside of an atom, where quantum mechanics plays a big role and gravity is negligible. Because gravity has such feeble influence on atomic-mass particles, we can use quantum mechanics and safely ignore gravity. Physicists can also make predictions about phenomena at large distances, such as inside galaxies, where gravity dominates predictions and quantum mechanics can be ignored.

However, we lack a theory that includes both quantum mechanics and gravity—and works at all possible energies and distances. In particular, we don't know how to calculate at enormously high energies and extremely short distances—comparable to the Planck energy or length. Because the influence of gravity is bigger for heavier and more energetic particles, gravity acting on Planck mass particles would play an essential role. And at the tiny Planck length, quantum mechanics would too.

Although this problem doesn't spoil any calculations for observable phenomena—certainly not those at the LHC—it does mean theoretical physics is incomplete. Physicists don't yet know how to consistently include quantum mechanics and gravity at extremely high energies or short distances where both have comparable importance for predictions and neither can be neglected. This important gap in our understanding could potentially point the way forward. Many think string theory could be the resolution.

The name "string theory" derives from the fundamental oscillating string that formed the core of the initial formulation. Particles exist in string theory, but they arise from the vibrations of a string. Different particles correspond to different oscillations, much as different notes arise from a vibrating violin string. In principle, experimental evidence for string theory should consist of new particles that would correspond to the many additional vibrational modes that a string can produce.

However, most such particles are likely to be much too heavy to ever observe, and that's why it's so difficult to experimentally verify whether string theory is realized in nature. String theory's equations describe objects that are so incredibly tiny and that possess such extraordinarily high energy that any detector we could even imagine would be unlikely to ever see them. It is defined at an energy scale that is about 10 million billion times larger than those we can experimentally explore with current instruments. At present, we still don't even know what will happen when the energy of particle colliders increases by a factor of 10.

String theorists can't uniquely predict what happens at experimentally accessible energies since the particle content and other properties depends on the as yet undetermined configuration of fundamental ingredients in the theory. String theory's consequences in nature depend on how the elements arrange themselves. As it is currently formulated, string theory contains more particles, more forces, and more dimensions than we see in our world. What is it that distinguishes those particles, forces, and dimensions that are visible from those that are not?

For example, space in string theory is not necessarily the space we see around us—space with three dimensions. Instead, string theory's gravity describes six or seven additional dimensions of space. A workable version of string theory has to explain how the invisible extra dimensions are different from the three we know. As fascinating and remarkable as string theory is, puzzling features like its extra dimensions obscure its connection to the visible universe.

To get from the high energy at which string theory is defined to predictions about measurable energies, we need to deduce what the original theory will look like with the heavier particles removed. However, there are many possible manifestations of string theory at accessible energies, and we don't yet know how to distinguish among the enormous range of possibilities, or even how to find the one that looks like our world. The problem is that we don't yet understand string theory sufficiently well to derive its consequences at the energies we see. The theory's predictions are hindered by its complexity. Not only is the challenge mathematically difficult, it is not even always clear how to organize string theory's ingredients and determine which mathematical problem to solve.

On top of that, we now know that string theory is much more complex than physicists originally thought and involves many other ingredients with different dimensionalities—notably branes. The name string theory still generally survives, but physicists also talk about M-theory, although no one really knows what the "M" stands for.

String theory is a magnificent theory that has already led to profound mathematical and physical insights, and it might well contain the correct ingredients to ultimately describe nature. Unfortunately, an enormous theoretical gulf separates the theory as it is currently understood from predictions that describe our world.

Ultimately, if string theory is correct, all the models that describe real-world phenomena should be derivable from its fundamental premises. But its initial formulation is abstract, and its connection to observable phenomena is remote. We would have to be very lucky to find all the correct physical principles that will make string theoretical predictions match our world. That is string theory's ultimate goal, but it is a daunting task.

Although elegance and simplicity can be the hallmarks of a correct theory, we can only really judge a theory's beauty when we have a reasonably comprehensive understanding of how it works. Discovering how and why nature hides string theory's extra dimensions would be a stunning achievement. Physicists want to figure out how this occurs.

THE LANDSCAPE

As I joked in *Warped Passages,* most attempts to make string theory realistic have had something of the flavor of cosmetic surgery. In order to make string theory conform to our world, theorists have to find ways to hide the pieces that shouldn't be there, removing particles from view and tucking dimensions demurely away. But although the resulting sets of particles come tantalizingly close to the correct set, you can nonetheless tell that they aren't quite right.

More recent attempts to make string theory realistic have something of the flavor of a casting call. Although most ingenues can't act very well and some have frozen faces that don't let them emote, with enough auditions, a beautiful talented actor might show up.

Similarly, some ideas about string theory also rely on our universe being the rare but ideal configuration of its ingredients. Even if string theory does ultimately unify all the known forces and particles, it might contain a single stable basin representing a particular set of particles, forces, and interactions, or more likely, a more complicated landscape with many possible hills and valleys and a variety of possible implications.

According to recent research, string theory can manifest itself in many possible universes in a scenario corresponding to a *multiverse.* The different universes can be so far apart that they never interact—even through gravity—over their lifetimes. In that case, completely different evolution can occur in each of these universes, and we would end up in only one of them.

If these universes existed and there were no way of populating them, we would be justified in ignoring all but our own. But cosmological evolution provides ways to create all of them. And the different universes can have significantly different properties, with different matter, forces, or energy.

Some physicists employ the idea of the landscape in conjunction with the *anthropic principle* to try to address the particularly thorny questions in string theory and particle physics. The anthropic principle tells us that since we live in a universe that permits galaxies and life, certain parameters must take values at or near the values they do—or we would never be here to ask the question. For example, the universe couldn't have so much energy that it would expand at a rate too quickly for matter to collapse into cosmic structures.

If this is the case, we need to determine what physical features, if any, favor one configuration of particles and forces and energy over another. We don't even know which properties should be predictable and which are simply necessary for us to be sitting around discussing science in the first place. Which properties have fundamental explanations and which are an accident of location?

Personally, I believe a landscape of many possible configurations where we might reside is reasonably likely since there are many possible solutions to any set of equations for gravity we write down, and I don't see any reason why what we observe should be all there is. But I find the anthropic principle as a way of explaining observed phenomena unsatisfying. The problem is we never know whether the anthropic principle suffices. Which phenomena should we be able to uniquely predict and which are determined by "just so" stories? On top of that, an anthropic explanation cannot be tested. It might turn out to be correct. But it will certainly be abandoned if a more fundamental explanation from first principles comes along.

BACK ON SOLID GROUND

String theory very likely contains some deep and promising ideas. It has already given us insights into quantum gravity and mathematics and provided interesting ingredients for model builders to pursue. But it will most likely be a long time before we can solve the theory sufficiently to answer the questions we would most like to solve. Deriving string theory's consequences for the real world directly from scratch might just be too difficult. Even if successful models ultimately arise from string theory, the clutter of superfluous elements makes them very difficult to find.

The model-building approach in physics is fueled by the instinct that the energies at which string theory makes definite predictions are too remote from those we can observe. As with many phenomena that have different descriptions on different scales, it could be that the mechanisms that address questions in particle physics are best studied at the relevant energies.

Physicists share common goals, but we have different expectations about how best to achieve them. I prefer the model-building approach because it is more likely to receive experimental guidance in the near future. My colleagues and I might use ideas from string theory, and some of our research might have string theory implications, but applying string theory is not my primary goal. Understanding testable phenomena is. Models can be described and subjected to experimental tests, even before any connection to a more fundamental theory is made.

Model builders pragmatically admit that we can't derive everything at once. A model's assumptions could be part of the ultimate underlying theory, or they might simply illuminate new relationships that have still deeper theoretical underpinnings. Models are effective theories. Once a model proves correct, it can provide direction for string theorists, or anyone attempting a more top-down approach. And models already benefit from the rich set of ideas that string theory provides. But models primarily focus on lower energies, and experiments that apply at these scales.

Models that go beyond the Standard Model incorporate its ingredients as well as the results at energies that have already been explored, but they also contain new forces, new particles, and new interactions that can be seen only at shorter distances. Even so, fitting everything we know is difficult, and the resulting precise model that I or anyone else works on often loses much of its initial elegance. For this reason, model builders need to have open minds.

People are often puzzled when I tell them that I work on many different models when I know that they can't all be correct and that the LHC should tell us more about which could be right. They are even more surprised when I explain that I don't necessarily assign huge probabilities to any particular model I am thinking about. Nonetheless, I choose projects that illuminate a genuinely new explanatory principle or new type of experimental search. The models I consider generally have some interesting feature or mechanism that provides interesting potential explanations for mysterious phenomena. Given the many unknowns—and uncertain criteria for progress—predicting and interpreting reality poses formidable challenges. It would be miraculous to get it all right from the get-go.

One of the beautiful aspects of the extra-dimensional theories is that ideas from both the top-down and bottom-up camps converged to produce them. String theorists recognized the critical role of branes in their theoretical formulations. And model builders realized that by reinterpreting the hierarchy problem as a question about gravity, they could find alternative solutions.

The Large Hadron Collider is now testing such ideas. Whatever the LHC discovers will guide and constrain model building in the future. With its higher-energy experimental results, we'll be able to piece together observations to determine what is right. Even if observations don't conform to any one particular proposal, the lessons we learned from constructing those models will help narrow down the possibilities for which theory is ultimately correct.

Model building helps us recognize the possibilities, suggest experimental searches, and interpret data once they are available. We might be lucky and get it right. But model building also gives us insights into what to look for. Even if no particular model's predictions turn out to be completely correct, they will help us deduce the implications of any new experimental result. The results will distinguish among the many ideas and determine which—if any—of the specific implementations correctly describes reality. If no current proposal works, data will nonetheless help determine what the right model might be.

High-energy experiments are not merely searching for new particles. They are searching for the structure of underlying physical laws with even greater explanatory power. Until experiments help determine the answers, we are all just making guesses. For now we'll apply aesthetic criteria (or prejudice) to favor certain models. But when experiments reach the energies or distances and statistics necessary to distinguish among models, we will know much more. Experimental results, such as those we hope that the LHC will provide, will determine which of our conjectures are correct and help us establish the underlying nature of reality.

Part V:
SCALING THE UNIVERSE

CHAPTER NINETEEN

INSIDE OUT

Back when I was in elementary school, I woke up one morning to read the bewildering news that the universe (at least in our understanding) had suddenly aged by a factor of two. I was astonished by this revision. How could something as important as the universe's age be at liberty to change so radically without destroying everything else about it that we knew?

Today my surprise works in the opposite direction. I am stunned by how much we can precisely measure now about the universe and its history. Not only do we know the universe's age much more accurately than ever before, but we know how the universe grew with time, how nuclei were formed, and how galaxies and clusters of galaxies began their evolution. Before, we had a qualitative picture of what had happened. Now we have an accurate scientific picture.

Cosmology has recently entered a remarkable era in which revolutionary advances, both experimental and theoretical, have precipitated a more extensive and detailed description than anyone would have believed possible even 20 years ago. By combining improved experimental methods with calculations rooted in general relativity and particle physics, physicists have established a detailed picture of what the universe looked like in its earlier stages and how it evolved into its form today.

So far, this book has focused primarily on smaller scales at which we examine the inner nature of matter. Having reached the current limit of our inward journey, let's now complete the tour over distance scales we began in Chapter 5 and turn our attention outward to consider the sizes of objects in the outer universe.

We need to be wary of one big difference in this journey to cosmic scales since we can't neatly characterize all aspects of the universe according to size alone. Observations don't just record the universe today. Because of the finite speed of light, they also look back in time. Structures we observe today can be early universe occupants whose light reached our telescopes only billions of years after being emitted. The size of the current greatly expanded universe we now see encompasses many times the size of the universe earlier on.

Size nevertheless plays a critical role in characterizing our observations—both of the current universe and its history over time, and this chapter explores both. In the second half, we'll consider the evolution of the universe as a whole, from its tiny initial size to the vast structure we now observe. But first we'll look out at the universe as it appears today in order to familiarize ourselves with some of the lengths that characterize what surrounds us. We'll work our way up in scales to consider larger sizes and more distant objects—on Earth and in the cosmos—to get a feeling for the bigger types of structures that are out there to explore. This tour of large scales will be briefer than our earlier tour of matter's interior. Despite the richness of structure in the universe, most of what we see can be explained with known physical laws—not fundamental, new ones. Star and galaxy formation rely on known laws of chemistry and electromagnetism—science rooted in the small scales we have already discussed. Gravity, however, now plays a critical role as well, and the best description will depend on the speed and density of the objects it is acting on, leading to varying theoretical descriptions in this case too.

TOUR OF THE UNIVERSE

The book and film *Powers of Ten,*[67] one of the iconic tours of distance scales, starts and ends with a couple sitting in Grant Park in Chicago— as good a place to begin our journey as any. Let's momentarily pause on

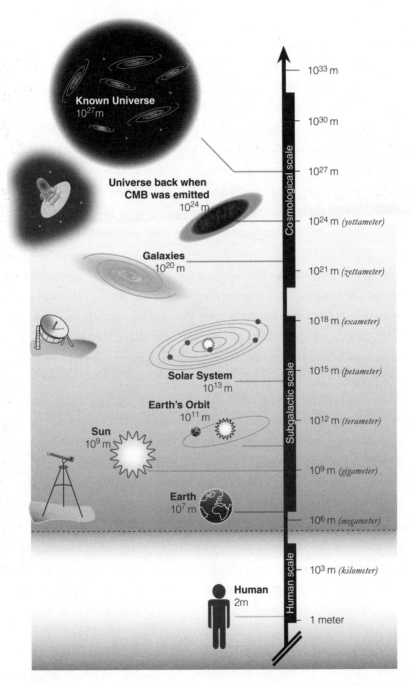

Known Universe
10^{27} m

10^{33} m

10^{30} m

10^{27} m

Cosmological scale

Universe back when CMB was emitted
10^{24} m

10^{24} m *(yottameter)*

Galaxies
10^{20} m

10^{21} m *(zettameter)*

10^{18} m *(exameter)*

Solar System
10^{13} m

10^{15} m *(petameter)*

Subgalactic scale

Earth's Orbit
10^{11} m

10^{12} m *(terameter)*

Sun
10^{9} m

10^{9} m *(gigameter)*

Earth
10^{7} m

10^{6} m *(megameter)*

10^{3} m *(kilometer)*

Human scale

Human
2m

1 meter

[FIGURE 70] A tour of large scales, and the length units that are used to describe them.

(what we now know to be largely empty) solid ground to view the familiar lengths and sizes around us. After momentarily reflecting on their human scale of about a couple of meters' height, let's take leave of this comfortable resting place and ascend to greater heights and sizes. (Refer to Figure 70 for a sampling of the scales this chapter explores.)

One of the more spectacular demonstrations of human response to height that I've seen occurred during a performance of Elizabeth Streb's dance company. Her dancers (or "action engineers") fall onto their stomachs from a rail raised higher and higher until the final dancer falls a full 30 feet. That is definitely beyond our comfort zone as the many gasps in the audience make abundantly clear. People shouldn't fall from that height—certainly not onto their faces.

Though maybe not so dramatic, most tall buildings inspire strong reactions too, ranging from awe to alienation. One of the challenges architects face is to humanize structures that are so much bigger than we are. Buildings and structures vary in size and shape, but our response to them inevitably reflects our psychological and physiological attitudes toward size.

The world's tallest man-made structure is Burj Khalifa in Dubai, United Arab Emirates, which stands 828 meters (2,717 ft.) tall. That is dauntingly high, but it's largely empty and the movie *Mission Impossible 4* probably won't confer on it the same cultural status that *King Kong* gave to the Empire State Building. New York's iconic 381-meter building stands at less than half the height of Burj Khalifa. However, to its credit, it has a much higher occupancy rate.

We live in a world surrounded by much larger natural entities, many of which inspire awe. In the vertical direction, Mt. Everest, at 8.8 kilometers, is the highest peak on Earth. Mt. Blanc, the tallest mountain in Europe (at least if you're not from the country of Georgia), is only about half as high—but I was still pretty happy years ago when I made it to the summit—though my friend and I look pretty miserable in the photo we took at the top. At 11 kilometers deep, the Mariana Trench is the deepest known place in the ocean, and the lowest elevation of the Earth's surface crust. This otherworldly trench was the director James Cameron's destination once he had successfully conquered three-dimensional imagery with his successful movie *Avatar*.

Natural bodies spread on the Earth's surface over far more extended regions. The Pacific Ocean, for example, is about 20 million meters wide, while Russia—at nearly eight million meters across—is almost half the extent of that. The nearly spherical Earth itself is some 12 million meters in diameter, with a circumference about three times as big. The United States, at 4.2 million meters across, is about a tenth this wide, but is still bigger than the diameter of the Moon, which measures about 3.6 million meters.

Objects in outer space have a large range of sizes as well. Asteroids, for example, vary quite a bit—tiny ones can be as small as pebbles, while bigger ones are far greater than any feature on Earth. At approximately a billion meters across, the Sun is about 100 times the size of the Earth. And the solar system, which I'll take to be roughly the distance from the Sun to Pluto (which is in the solar system whether or not it merits planet status) is about 7,000 times the radius of the Sun.

The distance from the Earth to the Sun is considerably smaller—a mere 100 billion meters—a hundredth of a thousandth of a light-year. A light-year is the distance light can travel in a year—the product of 300 million meters/second (the speed of light) and 30 million seconds (the number of seconds in a year). Because of this finite speed of light, the illumination we see from the Sun is already about eight minutes old.

Many visible structures, of varying shapes and sizes, exist within our vast universe. Astronomers have organized most astral bodies according to type. To set some scales, galaxies are typically about 30,000 light-years or 3×10^{20} meters across. That includes our galaxy—the Milky Way—which is about three times that size. Galaxy clusters, which contain from tens to thousands of galaxies, are about 10^{23} meters in size, or 10 million light-years big. Light takes about 10 million years to traverse from one end of a galaxy cluster to the other.

Yet despite the huge range of sizes, most of these bodies act in accordance with Newton's laws. The orbit of the Moon, like the orbit of Pluto, or the orbit of the Earth itself, can be explained in terms of Newtonian gravity. Based on the planet's distances from the Sun, its orbit can be predicted with Newton's gravitational force law. That's the same law that caused Newton's apple to fall to Earth.

Nonetheless, more precise measurements of planetary orbits revealed that Newton's laws were not the final word. General relativity was needed to explain the precession of the perihelion of Mercury, which is the observed change in its orbit around the Sun over time. General relativity is a more comprehensive theory that includes Newton's laws when densities are low and speeds are small, but also works outside these restrictions.

General relativity isn't needed to describe most objects however. But its effects can accumulate over time, and are prominent when objects are sufficiently dense, as with black holes. The black hole at the center of our galaxy is about 10 trillion (10^{13}) meters in radius. The enclosed mass is very large—about 4 million times the mass of the Sun—and, as with all other black holes, requires general relativity to describe its gravitational properties.

The entire visible universe is currently 100 billion light-years across—10^{27} meters, a million times the size of our galaxy. That is enormous and superficially surprising since it is bigger than the distance we can actually observe, 13.75 billion years since the time of the Big Bang. Nothing is supposed to travel faster than light speed so with the universe only 13.75 billion years old, this size might seem impossible.

However, no such contradiction exists. The reason the universe as a whole is bigger than the distance a signal could have traveled given its age is that space itself has expanded. General relativity plays a big role in understanding this phenomenon. Its equations tell us that the very fabric of space has expanded. We can observe places in the universe that are that far apart, even though they cannot see each other.

Given the finite speed of light and the finite age of the universe, this section has now taken us to the limit of observable sizes. The visible universe is all our telescopes can access. Nonetheless, the size of the universe is almost certainly not limited to what we can see. As with small scales, where we can make conjectures that extend beyond current experimental constraints, we can also consider what exists beyond the observable universe. The only limit to the largest sizes we can think about is our imagination, and our patience for contemplating structure that we can't hope to observe.

We really don't know what exists beyond the *horizon*—the boundary of the observable universe. The limits to our observations allow for the possibility of new and exotic phenomena beyond. Different structures, different dimensions, and even different laws of physics can in principle apply so long as they don't contradict anything that has been observed. That doesn't mean every possibility is realized in nature, as my astrophysics colleague Max Tegmark sometimes asserts. However, it does mean there are many possibilities for what can be out there.

We don't yet know if other dimensions or other universes exist. Really, we can't even say with certainty whether the universe as a whole is finite or infinite, though most of us think it's likely to be the latter. No measurement shows any sign of its ending, but measurements only reach so far. In principle, the universe could end, or even have the shape of a ball or balloon. But no theoretical or experimental clue leads us in that direction at present.

Most physicists prefer not to think too much about the regime beyond the visible universe, since we are unlikely to ever know what is there. However, any theory of gravity or quantum gravity gives us the mathematical tools to contemplate the geometry of what might exist. Based on theoretical methods and ideas about extra dimensions of space, physicists sometimes consider exotic other universes, which are not in contact with us over the lifetime of our universe or are only in contact via gravity. As discussed in Chapter 18, string theorists and others contemplate the existence of a multiverse that contains many disconnected independent universes that are consistent with string theory's equations, sometimes combining these ideas with the anthropic principle that exploits the possible riches of universes that might exist. Some even try to find observable signatures of such multiverses for the future. As we saw in Chapter 17, in one distinct scenario, a two-brane "multiverse" might even help us understand questions in particle physics and in that case have testable consequences. But most additional universes, though conceivable and maybe even likely, will probably remain beyond the realm of experimental testability for the foreseeable future. They will then remain theoretical abstract possibilities.

THE BIG BANG:
FROM SMALL TO LARGE THROUGH TIME

Now that we've ventured out to the largest sizes we can observe or discuss in the context of the observable universe, and reached the outer limits as to what we can see (and contemplate with our imagination), let's explore how the universe we do live in and observe evolved over time to create the enormous structures we see today. The Big Bang theory tells us how the universe grew during its 13.75-billion-year life span from its small initial size to the current extent, 100 billion light-years across. Fred Hoyle facetiously (and skeptically) named the theory after the initial explosion when a hot dense fireball began to expand into the massive extent of stars and structures we now observe: growing, diluting matter, and cooling as it evolved.

However, the one thing we certainly don't know is what banged in the beginning and how it happened—or even the precise size it had been when it did. Despite our understanding of the universe's late evolution, its beginnings remain shrouded in mystery. Nonetheless, although the Big Bang theory does not tell us anything about the universe's initial moment, it is a very successful theory that tells us much about its subsequent history. Current observations combined with the Big Bang theory teach us quite a lot about how the universe has evolved.

No one knew the universe was expanding when the twentieth century began. At the time that Edwin Hubble first peered into the sky, very little was known. Harlow Shapley had measured the size of the Milky Way to be 300,000 light-years across, but he was convinced that the Milky Way was all that the universe contained. In the 1920s, Hubble realized that some of the nebula that Shapley had thought were clouds of dust—which did indeed merit this uninspiring name—were in fact galaxies, millions of light-years away.

Once he identified galaxies, Hubble made his second stunning discovery—the universe's expansion. In 1929, he observed that galaxies red-shifted, which is to say there was a Doppler effect in which light waves shifted to longer wavelengths for more distant objects. This red shift demonstrated that galaxies were receding, much as the high-pitched wail of a siren decreases in frequency as an ambulance speeds away. (See Figure 71.) The galaxies he had identified were not stationary with respect to our location, but were all expanding away from us. This was evidence that we live in an expanding universe, in which galaxies are growing farther apart.

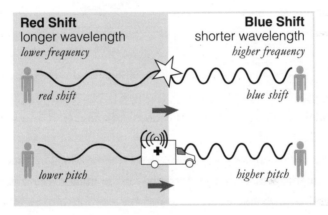

[**FIGURE 71**] The light from an object moving away from us is shifted to lower frequencies—or shifted toward the red end of the spectrum—whereas light from objects moving away is shifted to higher frequencies, or blue shifted. This is analogous to the noise from a siren that is lower pitched when an ambulance moves away and higher pitched when it approaches.

The universe's expansion is different from the pictures we might first imagine since the universe doesn't expand into some preexisting space. The universe is all there is. Nothing is present for it to expand into. The universe, as well as space itself, expands. Any two points within it grow farther apart as time progresses. Other galaxies move farther away from us, but our location is not special—they move farther from each other as well.

One way to picture this is to imagine the universe as the surface of a balloon. Suppose you had marked two points on the balloon's surface. As the balloon blows up, the surface becomes stretched and those two points grow farther apart. (See Figure 72.) This is in fact what happens to any two points in the universe as it expands. The distance between any two points—or any two galaxies—increases.

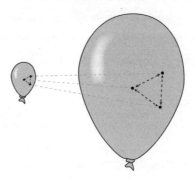

[FIGURE 72] The "ballooniverse" illustrates how all points move away from one another as the balloon (universe) expands.

Notice in our analogy that the points themselves don't necessarily expand—just the space between them. This is in fact what happens in the expanding universe as well. Atoms, for example, are tightly bound together via electromagnetic forces. They don't get any bigger. Neither do relatively dense strongly bound structures such as galaxies. The force driving the expansion acts on them too, but because other force contributions are at work, the galaxies don't themselves grow with the overall expansion of the universe. They feel such strong attractive forces that they remain the same size while their relative distance from each other gets bigger.

Of course, this balloon analogy is not perfect. The universe has three spatial dimensions, not two. Furthermore, the universe is large and probably infinite in size, and not small and curved like the balloon's surface. On top of that, the balloon exists in our universe and expands into existing space, unlike the universe, which permeates space and doesn't expand into something else. But even with these caveats, the surface of a balloon illustrates quite nicely what it means for space to expand. Every point moves away from every other point at the same time.

A balloon analogy—this time referring to the interior—is also helpful for understanding how the universe cooled from its initial hot dense fireball existence. Imagine an extremely hot balloon that you allow to expand to a very big size. Though it might have been too hot to handle at first, the expanded balloon will contain much cooler air that would no longer be alien to human contact. The Big Bang theory predicts that the initial hot dense universe expanded, all the while cooling as it did so.

Einstein had actually derived an expanding universe from his equations of general relativity. At that time, however, no one had yet measured the universe's expansion, so he didn't trust his prediction. Einstein introduced a new source of energy in an attempt to reconcile his theory with a static universe. After Hubble's measurements, Einstein dispensed with the fudge he had made, calling it "his biggest blunder." This modification was not entirely erroneous, however. We will soon see that more recent measurements show that the cosmological constant term he added is actually necessary to account for recent observations—although the measured magnitude, which accounts for the recently established acceleration of its expansion, is about an order of magnitude bigger than the one Einstein proposed to merely stall it.

The expansion of the universe was a nice example of a convergence of top-down and bottom-up physics. Einstein's theory of gravity implies that the universe expands, yet only with the discovery of the expansion did physicists feel confident they were on the right track.

Today, we refer to the number that determines the rate at which the universe expands at present as the *Hubble constant*. It is a constant in the sense that the fractional expansion everywhere in space is the same. However, the Hubble parameter is not constant over time. At an earlier time, when the universe was hotter and denser and gravitational effects were stronger, it expanded at a far more rapid rate.

Measuring the Hubble constant precisely is difficult, since we face exactly the problem we raised earlier of disentangling the past from the present. We need to know how far away the red-shifting galaxies are, since the red shift depends both on the Hubble parameter and distance. This imprecise measurement was the source of the factor-of-two uncertainty in the age of the universe that I mentioned at the beginning of this chapter. If the Hubble parameter measurements were uncertain by a factor of two, so too would be the universe's age.

That controversy is now pretty much resolved. The Hubble parameter has been measured by Wendy Freedman of the Smithsonian Astronomical Observatories and her collaborators and others, and the expansion rate is about 22 kilometers per second for a galaxy a million light-years away. Based on this value, we now know the universe is about 13.75 billion years old. This might under- or overestimate the age by 200 million years, but not by a factor of two. Although this might still sound like a good deal of uncertainty, the range is too small to make any great difference in our understanding today.

Two other key observations that agreed nicely with predictions further confirmed the Big Bang theory. One class of measurement that relied on both particle physics and general relativity predictions and therefore confirmed both was the density of various elements in the cosmos, such as helium and lithium. The amount of these elements that the Big Bang theory predicts agrees with measurements. This is in some respects indirect proof, and detailed calculations based on nuclear physics and cosmology are required to compute these values. Even so, this agreement of many different element abundances with predictions would be an unlikely coincidence unless physicists and astronomers were on the right track.

When the American Robert Wilson and the German-born Arno Penzias accidentally discovered the 2.7-degree microwave background in 1964, it was further confirmation of the Big Bang theory. To put this temperature in perspective, nothing is colder than absolute zero, which is zero degrees kelvin. The universe's radiation is less than three degrees warmer than this absolute limit to how cold anything can be.

The collaboration and adventure of Robert Wilson and Arno Penzias (for which they won the 1978 Nobel Prize) was a superb example of how science and technology sometimes work in concert to achieve results beyond what anyone had imagined. Back when AT&T was a phone monopoly, it did something rather wonderful, which was to create Bell Laboratories, a spectacular research environment where pure and applied research proceeded side by side.

Robert Wilson, who was a detail-oriented gadget technology geek, and Arno Penzias, who was more of a big picture scientist, both worked there, and together used and developed radio telescopes. Wilson and Penzias were interested in science and technology, while AT&T was understandably interested in communications, so radio waves in the sky were important to everyone involved.

While pursuing a specific radio astronomy goal, Wilson and Penzias found what they initially considered a mysterious nuisance that they simply couldn't explain. It seemed to be uniform background noise—essentially static. It wasn't coming from the Sun, and it wasn't related to a nuclear test from the previous year. They tried every explanation they could think of, most famously pigeon droppings, in their nine-month attempt to figure out what was going on. After considering all imaginable possibilities, cleaning out the pigeon droppings (or "white dielectric material" as Penzias called it), and even shooting the pigeons, the noise still didn't go away.

Wilson told me how lucky they were in the timing of their discovery. They didn't know about the Big Bang, but Robert Dicke and Jim Peebles at Princeton University did. The physicists there had just realized that one implication of the theory would be a relic microwave radiation. They were in the process of designing an experiment to measure this radiation when they discovered they had been scooped—by the Bell Lab scientists who hadn't yet realized what they had discovered. Luckily for Penzias and Wilson, the MIT astronomer Bernie Burke, who Robert Wilson described to me as the early version of the Internet, knew about the Princeton research and also the Penzias and Wilson discovery. He put two and two together and brought the connection to fruition by bringing the relevant players into contact.

This was a lovely example of science in action. The research was done for a specific scientific purpose that could also have ancillary technological and scientific benefits. The astronomers weren't looking for what they found, but they were extremely technologically and scientifically skilled. When they discovered something, they knew not to dismiss it. Their research—while looking for relatively small phenomena—resulted in a discovery with tremendously deep implications, which they found because they and others were thinking about the big picture at the same time. The discovery by the Bell Lab scientists was accidental, but it forever changed the science of cosmology.

The cosmic radiation has proved to be a tremendous tool—not just for confirming the Big Bang but for turning cosmology into a detailed science. The cosmic microwave background (CMB) radiation gives us a very different way of observing the past than traditional astronomy measurements.

In the past, astronomers would observe objects in the sky, try to determine their age, and attempt to deduce the evolutionary history that produced them. With the CMB, scientists can now also look directly back in time before structure such as stars and galaxies were even formed. The light they observe was emitted long ago—very early in the universe's evolution. When the microwave background we now observe was emitted, the universe was only about one-thousandth its current size.

Although the universe was originally filled with all types of particles—both charged and uncharged—once it cooled sufficiently, 400,000 years into its evolution, charged particles combined together into neutral atoms. Once this happened, light no longer scattered. Observed CMB radiation therefore arrives directly from about four hundred thousand years into the universe's evolution—unhindered and uninterrupted—to telescopes on Earth and on satellites. The background radiation Penzias and Wilson discovered was the same radiation present in the earlier stages of the universe's history, but it has been diluted and cooled through its expansion. The radiation traveled directly to the telescopes that detected it with no hindrances from scattering off any intervening charged particles en route. This light gives us a direct and precise window into the past.

The Cosmic Microwave Background Explorer (COBE), a four-year-long satellite mission launched in 1989, measured this background radiation extremely accurately, and the mission scientists found that their measurements agreed with predictions to better than one part in 1,000. But COBE measured something new as well. By far, the most interesting thing that COBE measured was a tiny bit of nonuniformity in temperature across the sky. Although the universe is extremely smooth, tiny inhomogeneities at the level of less than one in 10,000 in the early universe grew bigger and were essential to the development of structure. The inhomogeneities originated on minuscule length scales, but were stretched to sizes relevant to astrophysical measurements and structure. Gravity caused the denser regions where the perturbations were especially large to become more concentrated and form the massive objects we currently observe. The stars, galaxies, and clusters of galaxies that we discussed earlier are all the result of these initial tiny quantum mechanical fluctuations and their evolution through the gravitational force.

The microwave background measurement continues to be critical to our understanding of the universe's evolution. It's role as a direct window into the early universe cannot be underestimated. More recently, along with more traditional methods, CMB measurements have provided experimental insights into several other more mysterious phenomena—cosmological inflation, dark matter, and dark energy—subjects that we turn to next.

CHAPTER TWENTY

WHAT'S SO LARGE TO YOU
IS SO SMALL TO ME

When I was an MIT professor, the department ran out of office space on the third floor where the particle physicists worked. So I relocated to the open office next door to Alan Guth's on the floor below, which at the time housed theoretical astronomers and cosmologists. Although Alan started his career as a particle physicist, he is known today as one of the best cosmologists around. At the time of my office move, I had already explored some connections between particle physics and cosmology. But it's a lot easier to continue such research when your neighbor shares those interests—and is as messy as you are so that in his office you feel right at home.

Many particle physicists have gone further afield than a single floor and crossed over into a wide variety of other research areas. Wally Gilbert, a cofounder of Biogen, started life as a particle physicist but left to do biology and Nobel Prize–winning chemistry research. Many since have followed in his footsteps. On the other hand, many of my graduate student friends left particle physics to be "quants" on Wall Street where they could bet on changes in future markets. They chose just the right time to make such a move since the new financial instruments to hedge such bets were only just being developed at the time. In the crossover to biology, some ways of thinking and organizing problems carried over, whereas in finance some of the methods and equations did.

But the overlap between particle physics and cosmology is of course far deeper and richer than either of the above. Close examination of the universe on different scales has exposed the many connections between elementary particles on the smallest scales and the universe itself at the largest. After all, the universe is by definition unique and encompasses everything within it. Particle physicists, who look inward, ask what type of fundamental matter exists at the core of matter, and cosmologists, who look outward, study how whatever it is that is out there has evolved. The universe's mysteries—most notably what it is made of—matter to cosmologists and particle physicists alike.

Both types of researchers investigate basic structure and employ fundamental physical laws. Each needs to take into account the results of the other. The content of the universe that is studied by particle physicists is an important research subject for cosmologists too. Furthermore, the laws of nature that incorporate both general relativity and particle physics describe the universe's evolution, as they must if both theories are correct and apply to a single cosmos. At the same time, the known evolution of the universe constrains what properties matter can have if it is to avoid disrupting the observed history. The universe was in some respects the first and most powerful particle accelerator. Energies and temperatures were very high in the early stages of its evolution, and the high energies that accelerators currently achieve aim to reproduce some aspects of those conditions today on Earth.

Recent attention to this convergence of interests has led to many fruitful investigations and major insights and will hopefully continue to do so. This chapter considers some of the big open questions in cosmology that particle physicists and cosmologists both explore. The overlapping arenas include cosmological inflation, dark matter, and dark energy. We'll consider aspects we understand about each of these phenomena and—more important for active research—those that we don't.

COSMOLOGICAL INFLATION

Even though we can't yet say what happened at the very beginning of the universe, since we would need a comprehensive theory that incorporates both quantum mechanics and gravity, we can assert with reasonable certainty that at some time very early on (perhaps as early as 10^{-39} seconds into the universe's evolution), a phenomenon called *cosmological inflation* occurred.

In 1980, Alan Guth first suggested this scenario, which says that the very early universe essentially exploded outward. Interestingly, he was initially trying to solve a problem for particle physics involving the cosmological consequences of Grand Unified Theories. Coming from a particle background, he used methods rooted in field theory—the theory combining special relativity and quantum mechanics that particle physicists employ for our calculations. But he ended up deriving a theory that revolutionized our thinking about cosmology. How and when inflation occurred is still a matter of speculation. But a universe that underwent this explosive expansion would leave clear evidence, and much of it has now been found.

In the standard Big Bang scenario, the early universe grew calmly and steadily—for example, doubling in size when its age increased by a factor of four. But in an inflationary epoch, a patch of the sky underwent a phase of incredibly rapid expansion, growing exponentially with time. The universe doubled in size in a fixed time and then doubled again in that same time and then kept doubling at least 90 times in a row until the inflationary epoch ended and the universe was as smooth as we see it today. This exponential expansion means, for example, that when the universe's age had multiplied by 60 times, the size of the universe would have increased by more than a trillion trillion trillions in size. Without inflation, it would have increased by a mere factor of eight. In some sense, inflation was the beginning of our story of evolving from the small to the large—at least the part that we can potentially understand through observations. The initial enormous inflationary expansion would have diluted the matter and radiation content of the universe to practically nothing. Everything we observe today in the universe must therefore have arisen right after inflation, when the energy that drove the inflationary explosion converted into matter and radiation. At this point in time, conventional Big Bang evolution took over—and the universe began its further expansion into the huge structure we see today.

We can think of the inflationary explosion as the "bang" that was the precursor to the universe's evolving according to the standard Big Bang theory. It's not truly the beginning—we don't know what happened when quantum gravity played a role—but it's when the Big Bang stage of evolution, with matter cooling and eventually aggregating, began.

Inflation also partially answers why there is something rather than nothing. Some of the enormous energy density stored during inflation was converted (consistently with $E = mc^2$) to matter, and that is the matter that evolves to what we see today. As I discuss at the close of this chapter, we physicists still would like to know why there is more matter than antimatter in the universe. But whatever the answer to that question, the matter we know began evolving according to Big Bang theory predictions as soon as cosmological inflation ended.

Inflation was derived as a bottom-up theory. It solved important problems for the conventional Big Bang explosion, but only a few really believed any of the actual models for how it came about. No compelling high-energy theory seemed to obviously imply inflation. Since it was so challenging to make a credible model, many physicists (including those at Harvard when I was a graduate student) doubted the idea could be right. On the other hand, Andrei Linde, a Russian-born physicist now at Stanford, and one of the first to work on inflation, thought it had to be correct simply because no one had found any other solution to the puzzles about the size, shape, and uniformity of the universe that inflation addressed.

Inflation was an interesting example of the truth-beauty connection—or lack thereof. Whereas the exponential expansion of the universe beautifully and succinctly explained many phenomena about how the universe started, the search for a theory that naturally yields the exponential expansion led to many not-so-pretty models.

Recently, however, most physicists—even though not yet satisfied with most models—have become convinced that inflation, or something very similar to inflation, did occur. Observations of the last several years have confirmed the cosmological picture of Big Bang cosmology preceded by inflation. Many physicists now trust that Big Bang evolution and inflation have occurred because predictions based on these theories have been confirmed with impressive precision. The true model underlying inflation is still an open question. But the exponential expansion has a lot of evidence supporting it at this point.

One type of evidence for cosmological inflation has to do with the deviations from perfect uniformity in the cosmic microwave background radiation that the previous chapter introduced. The background radiation tells us much more than just that the Big Bang occurred. The beauty of it is that because it is essentially a snapshot of the universe very early on—before stars had time to form—it lets us look back directly into the beginnings of structure at the time when the universe was still very smooth. Cosmic microwave background (CMB) measurements also revealed tiny departures from perfect homogeneity. Inflation predicts this because quantum mechanical fluctuations caused inflation to end at slightly different times in different regions of the universe, giving rise to tiny deviations from absolute uniformity. The satellite-based Wilkinson Microwave Anisotropy Probe (WMAP), named for the Princeton physicist David Wilkinson who pioneered the project, made detailed measurements that distinguished inflationary predictions from other possibilities. Despite the fact that inflation happened long ago at incredibly high temperatures, theory based on inflationary cosmology nonetheless predicts the exact statistical properties of the pattern of temperature variations that should be imprinted on the radiation in the sky today. WMAP measured the small inhomogeneities in temperature and energy density with more accuracy and on smaller angular scales than had been done before, and the pattern conformed to inflationary expectations.

The chief confirmation of inflation that WMAP gave us was the measurement of the universe's extreme flatness. Einstein taught us that space can be curved. (See Figure 73 for examples of curved two-dimensional surfaces.) The curvature depends on the energy density of the universe. At the time when inflation was first proposed, it was known that the universe was far flatter than naive expectations would suggest, but the measurements were far too imprecise to test the inflationary prediction that the universe would expand so much that any curvature would be stretched away. Microwave background measurements have now demonstrated that the universe is flat at the level of one percent, which would be extremely difficult to understand without some underlying physical explanation.

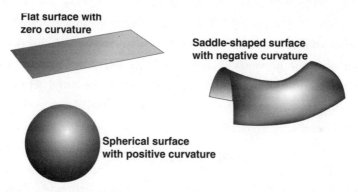

[FIGURE 73] Zero, positive, and negative curvature on two-dimensional surfaces. The universe, too, can be curved, but in four-dimensional spacetime that is difficult to draw.

This flatness of the universe was a huge victory for inflationary cosmology. Had it not been true, inflation would have been ruled out. The WMAP measurements were also a victory for science. When theorists first proposed the detailed measurements of the microwave background that would eventually tell us about the geometry of the universe, everyone thought it interesting enough to throw out to the science community, but far too difficult technically to achieve any time soon. Within the decade, confounding all expectation, observational cosmologists made the necessary measurements and gave us amazing insights into how the universe has evolved. WMAP is still providing new results, performing detailed measurements of the variation in temperature across the sky. The Planck satellite in operation today is measuring these fluctuations more precisely still. The CMB measurements have proven to be a prime resource of insight into the early universe and will most likely continue to be so.

Recent detailed studies of the cosmic radiation left throughout the sky have led to other enormous leaps in our quantitative knowledge of the universe and its evolution. The details of the radiation have provided rich information about the matter and energy that surrounds us. In addition to telling us the conditions when the light first started heading toward us, the CMB tells us about the universe through which the light had to travel. If the universe had changed in the last 13.75 billion years, or if its energy were different than expected, relativity tells us that it would have affected the path that the light-ray took and consequently the measured properties of the radiation that was measured. Since it is such a sensitive probe of the energy content of the universe today, the microwave background gives information about what the universe contains. This includes the dark matter and dark energy we will now consider.

HEART OF DARKNESS

In addition to successfully confirming inflationary theory, CMB measurements presented a few major mysteries that cosmologists, astronomers, and particle physicists now want to address. Inflation tells us that the universe should be flat but it doesn't tell us where the energy required to make it flat now resides. Nonetheless, based on Einstein's equations of general relativity, we can calculate the energy needed for the universe to be flat today. It turns out that known visible matter alone provides a mere four percent of the energy required.

An additional puzzle that had already indicated the need for something new concerned the tininess of the fluctuations in temperature and density that COBE had measured. With only visible matter and such tiny perturbations, the universe wouldn't have lasted long enough for the perturbations to have grown large enough for structure to have formed. The existence of galaxies and clusters of galaxies in conjunction with the tininess of the measured fluctuations pointed to the existence of matter that no one had yet directly seen.

In fact, scientists had already known that a new type of matter known as dark matter should exist well before COBE's microwave radiation results. Other observations that we will get to soon that had already

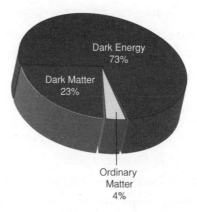

[**FIGURE 74**] Pie chart illustrating the relative amounts of visible matter, dark matter, and dark energy of which the universe is composed.

indicated additional unseen matter must exist. This mysterious stuff, which became known as dark matter, exerts gravitational forces, but it doesn't interact with light. Because it neither emits nor absorbs light, it is invisible—not dark. Dark matter (we'll keep using the term) has so far provided few tangible identifying features other than its gravitational influence and that it is so feebly interacting.

Furthermore, gravitational influence and measurements indicate the presence of something even more mysterious than dark matter, known as dark energy. This is energy that permeates the universe, but doesn't clump like ordinary matter or dilute as it expands. It is very much like the energy that precipitated inflation, but its density today is much smaller than it was back then.

Although we now live in a renaissance era of cosmology, in which theories and observations have advanced to the stage where ideas can be precisely tested, we also live in the dark ages. About 23 percent of the universe's energy is carried by dark matter, and approximately another 73 percent is carried by the mysterious dark energy, as is illustrated in the pie chart. (See Figure 74.)

The last time something was called "dark" in physics was in the mid-1800s, when Urbain Jean Joseph Le Verrier of France proposed an unseen dark planet, which he named Vulcan. Leverrier's goal was to explain the peculiar trajectory of the planet Mercury. Leverrier, along with John C. Adams of England, had previously deduced the existence of Neptune based on its effects on the planet Uranus. Yet he was wrong about Mercury. It turned out that the reason for Mercury's strange orbit was much more dramatic than the existence of another planet. The explanation could be found only with Einstein's theory of relativity. The first confirmation that his theory of general relativity was correct was that he could use it to accurately predict Mercury's orbit.

It could turn out that dark matter and dark energy are a consequence of known theories. But it might also be that these missing elements of the universe presage a similar significant change of paradigm. Only time will tell which of these options will resolve the dark matter and dark energy problems.

Even so, I'd say that dark matter is very likely to have a more conventional explanation, consistent with the type of physical laws we now know. After all, even if novel matter acts in accordance with force laws similar to those we know, why should all matter behave exactly like familiar matter? To put it more succinctly, why should all matter interact with light? If the history of science has taught us anything, it should be the shortsightedness of believing that what we see is all there is.

Many people think differently. They find dark matter's existence very mysterious and ask how it can possibly be that most matter—about six times the amount we see—is something we can't detect with conventional telescopes. Some are even suspicious that dark matter is really some sort of mistake. Personally, I think quite the opposite (though admittedly not even all physicists see it this way). It would perhaps be even more mysterious if the matter we can see with our eyes is all the matter that exists. Why should we have perfect senses that can directly perceive everything? Again, the lesson of physics over the centuries is how much is hidden from view. From this perspective, it's mysterious why the stuff we do know should constitute even as much as 1/6 of the energy of all matter, an apparent coincidence that my colleagues and I are currently trying to understand.

We know something with dark matter's properties has to be there. Although we don't exactly "see" it, we do detect dark matter's gravitational influence. We know dark matter exists due to the extensive observational evidence of its gravitational effects in the cosmos. The first clue that it existed came from the speed with which stars rotated in galaxy clusters. In 1933, Fritz Zwicky observed that galaxies in galaxy clusters orbited faster than could be accounted for by the visible mass, and Jan Oort soon after observed a similar phenomenon in the Milky Way. Zwicky was convinced enough by his work to conjecture the existence of dark matter that no one could directly see. But neither of these observations was conclusive. A faulty measurement or some other galaxy dynamics seemed like a far more plausible explanation than some invisible substance invented solely to provide additional gravitational attraction.

At the time Zwicky made his measurements, he didn't have the resolution to see individual stars. Much more solid evidence for dark matter came from Vera Rubin, an observational astronomer, who much later—in the late 1960s and early 1970s—made detailed quantitative measurements of stars rotating in galaxies. What first seemed to be a "boring" study of stars orbiting in a galaxy—a study Vera turned to since it provided less-well-trodden territory than other astronomical activities at the time—emerged as the first solid evidence of dark matter in the universe. Rubin's observations with Kent Ford yielded incontrovertible evidence that Zwicky's conclusion years earlier had been correct.

You might wonder how someone could look through a telescope and see something dark. The answer is that she could see its gravitational consequences. The properties of a galaxy, such as the rate at which its stars orbit around, are influenced by how much matter it contains. With only visible matter present, one would have expected those stars well beyond the galaxy to be rather insensitive to the galaxy's gravitational influence. Yet stars ten times farther away than the luminous central matter rotated with the same velocity as stars closer to the galaxy's center. This implied that the mass density did not fall off with distance, at least to distances as far from the galaxy's center as ten times the distance of the luminous matter. Astronomers concluded that galaxies consisted primarily of unseen dark matter. The luminous matter we see is a significant fraction, but most of the galaxy is invisible, at least in the ordinary sense of the word.

We now have a good deal of other supplementary evidence for dark matter's existence. Some of the most direct is from lensing, illustrated in Figure 75. Lensing is the phenomenon that occurs when light passes a massive object. Even if that object itself doesn't emit light, it does exert a gravitational force. And that gravitational force can cause light emitted by a nondark object behind (as seen from our vantage point) to bend. Because the light bends in different directions according to the path it takes around the dark object and because we automatically project straight lines for light, this lensing can produce multiple images of the original bright object in the sky. These multiple images allow us to "see" the dark object—or at least infer its existence and properties by deducing the gravity needed to bend the observed light.

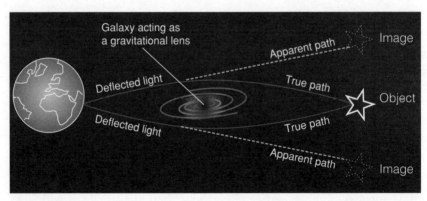

[FIGURE 75] Light passing a massive object can bend, which from the perspective of the observer appears to create multiple images of the original object.

Perhaps the strongest evidence to date that dark matter, rather than a modified gravitational theory, explains such phenomena comes from the *Bullet cluster,* which involved two colliding clusters of galaxies. (See Figure 76.) Their collision demonstrated that the clusters contain stars, gas, and dark matter. The hot gas in the cluster interacts strongly—so strongly that the gas remains concentrated in the central collision region. Dark matter, on the other hand, doesn't interact—at least not very much. So the dark matter just passed through. Lensing measurements showed that the dark matter was indeed separated from the hot gas in just the way implied by a model of very weakly interacting dark and strongly interacting ordinary matter.

[FIGURE 76] The Bullet cluster indicates that clusters of galaxies contain dark matter, and that their dynamics are unlikely to be explained by modified gravitational laws. That's because we can see a separation between the more strongly interacting ordinary matter that gets trapped in the middle when two clusters collide and the far more weakly interacting dark matter, which is detected by gravitational lensing, and evidently just passes through.

We have further evidence for dark matter's existence from the cosmic microwave background discussed earlier. Unlike lensing, the radiation measurements don't tell us anything about the distribution of dark matter. Instead they tell us the net energy content carried by dark matter—how big a piece of the cosmic pie is constituted by the energy it carries.

CMB measurements tell us a great deal about the early universe and give us detailed information about its properties. These measurements argue not only for dark matter. They also support the existence of dark energy. According to Einstein's equations of general relativity, the universe could only be flat with just the right amount of energy. Matter, even accounting for dark matter, simply didn't suffice to account for the flatness measured by WMAP and balloon-based detectors. Other energy had to exist. Dark energy is the only way to account for the universe's flatness—with no measurable curvature of three-dimensional space and agree with all other measurements to date.

Dark energy, which carries the bulk of the universe's energy —approximately 70 percent—is even more puzzling than dark matter. The evidence that convinced the physics community of dark energy's existence was the discovery that the expansion of the universe is currently accelerating—much as it did during inflation earlier on but at a very much slower rate. In the late 1990s, two independent research teams, the Supernova Cosmology Project and the High-z Supernova Team, surprised the physics community when they discovered that the rate of expansion of the universe is no longer slowing down, but is actually increasing.

Before the supernova measurements, a few hints had pointed to the existence of missing energy, but the evidence had been weak. But careful measurements in the 1990s showed that distant supernova were dimmer than expected. Since this particular type of supernova has fairly uniform and predictable emission, this could only be explained by something new. And that something new seems to be an accelerated expansion of the universe—that is, it is expanding at an increasingly faster rate.

This acceleration would not arise from ordinary matter, whose gravitational attraction would slow the universe's expansion. The only explanation could be a universe that acts like one that is inflating, but with far smaller energy than during the inflationary phase the universe had undergone much earlier on. This acceleration could be due only to something that acted like the cosmological constant that Einstein had introduced, or dark energy, as it has become known.

Unlike matter, dark energy exerts negative pressure on its environment. Ordinary positive pressure favors inward collapse, whereas negative pressure leads to accelerated expansion.[68] The most obvious candidate for negative pressure—one that agrees with measurements so far—is Einstein's cosmological constant, representing an energy and pressure that permeates the universe but is not carried by matter. Dark energy is the more general term we now use to allow for the possibility that the cosmological constant's assumed relationship between energy and pressure isn't precisely true but is only approximate.

Today dark energy is the dominant component of the universe's energy. This is all the more remarkable because the amount of dark energy density turns out to be extraordinarily small. Dark energy has dominated only for the last few billion years. Earlier in the universe's evolution, first radiation and then matter were dominant. But radiation and matter, which are shared over the volume of an ever-increasing universe, dilute. Dark energy density, on the other hand, remained constant, even when the universe grew. By the time the universe had lasted so long as it has, the energy density in radiation and matter had decreased so enormously that dark energy, which doesn't dissipate, eventually took over.

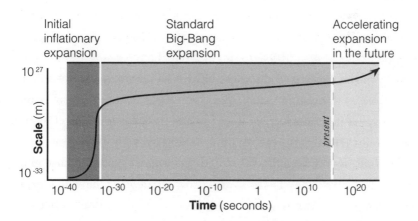

[FIGURE 77] The universe has expanded differently over time. During the inflationary phase it quickly expanded exponentially. The conventional Big Bang expansion took over when inflation ended. Dark energy now makes the expansion rate accelerate again.

Despite dark energy's incredibly tiny size, it was bound to eventually dominate. After 10 billion years of expanding at an increasingly slower rate, the impact of dark energy was finally felt and the universe sped up its expansion. Eventually, the universe will end up with nothing in it but vacuum energy and its expansion will accelerate accordingly. (See Figure 77.) The meek energy might not inherit the Earth, but it is in the process of inheriting the universe.

FURTHER MYSTERIES

The necessity for dark energy and dark matter tells us that we can't be as smug about our understanding of the evolution of the universe as the incredible agreement of cosmological theory with cosmological data might suggest. Most of the universe is stuff whose identity remains a mystery. Twenty years from now, people might smile at our ignorance.

And these are not the only puzzles evoked by the energy of the universe. The value of the dark energy, in particular, is actually the tail end of a much larger mystery: why is the energy that pervades the universe so small? Had the amount of dark energy been greater, it would have dominated matter and radiation much earlier in the evolution of the universe, and structure (and life) would not have had time to form. On top of that, no one knows what was responsible earlier on for the large energy density that triggered and fueled inflation. But the biggest problem with the energy of the universe is the *cosmological constant problem.*

Based on quantum mechanics, we would have expected a much larger value for dark energy—both during inflation and now. Quantum mechanics tells us that the vacuum—the state with no permanent particles present—is actually filled with ephemeral particles that pop in and out of existence. These short-lived particles can have any energy. They sometimes can have energy so large that gravitational effects can no longer be neglected. These highly energetic particles contribute an extremely large energy to the vacuum—much larger than the long evolution of our universe would allow. In order for the universe to look like the one we see, the value of the vacuum energy has to be an astonishing 120 orders of magnitude smaller than the energy that quantum mechanics would lead us to expect.

And there is yet a further challenge associated with this problem. Why do we happen to live in the time when the energy densities of matter, dark matter, and dark energy are comparable? Certainly dark energy dominates over matter, but it's by less than a factor of three. Given that these energies in principle have entirely different origins and any one of them could have overwhelmed the others, the fact that their densities are so close seems very mysterious. The peculiarity of this coincidence is especially notable because it is only (roughly speaking) in our time that this coincidence is true. Earlier in the universe, dark energy was a much smaller fraction of the whole. And later on it will be a much greater fraction. Only today are the three components—ordinary matter, dark matter, and dark energy—comparable.

The questions of why the energy density is so extraordinarily tiny and why these different energy sources contribute similar amounts today are entirely unsolved. In fact, some physicists believe that there is no true explanation. They think we live in a universe with such an incredibly unlikely value for the vacuum energy because any larger value would have prevented the formation of galaxies and structure—and us—in the universe. We wouldn't be here to ask about the value of the energy in any universe with a somewhat larger value of the cosmological constant. Those physicists believe that there are many universes, and each of these universes contains a different value of the dark energy. Out of the many possible universes, only the ones that could give rise to structure could possibly contain us. The value of the energy in this universe is ridiculously small, but we could exist only in a universe with just such a small value. This reasoning is the *anthropic principle* we considered in Chapter 18. As I said then, I'm not convinced. Nonetheless, neither I nor anyone else has a better answer. The explanation for the value of the dark energy is perhaps the most major mystery particle physicists and cosmologists face today.

In addition to puzzles about energy, we also have a further cosmological mystery about matter: Why is there matter in the universe at all? Our equations treat matter and antimatter on the same footing. They annihilate when they find each other, and both disappear. Neither matter nor antimatter should remain when the universe has cooled.

Whereas dark matter doesn't interact very much and therefore sticks around, ordinary matter interacts quite a lot through the strong nuclear force. Without an exotic addition to the Standard Model, almost all of our usual matter would have disappeared by the time the universe had cooled to its current temperature. The only reason matter can be left is that there is a predominance of matter over antimatter. This isn't built into the simplest versions of our theories. We need to find reasons that protons exist but can't find antiprotons with which they can annihilate. Somewhere a matter-antimatter asymmetry must be built in.

The amount of leftover matter is smaller than the amount of dark matter, but it is still a sizable chunk of the universe—not to mention the source of everything we know and love. How and when this matter-antimatter asymmetry was created is another big question that particle physicists and cosmologists very much want to tackle.

The question of what constitutes the dark matter of course remains critical as well. Perhaps eventually we will find that the underlying model connects the dark matter density to that of matter, as recent research suggests. In any case, we hope to soon learn a lot more about the dark matter question from experiments—a sampling of which we'll now explore.

CHAPTER TWENTY-ONE

VISITORS FROM THE DARK SIDE

When the LHC's chief engineer, Lyn Evans, spoke at the California LHC/Dark Matter conference in January 2010, he closed by teasing the audience about how for the last couple of decades, "You theorists have been thrashing around in the dark (sector)." He added the caveat, "Now I understand why I spent the last fifteen years building the LHC." Lyn's comments referred to the paucity of high-energy data over the previous years. But they were also hints about the possibility that LHC discoveries might shed light on dark matter.

Many connections exist between particle physics and cosmology, but one of the most intriguing is that dark matter might actually be made at the energies explored by the LHC. The remarkable fact is that if a stable particle species with weak scale mass exists, the amount of energy carried by particles of this type that survived from the early universe to today would be about right to account for dark matter. The result of calculations about the amount of dark matter that is left over by an initially hot—but cooling—universe demonstrate that this might be the case. That means that not only is dark matter literally right under our noses, its identity might prove to be too. If dark matter is indeed composed of such a weak mass particle, the LHC might not only give us insight into particle physics questions, it might also provide clues to what is out there in the universe and how it all began—questions that are incorporated into the science of cosmology.

But LHC experiments are not the only way to search for dark mat-

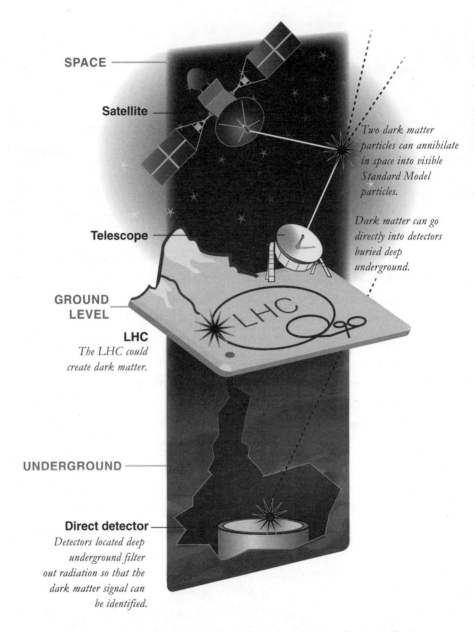

SPACE

Satellite

Two dark matter particles can annihilate in space into visible Standard Model particles.

Telescope

Dark matter can go directly into detectors buried deep underground.

GROUND LEVEL

LHC
The LHC could create dark matter.

UNDERGROUND

Direct detector
Detectors located deep underground filter out radiation so that the dark matter signal can be identified.

[FIGURE 78] Dark matter searches take a three-pronged approach. Underground detectors look for dark matter directly hitting target nuclei. The LHC might create dark matter that leaves evidence in its experimental apparatuses. And satellites or telescopes might find evidence of dark matter annihilating and producing visible matter out in space.

ter. The fact is that physics has now entered a potentially exciting era of data, not just for particle physics, but also for astronomy and cosmology. This chapter explains how experiments in the upcoming decade will search for dark matter using a three-pronged approach. It first explores why weak-scale-mass dark matter particles are favored, and after that, how the LHC might produce and identify dark matter particles if this hypothesis is right. We'll then consider how dedicated experiments that are specifically designed to search for dark matter particles look for their arrival to Earth and try to register their feeble but potentially detectable interactions. Finally, we'll consider the ways in which telescopes and detectors on the ground and in space look for products of dark matter particles annihilating in the sky. These three different ways of searching for dark matter are illustrated in Figure 78.

TRANSPARENT MATTER

We know the density of dark matter, that it is cold (which is to say, it moves slowly relative to the speed of light), and that it interacts at most extremely weakly—certainly with no significant interaction with light. And that's about it. Dark matter is transparent. We don't know its mass, if it has any non-gravitational interactions, or how it was created in the early universe. We know its average density, but there could be one proton mass per cubic centimeter in our galaxy or there could be one thousand trillion times the proton mass stored in a compact object that is distributed throughout the universe every kilometer cubed. Either gives the same average dark matter density, and either could have seeded the formation of structure.

So although we know it's out there, we don't yet know the nature of dark matter. It could be small black holes or objects from other dimensions. Most likely, it is simply a new elementary particle that doesn't have the usual Standard Model interactions—perhaps a stable neutral remnant of a soon-to-be-discovered physical theory that will appear at the weak mass scale. Even if that's the case, we would want to know what the properties of the dark matter particle are—its mass and its interactions and if it is part of some such larger sector of new particles.

One reason the elementary particle interpretation is currently favored is the point alluded to above—the abundance of dark matter, the fraction of energy it carries—supports this hypothesis. The surprising fact is that a stable particle whose mass is roughly the weak energy scale that the LHC will explore (again via $E = mc^2$) has a *relic density* today—the fraction of energy stored in the particles in the universe—in the right ballpark to be dark matter.

The logic goes as follows. As the universe evolved, the temperature decreased. Heavier particles that were abundant when the universe was hotter are much more dispersed in the later cooler universe since the energy at low temperature is insufficient to create them. Once the temperature dropped sufficiently, heavy particles efficiently annihilated with heavy antiparticles so that both of them disappeared, but the reverse process where they were created no longer occurred at any significant rate. Therefore, due to annihilation, the number density of heavy particles decreased very rapidly as the universe cooled down.

Of course, in order to annihilate, particles and antiparticles have to first find each other.[69] But as their number decreased and they became more diffuse, this became less likely. As a consequence, particles annihilated less efficiently later in the universe's evolution since it takes two of them in the same place to tango.

The result is that substantially more stable, weak-mass particles could remain today than a naive application of thermodynamics would suggest—at some point both particles and antiparticles became so dilute that they just couldn't find and eliminate each other. How many particles are left today depends on the mass and the interactions of the putative dark matter candidate. Physicists know how to calculate the relic abundance if we know these quantities. The intriguing and remarkable fact is that stable weak-mass particles happen to be such that they are left with about the right abundance to be the dark matter.

Of course, since we know neither the exact particle mass nor the precise interactions (not to mention the model of which this stable particle might be a part), we don't yet know if the numbers work out exactly. But the fortuitous, albeit rough, agreement between numbers associated with what on the surface appear to be two entirely different phenomena is intriguing, and might well be a signal that weak-scale physics accounts for the dark matter in the universe.

This type of dark matter candidate has become generically known as a *WIMP*, or a *Weakly Interacting Massive Particle*. The word "weak" here is a descriptive term and not a reference to the weak force—a WIMP would interact even more weakly than the Standard Model's weakly interacting neutrinos. Without more direct evidence for dark matter and its properties of the sort the LHC might reveal, we won't know whether dark matter indeed consists of WIMPs. This is why we need experimental searches such as those we now consider.

DARK MATTER AT THE LHC

The intriguing possibility of producing dark matter is one reason cosmologists are curious about the physics of the weak energy scale and what the LHC might find. The LHC has just the right energy to search for a WIMP. If dark matter is indeed composed of a particle associated with the weak energy scale as the above calculation suggests, it just might be created at the LHC.

Even if that's the case, however, the dark matter particle won't necessarily be discovered. After all, dark matter doesn't interact a lot. Due to their limited interactions with Standard Model matter, dark matter particles certainly won't be produced directly or found in a detector. Even if produced, they will just fly through. Nonetheless, all is not lost (even if the dark matter particle will be). Any solution to the hierarchy problem will contain other particles—most of which have stronger interactions. Some of these might be copiously produced and subsequently decay into dark matter that will then carry away undetected momentum and energy.

Supersymmetric models are the most well-studied weak scale models of this type that naturally contain a viable dark matter candidate. If supersymmetry applies in the world, the lightest supersymmetric particle

(LSP) might constitute the dark matter. This lightest particle, which carries zero electric charge, interacts too weakly to be produced on its own sufficiently often to find. However, gluinos—supersymmetric partners of the strong-force-communicating gluons, and squarks—supersymmetric partners of quarks—would be created if they exist and are in the right mass range. And, as was discussed in Chapter 17, both of these supersymmetric particles would eventually decay into the LSP. So even though a dark matter particle wouldn't be produced directly, decays of other more prolifically created particles could conceivably create LSPs at an observable rate.

Other weak-scale dark matter scenarios that have testable consequences would have to be produced and "detected" in much the same way. The mass of the dark matter particle should be around the weak scale energy that the LHC will study. Those particles won't be produced directly because of their feeble interaction strength, but many models contain other new particles that could decay into them. We might then learn of the dark matter particle's existence and possibly its mass through the missing momentum it carries away.

Finding dark matter at the LHC would certainly be a major accomplishment. If it is found there, experimenters could even study some of its properties in detail. However, to really establish that a particle found at the LHC indeed constitutes the dark matter would require supplementary evidence. That is what detectors on the ground and in space might provide.

DIRECT DETECTION DARK MATTER EXPERIMENTS

The LHC's potential to create dark matter is certainly intriguing. But most cosmology experiments don't take place at accelerators. Experiments on Earth and in space that are dedicated to astronomical and dark matter searches are primarily responsible for addressing and advancing our understanding of potential solutions to cosmological questions.

Of course, dark matter's interactions with matter are very weak, so current searches rely on a leap of faith that dark matter, despite its near invisibility, nonetheless interacts feebly—but not impossibly so—with matter we know (and can build detectors out of). This isn't merely a wishful guess. It's based on the same relic density calculation mentioned above that shows that if dark matter is related to models proposed to explain the hierarchy problem, then the density of particles that remains is the correct amount to account for dark matter observations. Many of the WIMP dark matter candidates suggested by this calculation interact with Standard Model particles at rates that might well be detectable with the current generation of dark matter detectors.

Even so, because of dark matter's feeble interactions, the search requires either enormous detectors on the ground or, alternatively, very sensitive detectors that look for the products of dark matter meeting, annihilating, and creating new particles and their antiparticles on Earth or in space. You probably wouldn't win the lottery if you bought only a single ticket, but if you could buy more than half of what's available, then your odds would be pretty good. Similarly, very large detectors have a reasonable chance of finding dark matter, even though dark matter's interaction with any single nucleon in the detector is extremely small.

The challenging task for dark matter detectors is to detect the neutral—uncharged—dark matter particles, and afterward distinguish them from cosmic rays or other background radiation. Particles with no charges don't interact with detectors in conventional ways. The only trace of a dark matter particle passing through a detector would be the consequences of hitting nuclei in the detector and changing its energy by a minuscule amount. Because this is the only observable consequence, dark matter detectors have no choice but to search for evidence of the tiny amounts of heat or recoil energy that get created when dark matter particles pass through. Detectors are therefore designed to be either very cold or very sensitive in order to record the small heat or energy deposits from dark matter particles subtly ricocheting off.

The very cold devices, known as *cryogenic detectors,* detect the small amount of heat emitted when a dark matter particle enters the apparatus. A small amount of heat added to an already hot detector would be too difficult to notice, but with specially designed cold detectors, the tiny heat deposit can be absorbed and recorded. Cryogenic detectors are made with a crystalline absorber such as germanium. Experiments of this sort include the Cryogenic Dark Matter Search (CDMS), CRESST, and EDELWEISS.

The other class of direct detection experiments involves noble liquid detectors. Even though dark matter doesn't directly interact with light, the energy added to an atom of xenon or argon when a dark matter particle hits it can lead to a flash of characteristic scintillation. Experiments with xenon include XENON100 and LUX, and the other noble liquid experiments, ZEPLIN and ArDM.

Everyone in the theoretical and experimental communities is eager to know what the new results from these experiments will be. I was fortunate to be present at a dark matter conference at the KITP in Santa Barbara organized in December 2009, by two leading dark matter experts, Doug Finkbeiner and Neal Weiner, when CDMS, one of the most sensitive dark matter detection experiments, was about to release new results. In addition to being young and tall contemporaries who had done their PhDs together at Berkeley, Doug and Neal both had a great understanding of dark matter experiments and what their implications might be. Neal had more of a particle physics background, and Doug had done more astrophysics research, but they converged on the topic of dark matter when it became clear that dark matter studies would involve both. At the conference, they had collected leading theoretical and experimental expertise on the subject.

The most riveting talk of the day occurred the morning I arrived. Harry Nelson, who is a professor at the University of California Santa Barbara, talked about year-old CDMS results. You might wonder why a talk about old results should receive so much attention. The reason was that everyone at the conference knew that only three days later the experiment would release new data. And rumors were flying that scientists at the CDMS experiment had actually seen compelling evidence of a discovery, so everyone wanted to understand the experiment better. For years theorists had listened to talks about dark matter detection but had listened primarily to their results and had paid only superficial attention to the details. But with imminent dark matter detection conceivable, theorists were eager to learn more. Later in the week, the results were released and disappointed the audience's greatly exaggerated expectations. But at the time of the talk, everyone was absorbed. Harry steadfastly managed to give his talk despite the many probing questions about the soon-to-be-released results.

Because it was a two-hour informal presentation, those of us in attendance could interrupt whenever necessary to understand as much as possible. The talk nicely addressed questions that the audience, which consisted mostly of particle physicists, would find confusing. Harry, who was trained as a particle physicist—not as an astronomer—spoke the same language we did.

With these extraordinarily difficult dark matter experiments, the devil is in the details. Harry made that abundantly clear. The CDMS experiment is based on advanced low-energy physics technology—the kind more conventionally associated with so-called condensed matter or solid state physicists. Harry told us how before joining the collaboration he would never have believed such delicate detections could possibly work, joking that his experimental colleagues should be grateful he wasn't a referee on the original proposal.

CDMS works very differently from scintillating xenon and sodium iodide detection experiments. It has hockey-puck-size pieces of germanium or silicon topped by a delicate recording device, which is a phonon sensor. The detector operates at very low temperature—low enough to be just at the border between superconducting and non-superconducting. If even a small amount of energy from phonons, the sound units that carry the energy through the germanium or silicon, much like photons are the units of light—hit the detector, it can be enough to make the device lose superconductivity and register a potential dark matter event through a device called a *superconducting quantum interference device (SQUID)*. These devices are extraordinarily sensitive and measure the energy deposition extremely well.

But recording an event isn't the end of the story. The experimenters need to establish that the detector is recording dark matter—not just background radiation. The problem is that everything radiates. We radiate. The computer I'm typing on radiates. The book (or electronic device) you're reading radiates. The sweat from a single experimentalist's finger is enough to swamp any dark matter signal. And that doesn't even take into account all the primordial and man-made radioactive substances. The environment and the air as well as the detector itself carry radiation. Cosmic rays can hit the detector. Low-energy neutrons in the rock can mimic dark matter. Cosmic ray muons can hit rock and create a splash of material, including neutrons that can mimic dark matter too. There are about 1,000 times as many background electromagnetic events as predicted signal events, even with reasonably optimistic assumptions about the mass and interaction strength of the dark matter particles.

So the name of the game for dark matter experiments is *shielding and discrimination.* (This is the astrophysicists' term. Particle physicists use the more PC term *particle ID,* though these days I'm not sure that's so great either.) Experimenters need to shield their detector as much as possible to keep radiation out and discriminate potential dark matter events from uninteresting radiation scattering in the detectors. Shielding is accomplished in part by performing the experiments deep in mines. The idea is that cosmic rays will hit the rock surrounding the detector before they hit the detector itself. Dark matter, which has far fewer interactions, will make it to the detector unimpeded.

Fortunately for dark matter detection, plenty of mines and tunnels exist. The DAMA experiment, along with experiments called XENON10 and the bigger version XENON100—as well as CRESST, a detector that uses tungsten—take place in the Gran Sasso laboratory, situated in a tunnel in Italy about 3,000 meters underground. A 1,500-meter-deep cavern in the Homestake mine in South Dakota, originally built for gold excavation, will be home to another xenon-based experiment known as LUX. This experiment will take place in the very same cavity where Ray Davis discovered neutrinos from nuclear reactions taking place in the Sun. The CDMS experiment is in the Soudan mine, about 750 meters underground.

Still, all that rock above the mines and tunnels is not enough to guarantee that the detectors are radiation-free. The experiments further shield the actual detectors in a variety of ways. CDMS has a layer of surrounding polyethylene that will light up if something too strongly interacting to be dark matter comes through from the outside. Even more memorable is the surrounding lead from an eighteenth-century sunken French galleon. Older lead that has been underwater for centuries has had time to shed its radioactivity. It is a dense absorbing material that is perfect for shielding the detector from incoming radiation.

Even with all these precautions, a lot of electromagnetic radiation still survives. Distinguishing radiation from potential dark matter candidates requires further discrimination. Dark matter interactions resemble nuclear reactions that occur when a neutron hits the target. So opposite the phonon readout system is a more conventional particle physics detector that measures the ionization created when the alleged dark matter particle passed through the germanium or silicon. Together, the two measurements, ionization and phonon energy, distinguish nuclear events—the good processes that might be the result of dark matter—from events due to electrons, which are just radioactivity induced.

Other beautiful features of the CDMS experiment include the excellent position and timing measurements that it can make. This is nice because although the position is only directly measured in two directions, the timing of the phonons gives the position in the third coordinate. So experimenters can locate exactly where the event happened and discard background surface events. Another nice feature is that the experiment is segmented into the stacked hockey-puck-size detectors. A true event will occur in only one of these detectors. Locally induced radiation, on the other hand, won't necessarily be confined to a single detector. With all these features and an even better design to come, CDMS has a good chance of finding dark matter.

Nonetheless, impressive as it is, CDMS is not the only dark matter detector and cryogenic devices are not the only type. Later on in the week, Elena Aprile, one of the xenon experiment pioneers, gave comparable details about her experiments (XENON10 and XENON100), as well as other experiments performed with noble liquids. Since these would soon be the most sensitive detectors for dark matter, the audience paid rapt attention to her talk too.

Xenon experiments record dark matter events through their scintillation. Liquid xenon is dense and homogenous, has a large mass per
atom (enhancing the dark matter interaction rate), scintillates well,
ionizes fairly readily when energy is deposited so that the two types
of signals described above can efficiently discriminate against electromagnetic events, and is relatively cheap compared to other potential
materials—although the price had fluctuated by a factor of six in the
course of the decade. Noble gas experiments of this type have become
a lot better as they have gotten bigger, and they should continue to do
so. With more material, not only is detection more likely, but the outer
part of the detector can shield the inner part of the detector more efficiently, helping assure the significance of a result.

By measuring both ionization and the initial scintillation, experimenters distinguish signal from background radiation. The XENON100 experiment uses very special phototubes that were designed to work in the
low-temperature, high-pressure environment of the detector to measure
the scintillation. Argon detectors might provide even better scintillation
information in the future through their use of the detailed shape of the
scintillation pulse as a function of time, and that will also help separate
the wheat from the chaff.

The strange state of affairs today (although this might soon change)
is that one scintillation experiment—the DAMA experiment in the
Gran Sasso Laboratory in Italy—has actually seen a signal. DAMA,
unlike the experiments I just described, has no internal discrimination
between signal and background. Instead it relies on identifying dark
matter signal events solely by their time dependence, using the distinctive velocity dependence coming from the Earth's orbit around the Sun.

The reason the velocity of incoming dark matter particles is relevant
is that it determines how much energy is deposited in the detector. If the
energy is too low, the experiment won't be sensitive enough to know if
anything was there. More energy means the experiment is more likely to
record the event. Due to the Earth's orbital velocity, the speed of dark
matter relative to us (and hence the energy deposited) depends on the
time of year—making it easier to see a signal at some times of year (sum-

mer) than at others (winter). The DAMA experiment looks for an annual modulation in the event rate that accords with this prediction. And their data indicates they have found such a signal. (See Figure 79 for the oscillating DAMA data.)

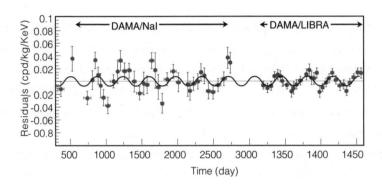

[**FIGURE 79**] Data from the DAMA experiment showing the modulation of the signal over time.

No one yet knows for sure whether the DAMA signal represents dark matter or is due to some possible misunderstanding about the detector or its environment. People are skeptical because no other experiment has yet seen anything. This absence of other signals is inconsistent with the predictions of most dark matter models.

Although confusing for the time being, this is the sort of thing that makes science interesting. The result encourages us to think about what different types of dark matter might exist and whether dark matter might have properties that make it easier for DAMA to see it than other dark matter detection experiments. Such results also force us to better understand the detectors so that we can identify spurious signals and tell if the data mean what the experimenters claim.

Other experiments all over the globe are working to achieve greater sensitivity. They could either rule out or confirm the DAMA dark matter discovery. Or they might independently discover a different type of dark matter on their own. Everyone would agree that dark matter had been discovered if even one other experiment confirms what DAMA has seen, but this has not yet occurred. Nonetheless, answers should be available soon. Even if the results just presented are out of date when you read this, the nature of the experiments most likely won't be.

INDIRECT DARK MATTER DETECTION

LHC experiments and ground-based cryogenic or noble liquid detectors are two ways to determine the nature of dark matter. The third and final way is through *indirect detection* of dark matter in the sky or on Earth.

Dark matter is dilute, but nonetheless occasionally annihilates with itself or with its antiparticle. This doesn't happen enough to significantly affect the overall density, but it might be enough to produce a measurable signal. That's because when dark matter particles annihilate, new particles get produced that carry away their energy. Depending on its nature, dark matter annihilation could sometimes yield detectable Standard Model particles and antiparticles, such as electrons and positrons, or pairs of photons. Astrophysical detectors that measure antiparticles or photons might then see signs of these annihilations.

The instruments that search for these Standard Model products of dark matter annihilation weren't initially designed with this goal. They were conceived as telescopes or detectors out in space or on the ground to detect light or particles in order to better understand what is in the sky. By looking at what types of stuff gets emitted by stars and galaxies and exotic objects that lie within them, astronomers can learn about the chemical composition of astronomical objects and deduce the properties and nature of stars.

The philosopher Auguste Comte in 1835 mistakenly said about stars, "We can never by any means investigate their chemical composition," which he thought beyond the boundary of attainable knowledge. Yet not too long after he said those words, the discovery and interpretation of the spectra of the Sun—the light that was emitted or absorbed—taught us about the composition of the Sun and proved him decidedly wrong.

Experiments today continue this mission when they try to deduce the composition of other celestial objects. Today's telescopes are very sensitive, and every few months we learn more about what is out there.

Fortunately for dark matter searches, the observations of light and particles that these experiments are already engaged in might also illuminate the nature of dark matter. Since antiparticles are relatively rare in the universe and the distribution of photon energies could exhibit distinctive and identifiable properties, such detection could eventually be associated with dark matter. The spatial distribution of these particles might also help distinguish such annihilation products from more common astrophysical backgrounds

HESS, the High Energy Stereoscopic System located in Namibia, and VERITAS, the Very Energetic Radiation Imaging Telescopic Array System in Arizona, are large arrays of telescopes on Earth that look for high-energy photons from the center of the galaxy. And the next generation of very high-energy gamma-ray observatory, the Cherenkov Telescope Array (CTA), promises to be even more sensitive. The Fermi Gamma-ray Space Telescope, on the other hand, orbits the sky 550 kilometers above the Earth every 95 minutes on a satellite that was launched at the beginning of 2008. Photon detectors on Earth have the advantage of having enormous collecting areas, whereas the very precise instruments on the Fermi satellite have better energy resolution and directional information, are sensitive to photons with lower energies, and have about 200 times the field of view.

Either of these types of experiments could see photons from annihilating dark matter, or from radiation produced by electrons and positrons resulting from dark matter annihilation. If we see either, we stand to learn a lot about the identity and properties of dark matter.

Other detectors look primarily for positrons, the antiparticles of electrons. Physicists working on an Italian-led satellite experiment called PAMELA have already reported their findings, and they look nothing like what was predicted. (See Figure 80 for PAMELA results.) The acronym in this case stands for the mouthful "Payload for Antimatter Matter Exploration and Light-nuclei Astrophysics," which is somewhat mitigated by the nice way PAMELA sounds when spoken with an Italian accent.

We don't yet know if the PAMELA excess events are due to dark matter or to misestimations of astronomical objects such as pulsars. But either way, the results have absorbed the attention of astrophysicists and particle physicists alike.

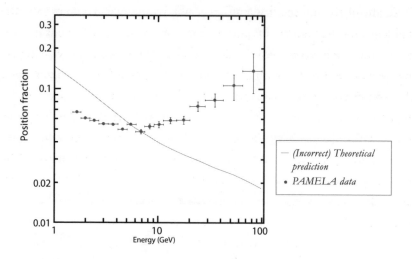

[**FIGURE 80**] Data from the PAMELA experiment, showing how badly experimental data (the crosses) agreed with theoretical predictions (the dotted curve).

Dark matter can also annihilate into protons and antiprotons. In fact, many models predict that this happens most frequently if dark matter particles do indeed find each other and annihilate. However, large numbers of antiprotons lurking in the galaxy due to known astronomical processes can mask the dark matter signal. Still, we might have a chance of seeing such dark matter through antideuterons, which are very weakly bound states of an antiproton and an antineutron, which might also be formed when dark matter annihilates. The Alpha Magnetic Spectrometer (AMS-02), now on the International Space Station, as well as dedicated satellite experiments, such as the General Antiparticle Spectrometer (GAPS), might ultimately find these antideuterons and thereby discover dark matter.

Finally, the uncharged particles called neutrinos that interact only via the weak force could be the key to the indirect detection of dark matter. Dark matter might get trapped in the center of the Sun or the Earth. The only signal that could get out in that case would be neutrinos, since unlike other particles, they won't be stopped by their interactions as they escape. Detectors named AMANDA, IceCube, and ANTARES are looking for these high-energy neutrinos.

If any of the above signals is observed—or even if they are not—we will learn more about the nature of dark matter—its interactions and its mass. In the meantime, physicists have been thinking about what signal to expect according to predictions from various possible dark matter models. And of course we ask about what any existing measurements might imply. Dark matter is tricky, since it interacts so weakly. But the hope is that with the many different types of dark matter experiments currently in operation, dark matter detection may be within imminent reach, and along with results from the LHC and elsewhere will provide a better sense of what is out there in the universe and how it all fits together.

Part VI: ROUNDUP

CHAPTER TWENTY-TWO

THINK GLOBALLY AND ACT LOCALLY

This book has presented glimpses of how the human mind can explore to the outer limits of the cosmos as well as into the internal structure of matter. In both pursuits, the late Harvard professor Sidney Coleman was considered one of the wisest physicists around. The story students told was that when Sidney applied for a postdoctoral fellowship after finishing graduate school, all except one of his letters of recommendation described him as the smartest physicist they had known—apart from Richard Feynman. The remaining letter was from Richard Feynman, who wrote that Sidney was the best physicist around—though he wasn't counting himself.

At Sidney's sixtieth birthday Festschrift celebration—a conference organized in his honor—many of the most notable physicists of his generation spoke. Howard Georgi, Sidney's Harvard colleague for many years and a fine particle physicist himself, observed that what struck him in watching the succession of talks by these very successful theoretical physicists was how differently they all think.

He was right. Each speaker had a particular way of approaching science and had made significant contributions through his (indeed they were all male) distinctive skills. Some were visual, some were mathematically gifted, and some simply had a prodigious capacity to absorb and evaluate information. Both top-down and bottom-up styles were represented among those present, whose accomplishments ranged from understanding the strong nuclear force in the interior of matter to the mathematics that could be derived using string theory as a tool.

Pushkin was right when he wrote, "Inspiration is needed in geometry, just as much as in poetry." Creativity is essential to particle physics, cosmology, and to mathematics, and to other fields of science, just as it is to its more widely acknowledged beneficiaries—the arts and humanities. Science epitomizes the extra richness that can enhance creative endeavors that take place in constrained settings. The inspiration and imagination involved are easily overlooked amid the logical rules. However, math and technology were themselves discovered and formulated by people who were thinking creatively about how to synthesize ideas— and by those who accidentally came upon an interesting result and had the creative alertness to recognize its value.

In the past few years, I've been fortunate to have had a variety of opportunities to meet and work with creative people in different walks of life, and it's interesting to reflect on what they share. Scientists, writers, artists, and musicians might seem very different on the surface, but the nature of skills, talents, and temperaments is not always as distinctive as you might expect. I'll now round up our story of science and scientific thinking with some of the qualities I've found most striking.

OUTLYING TALENT

Neither scientists nor artists are likely to be thinking about creativity per se when they do something significant. Few (if any) successful people sit down at their desks and decide, "I will be creative today." Instead, they are focused on a problem. And when I say focused, I mean single-mindedly, can't-help-but-think-about-it, intently-concentrated-on-their-work focused.

We usually see the end product of creative endeavors without witnessing the enormous dedication and technical expertise that underlie them. When I saw the 2008 film *Man on Wire*, which celebrated Philippe Petit's 1974 high-wire walk a quarter of a mile up in the air between the twin towers of the World Trade Center—a feat that at the time captured the attention of most New Yorkers like myself, but also many others around the world—I appreciated his sense of adventure and play and skill. But Philippe doesn't just bolt a tightrope into two walls and wiggle it around. The choreographer Elizabeth Streb showed me the inch-thick book with the many drawings and calculations he did before he installed a wire in her studio. Only then did I understand the preparation and focus that guaranteed the stability of his enterprise. Philippe was a "self-taught engineer," as he playfully described himself. Only after careful study and application of known laws of physics to understanding his materials' mechanical properties was he prepared to walk his tightrope. Of course until he actually did it, Philippe couldn't be absolutely sure he had taken everything into account—merely everything he could anticipate, which, not surprisingly, was enough.

If you find this level of absorption hard to believe, look around. People are frequently transfixed by their activities—whether of small or great significance. Your neighbor does crossword puzzles, your friends sit mesmerized watching sports on TV, someone on the subway is so absorbed in a book she misses her stop—not to mention the countless hours you might spend playing video games.

Those who are preoccupied by research are in the fortunate situation where what they do for a living coincides with what they love—or at the very least can't bear to neglect. Professionals in this category generally have the comforting idea (albeit possibly illusory) that what they do might have lasting significance. Scientists like to think we are part of a bigger mission to determine truths about the world. We might not have time for a crossword puzzle on a particular day but we will very likely want more time to spend on a research project—especially one connected to a bigger picture and larger goal. The actual act might involve the same sort of absorption as engaging in a game or even watching sports on TV.[70] But a scientist is likely to continue thinking about research when driving a car or falling asleep at night. The ability to stay committed to the project for days or months or years is certainly connected to the belief that the search is important—even if only a few might understand it (at least at first)—and even if the trajectory might ultimately prove to be wrong.

Lately it has become fashionable to question innate creativity and talent and attribute success solely to early exposure and practice. In a *New York Times* column, David Brooks summarized a couple of recent books on the subject this way: "What Mozart had, we now believe, was the same thing Tiger Woods had—the ability to focus for long periods of time and a father intent on improving his skills."[71] Picasso was another example he used. Picasso was the son of a classical artist and in his privileged environment was already making brilliant paintings as a child. Bill Gates too had exceptional opportunities. In his recent book, *Outliers*,[72] Malcolm Gladwell tells how Bill Gates's Seattle high school was one of the few to have a computer club, and how Gates subsequently had the opportunity to use the computers at the University of Washington for hours on end. Gladwell goes on to suggest that Gates's opportunities were more important to his success than his drive and talent.

Indeed, focusing and practice at an early stage so that the methods and techniques become hard wired is unquestionably part of many creative backgrounds. If you have a difficult problem to solve, you want to spend as little time as possible on the basics. Once skills (or math or knowledge) become second nature, you can call them up much more easily when you need them. Such embedded skills often continue operating in the background—even before they push good ideas into your conscious mind. More than one person has solved a problem while asleep. Larry Page told me that the seed idea for Google came to him in a dream—but that was only after he had been absorbed by the problem for months. People often attribute insights to "intuition" without recognizing how much lead time of detailed studies lies behind the moment of revelation.

So Brooks and Gladwell undoubtedly are correct in some respects. Though skill and talent matter, they won't get you very far without the honing of skills and intensity that comes with dedication and practice. But opportunities at a young age and systematic preparation are not the whole story. This description neglects the fact that the ability to focus and practice so intently is a skill in itself. The exceptional people who learn from what they did before and who can hold the accumulated lessons in their heads are far more likely to benefit from study and repetition. This tenacity allows for concentration and focus that will eventually pay off—in scientific research or any other creative pursuit.

The name of Calvin Klein's original perfume, "Obsession," was no accident: he became successful because (in his own words) he was obsessed. Even if golf pros perfect their swing over countless repeated attempts, I don't believe everyone can hit a ball a thousand times without becoming exceedingly bored or frustrated. My climbing friend, Kai Zinn, who works on difficult routes—hard 5.13s for those in the know—remembers the details and moves much better than I do. When he does a route ten times, he therefore benefits far more. This in turn makes him much more likely to persevere. I'll get bored and move on and remain a midlevel climber while Kai, who knows how to learn efficiently from repetition, will continually improve. Georges-Louis Leclerc, the eighteenth-century naturalist, mathematician, and author, succinctly summarized this ability: "Genius is only a greater aptitude for patience." Though I'll add that it's also rooted in impatience with lack of improvement.

SCALING A HILL OF BEANS

Practice, technical training, and drive are essential to scientific research. But they are not all that is required. Autistics—not to mention some academics and far too many bureaucrats—frequently demonstrate high-level technical skills yet lack creativity and imagination. All it takes is a trip to the movies these days to witness the limitations of drive and technical achievements without the support of these other qualities. Scenes in which animated creatures fight other animated creatures in hard-to-follow sequences might be impressive accomplishments in themselves, but they rarely possess the creative energy needed to fully engage many of us—even with the light and noise, I frequently fall asleep.

For me, the most absorbing films are those that address big questions and real ideas but embody them in small examples that we can appreciate and comprehend. The movie *Casablanca* might be about patriotism and love and war and loyalty but even though Rick warns Ilsa that "it doesn't take much to see that the problems of three little people don't amount to a hill of beans in this crazy world," those three people are the reason I'm captivated by the movie (plus, of course, Peter Lorre and Claude Rains).

In science, too, the right questions often come from having both the big and the small pictures in mind. There are grand questions that we all want to answer, and there are small problems that we believe to be tractable. Identifying the big questions is rarely sufficient, since it's often the solutions to the smaller ones that lead to progress. A grain of sand can indeed reveal an entire world, as the title of the Salt Lake City conference on scale (referred to in Chapter 3)—and the line of poetry by William Blake it refers to—remind us, and as Galileo understood so early on.

An almost indispensable skill for any creative person is the ability to pose the right questions. Creative people identify promising, exciting, and, most important, accessible routes to progress—and eventually formulate the questions correctly. The best science frequently combines an awareness of broad and significant problems with focus on an apparently small issue or detail that someone very much wants to solve or understand. Sometimes these little problems or inconsistencies turn out to be the clues to big advances.

Darwin's revolutionary ideas grew in part out of minute observations of birds and plants. The precession of the perihelion of Mercury wasn't a mistaken measurement—it was an indication that Newton's laws of physics were limited. This measurement turned out to be one of the confirmations of Einstein's gravity theory. The cracks and discrepancies that might seem too small or obscure for some can be the portal to new concepts and ideas for those who look at the problem the right way.

Einstein didn't even initially set out to understand gravity. He was trying to understand the implications of the theory of electromagnetism that had only recently been developed. He focused on aspects that were peculiar or even inconsistent with what everyone thought were the symmetries of space and time and ended up revolutionizing the way we think. Einstein believed it should all make sense, and he had the breadth of vision and persistence to extract how that was possible.

More recent research illustrates this interplay too. Understanding why certain interactions shouldn't occur in supersymmetric theories might seem like a detail to some. My colleague David B. Kaplan was mocked when he talked about such problems in Europe in the 1980s. But this problem turned out to be a rich source of new insights into supersymmetry and supersymmetry breaking, leading to new ideas that experimenters at the LHC are now prepared to test.

I'm a firm believer that the universe is consistent and any deviation implies something interesting yet to be discovered. After I made this point at a Creativity Foundation presentation in Washington, D.C., a blogger nicely interpreted this as my having high standards. But really, belief in the consistency of the universe is probably the principal driving force for many scientists when figuring out which questions to study.

Many of the creative people I know also have the ability to hold a number of questions and ideas in their heads at the same time. Anyone can look things up using Google, but unless you can put facts and ideas together in interesting ways, you aren't likely to find anything new. It is precisely the slightly jarring juxtaposition of ideas coming from different directions that often leads to new connections or insights or poetry (which was what the term creativity originally applied to).

A lot of people prefer to work linearly. But this means that once they are stuck or find that the path is uncertain, their pursuit is over. Like many writers and artists, scientists make progress in patches. It's not always a linear process. We might understand some pieces of a puzzle, but temporarily set aside others we don't yet understand, hoping to fill in these gaps later on. Only a few understand everything about a theory from a single continuous reading. We have to believe that we will eventually piece it all together so that we can afford to skip over something and then return, armed with more knowledge or a broader context. Papers or results might initially appear to be incomprehensible, but we'll keep reading anyway. When we find something we don't understand, we'll skip over it, get to the end, puzzle it out our own way, and then later on return to where we were mystified. We have to be absorbed enough to continue—working through what does and does not make sense.

Thomas Edison famously noted that, "Genius is one percent inspiration, ninety-nine percent perspiration." And—as Louis Pasteur once said—"In the fields of observation, chance favors the prepared mind." Dedicated scientists sometimes thereby find the answers they are looking for. But they might also find solutions to problems apart from the original target of investigation. Alexander Fleming didn't intend to find a cure for infectious diseases. He noticed a fungus had killed colonies of *Staphylococci* he'd been investigating and recognized its potential therapeutic benefits—though it took a decade and the involvement of others before penicillin was developed into a powerful world-changing medicine.

Subsidiary benefits often arise from a reserve of a broad base of questions. When Raman Sundrum and I worked on supersymmetry, we ended up finding a warped extra dimension that could solve the hierarchy problem. Afterward, by staring hard at the equations and putting them in a broader context, we also found that an infinite warped dimension of space could exist without contradicting any known observations or law of physics. We had been studying particle physics—a different topic altogether. But we had both the big and small pictures in mind. We were aware of the big questions about the nature of space even when concentrating on the more phenomenological issues such as understanding the hierarchy of mass scales in the Standard Model.

Another important feature of this particular work was that neither Raman nor I was a relativity expert, so we arrived at our research with open minds. Neither we (nor anyone else) would have conjectured that Einstein's theory of gravity permits an invisible infinite dimension unless the equations had shown us that it was possible. We doggedly pursued the consequences of our equations, unaware that an infinite extra dimension was supposed to be impossible.

Even so, we weren't immediately convinced we were right. And Raman and I hadn't dived into the radical idea of extra dimensions blindly. It was only after we and many others had tried employing more conventional ideas that it made sense to leave our spacetime box. Although an extra dimension is an exotic and novel suggestion, Einstein's theory of relativity still applies. Therefore, we had the equations and mathematical methods to understand what would happen in our hypothetical universe.

People subsequently used the results from this research assuming extra dimensions as launching points to discover new physical ideas that might apply in a universe with no such extra dimensions at all. By thinking about the problem in an orthogonal way (here, literally orthogonal), physicists recognized possibilities they had previously been entirely unaware of. It helped to think outside the box of three-dimensional space.

Anyone facing new ground has no choice but to live with the uncertainty that exists before a problem is completely solved. Even when starting from a nice solid platform of existing knowledge, someone investigating a new phenomenon inevitably encounters unknowns and the uncertainty that accompanies them—though admittedly with less risk to life and limb than a tightrope walker. Space adventurers, but artists and scientists, too, try to "boldly go where no one has gone before." But the boldness isn't random or haphazard and it doesn't ignore earlier achievements, even when the new territory involves new ideas or anticipates crazy-seeming experiments that appear to be unrealistic at first. Investigators do their best to be prepared. That's what rules, equations, and instincts about consistency are good for. These are the harnesses that protect us when traversing new domains.

In my colleague Marc Kamionkowski's words, it's "OK to be ambitious and futuristic." But the trick is still to determine realistic goals. An award-winning business student present at the Creativity Foundation event I participated in remarked that the recent successful economic growth that had escalated into an economic bubble stemmed in part from creativity. But he noted too that the lack of restraint also caused the bubble to burst.

Some of the most groundbreaking research of the past exemplifies the contradictory impulses of confidence and caution. The science writer Gary Taubes once said to me that academics are at the same time the most confident and the most insecure people he knows. That very contradiction drives them—the belief that they are moving forward coupled with the rigorous standards they apply to make sure they are right. Creative people have to believe that they are uniquely placed to make a contribution—while all the time keeping in mind the many reasons that others might have already thought of and dismissed similar ideas.

Scientists who were very adventuresome in their ideas could also be very cautious when presenting them. Two of the most influential, Isaac Newton and Charles Darwin, waited quite a while before sharing their great ideas with the outside world. Charles Darwin's research spanned many years, and he published the *Origin of Species* only after completing extensive observational research. Newton's *Principia* presented a theory of gravity that was well over a decade in development. He waited to publish until he had completed a satisfactory proof that bodies of arbitrary spatial extent (not just pointlike objects) obey an inverse square law. The proof of this law, which says gravity decreases as the square of the distance from the center of an object, led Newton to develop the mathematics of calculus.

It sometimes takes a new formulation of a problem to see it the right way and to redefine the boundaries so you can find a solution where, on the surface, none appears possible. Perseverance and faith often make a big difference to the outcome—not religious faith but faith that a solution exists. Successful scientists—and creative people of all kinds—refuse to get stuck in dead ends. If we can't solve a problem one way, we'll seek an alternative route. If we reach a roadblock, we'll dig a tunnel, find another direction, or fly over and get the lay of the land. Here's where imagination and superficially crazy ideas come in. We have to

[**FIGURE 81**] The nine-dots problem asks how to connect all the dots using only four segments without lifting your pen.

believe in the reality of an answer in order to continue, and to trust that ultimately the world has a consistent internal logic that we might eventually discover. If we think about something from the right perspective, we can often find connections that we would otherwise miss.

The expression "thinking outside the box" doesn't come from getting outside your work cubicle (as I once thought might be the case), but from the nine-dots problem, which asks how to connect nine dots with four lines without lifting your pen (see Figure 81). No solution to the nine-dots problem exists if you have to keep your pen inside the confines of the square, but no one told you that was a requirement. Going "outside the box" yields the solution (see Figure 82). At this point you might realize you can reformulate the problem in a number of other ways too. If you use thick dots, you can use three lines. If you fold the paper (or use a really thick line, as a young girl apparently suggested to the problem's creator), you can use just one line.

These solutions aren't cheating. They would be only if you have additional constraints. Education unfortunately sometimes encourages students not only to learn how to resolve problems, but also to second-guess the teacher's intention—narrowing the range of correct answers and potentially also the students' minds. In *The Quark and the Jaguar*,[73] Murray Gell-Mann cites Washington University physics professor Alexander Calandra's "Barometer Story,"[74] in which he tells of a teacher who wasn't

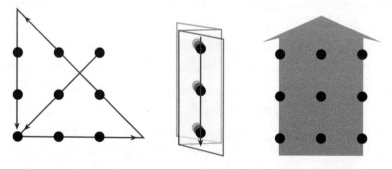

[**FIGURE 82**] Possible creative solutions to the nine-dots problem include "thinking outside the box," folding the paper so the dots align, or using a very thick pen.

sure he should give a student credit. The teacher had asked his students how they might use a barometer to determine the height of a building. This particular student answered that you could attach a string to the barometer, lower it to the ground, and find out how long the string was. When he was told to use physics, he suggested measuring the time it took for it to fall from the top of the building, or measuring the shadow at a known time of day. The student also volunteered the nonphysics solution of offering the superintendent the barometer in exchange for being told the height of the building. These answers might not have been what the teacher was looking for. But the student astutely—and humorously—recognized that the teacher's constraints weren't part of the problem.

When other physicists and I started thinking about extra dimensions of space in the 1990s, we not only went outside the box, we went outside three-dimensional space itself. We thought of a world in which the very stage in which we solved the problems was bigger than we had originally assumed. In doing so, we found potential solutions to problems that had plagued particle physicists for years.

Even so, research doesn't arise in a vacuum. It is enriched by the many ideas and insights that others have thought of before. Good scientists listen to one another. Sometimes we find the right problem or solution just by very carefully listening to, observing, or reading someone else's work. Often we collaborate to bring in different people's talents, and also to keep ourselves honest.

Even if everyone wants to be the first to solve an important problem, scientists still learn from and share with one another and work on common topics. Occasionally other scientists say things that contain the clues to interesting problems or solutions—even unwittingly. Scientists might have their own inspiration, but they will often also exchange ideas, work out the consequences, and make adjustments or start again if the original idea doesn't work. Imagining new ideas and keeping some while shooting others down is our bread and butter. That's how we advance. It's not bad. It's progress.

One of the most important roles I can play as an adviser to graduate students is to be alert to their good ideas, even when they haven't yet learned how to express them—and to listen when students find loop-holes in my suggestions. This back-and-forth is perhaps one of the best ways to teach—or at least foster—creativity.

Competition plays an important role as well—in science as well as in most any other creative endeavor. In a discussion of creativity, the artist Jeff Koons simply told those of us in the room that when he was young, his sister did art—and he realized that he could do it better. A young filmmaker explained how competition encourages him and his colleagues to absorb each other's techniques and ideas and thereby re-fine and develop their own. The chef David Chang expressed a similar thought a little more bluntly. His reaction after going to a new restaurant is, "That's delicious. Why didn't I think of that?"

Newton waited to publish until his results were complete. But he might also have been wary of his competitor Robert Hooke, who knew about the inverse square law as well—but lacked the calculus to support it. Nonetheless, Newton's publication seems to have been prompted in part by a question relayed to him about Hooke's overlapping research. Darwin, too, was clearly motivated to present his results by the knowl-edge that Alfred Russel Wallace was working on similar evolutionary ideas—and was likely to steal his thunder if he remained silent much longer. Both Darwin and Newton wanted to have their stories straight before presenting their revolutionary results, and developed them until they were extremely confident they were correct—or at least until they thought they might be scooped.

The universe repeatedly reveals itself to be cleverer than we are. Equations or observations open up ideas that no one would have dreamed of—and only creative open-minded inquiries will unearth such hidden phenomena in the future. Without incontrovertible evidence, no scientist would have invented quantum mechanics, and I suspect that anticipat-ing the precise structure of DNA and the myriad phenomena that make up life would have been pretty nearly impossible unless we were faced with the phenomena or equations that told us what was there. The Higgs mechanism is ingenious, as are the inner workings of the atom and the behavior of the particles that underlie everything we see.

Research is an organic process. We don't necessarily always know where we are headed, but experiments and theory serve as valuable guides. Preparation and skill, concentration and perseverance, asking the right questions, and cautiously trusting our imaginations will all help us in our search for understanding. So will open minds, conversations with others, wanting to do better than our predecessors or peers, and believing there are answers. No matter what the motivation, and independently of the particular skills that might come into play, scientists will continue to investigate inward and outward—and look forward to learning about the other ingenious mechanisms the universe has in store.

CONCLUSION

When I first looked at translations of German media reports on my physics research or my book, *Warped Passages*,[75] I was surprised by the repeated presence of the words "edge of the universe." The explanation of the plausible but seemingly random appearance of the phrase wasn't quite obvious at first—it turned out to be the computer's German translation of my last name.[76]

Yet we are indeed at the edge of the universe, both on small scales and on large ones. Scientists have experimentally explored distances from the weak scale of 10^{-17} centimeters to the size of the universe, 10^{30} centimeters. We can't be sure what the scales that demarcate true paradigm shifts in the future will be, but many scientific eyes are now focused on the weak scale, which the LHC and dark matter searches are experimentally exploring. At the same time, theoretical work continues to investigate scales ranging from the weak to the Planck energies, and to larger scales as well, as we attempt to fill in gaps in our understanding. It's hubris to think that what we've seen is all there is. New discoveries almost certainly await.

The era of modern science represents a mere blip on the timeline of history. But the remarkable insights gained through advances in technology and mathematics since its birth in the seventeenth century have taken us an impressively long way toward understanding the world.

This book has explored how high-energy physicists and cosmologists today determine their course and how a combination of theory and experiment could shed light on some deep and fundamental questions. The Big Bang theory describes the universe's current expansion, but it leaves open the questions of what happened earlier—and what is the nature of dark energy and dark matter. The Standard Model predicts elementary particle interactions, but leaves unresolved questions about why its properties are what they are. Dark matter and the Higgs boson could be around the corner—as could evidence for new spacetime symmetries or even new dimensions of space. We could be lucky and have answers soon. Or—if the relevant quantities are too heavy or too weakly interacting—it could take a while. We'll only know if we ask and look.

I've also presented speculations about some even more difficult-to-test ideas. Though they expand the imagination and might eventually connect to reality, they could also remain in the domain of philosophy or religion. Science won't disprove the landscape of multiple universes—or God for that matter—but it's unlikely to verify them either. Even so, some aspects of the multiverse—such as those that could explain the hierarchy—do have testable consequences. It's up to scientists to ferret these out.

The other major element of *Knocking on Heaven's Door* has been the concepts—such as scale, uncertainty, creativity, and rational critical reasoning—that inform scientific thought. We can believe that science will make progress toward reaching answers and that complexity can emerge over time even before we have a fully fleshed-out explanation. The answers might be complicated, but that doesn't justify abdicating faith in reason.

Understanding nature, life, and the universe poses extraordinarily difficult problems. We all would like to better understand who we are, where we came from, and where we are going—and to focus on things larger than ourselves and more permanent than the latest gadget or fashion. It's easy to see why some turn to religion for explanations. Without the facts and the inspired interpretations that demonstrated surprising connections, the answers scientists have arrived at so far would have been extremely difficult to guess. People who think scientifically advance our knowledge of the world. The challenge is to understand as much as we can, and curiosity—unconstrained by dogma—is what is required.

The line between legitimate inquiry and arrogance might be an issue for some, but ultimately critical scientific thinking is the only reliable way to answer questions about the makeup of the universe. Extremist anti-intellectual strands in some current religious movements are at odds with traditional Christian heritage—not to mention progress and science—but fortunately they don't represent all religious or intellectual perspectives. Many ways of thinking—even religious ones—incorporate challenges to existing paradigms and allow for the evolution of ideas. Progress for each of us involves replacing wrong ideas and building on the ones that are right.

I appreciated the sentiment when at a recent lecture, Bruce Alberts, former president of the National Academy of Sciences and current editor in chief of *Science* magazine, highlighted the need for the creativity, rationality, openness, and tolerance that are inherent to science—the robust combination of qualities that Jawaharlal Nehru, India's first prime minister, called "the scientific temper."[77] Scientific ways of thinking are critical in today's world, providing essential tools for dealing with many tough issues—social, practical, and political. I'd like to close with a few further reflections about the relevance of science and scientific thinking.

Some of today's complex challenges might be addressed with a combination of technology, information about large populations, and raw computing power. But many major advances—scientific or otherwise—simply require a lot of thought by isolated or small groups of inspired individuals working on hard problems for a long time. Although this book has focused on the nature and value of basic science, pure, curiosity-driven research has—along with advancing science itself—led to technological breakthroughs that have completely changed the way we live. In addition to giving us important ways of thinking about hard problems, basic science can lead to technological tools today that—when combined with more scientific thinking that absorbs the creativity and principles we've discussed—will help find solutions tomorrow.

The question now is how to address bigger questions in that context. How do we take technology beyond mere short-term goals? Even in a world of technology, we need both ideas and incentives. The company that makes a must-have gadget may be very successful, and it's easy to get caught up in the pursuit of a new one. But this can distract from the real issues we'd like technology to address. Although iPods are fun, the iPod lifestyle isn't going to solve the big problems of today's world.

Kevin Kelly, one of the founders of *Wired* magazine, said when we were on a panel together at a conference about technology and progress: "Technology is the greatest force in the universe." If that is indeed the case, science is responsible for the greatest force, since basic science was essential to the technology revolution. The electron was discovered with no ulterior motive, yet electronics has defined our world. Electricity too was a purely intellectual discovery, yet the planet is now pulsing with wires and cables. Even quantum mechanics, the esoteric theory of the atom, turned out to be the key to Bell Labs' scientists developing the transistor—the underlying hardware of the technology revolution. Yet none of the early investigators of the atom would have believed that the research they were doing would ever have any application, let alone one as grand as the computer and the information revolution. Both basic scientific knowledge and scientific ways of thinking were needed for the deep insights into the nature of reality that ultimately led to these breakthroughs.

No amount of computing power or social networking would have helped Einstein develop the theory of relativity any faster than he did. Scientists probably wouldn't have understood quantum mechanics any more rapidly either. This is not to deny that, once there is an idea or some new understanding of a phenomenon, technology expedites advances. And some problems simply do require sifting through large amounts of data. But usually a core idea is essential. The insights into the nature of reality that the practice of science gives us can ultimately lead to transformative breakthroughs that affect us in unpredictable ways. It is vital that we continue to pursue it.

It is now a given that technology is central. This is true in the sense that most new developments critically employ technology. But I would add that it is central in the sense of being neither the beginning nor the end, but rather a means of getting things done and communicating and connecting developments. What we want to use it for is our choice. And the insights that go into solving problems or new developments can arise from many forms of creative thought.

Technology also makes each of us the center of our own universe, as we see physically in MapQuest or metaphorically on any social networking site. But the problems of the world are far more extensive and global. Technology can enable solutions, but they are more likely to come when also prompted by clear and creative thinking—the kind we see in the best scientific work.

In the past, our nation's attention to science and technology— along with the recognition that we need to make long-term commitments and stick to them—has proved to be a successful strategy that kept us in the forefront of new developments and ideas. We now seem to be in danger of losing these values that have worked so well for us before. We need to recommit to these principles as we seek not just short-term advances but also to understand the costs and benefits for the long term.

Rational inquiry about the world deserves more credit, so that we can use it to address some of the serious challenges that lie ahead. Bruce Alberts in his lecture also advocated scientific thinking as a way of arming people against rants, simplified TV news, and overly subjective talk radio. We don't want people to drift away from the scientific method, since that method is essential to reaching meaningful conclusions about the many complex systems that societies today must deal with—among them the financial system, the environment, risk assessment, and health care.

One of the key elements in making advances and solving problems— whether scientific or otherwise—has been and will be an awareness of scale. Categorizing what has been observed and understood by scale has taken us very far in our understanding of physics and the world— whether the units are physical scales, population groups, or time frames. Not only scientists, but political, economic, and policy leaders too need to keep such concepts in mind.

Supreme Court Justice Anthony Kennedy, in a speech to the Ninth Judicial Circuit, referred not only to the significance of scientific thinking, but also to the important contrast between "micro" and "macro" thinking—words that apply as much to the small-scale and large-scale elements of the universe as to the detailed and global ways we think about the world. As we have seen in this book, one of the factors in addressing issues—scientific as well as practical and political—is the interplay between the two scales of thought. The awareness of both is one of the factors that contributes to creative ideas.

Justice Kennedy also noted that among the elements of science that he likes are "the ridiculous solutions [that] often turn out to be the ones that are true." And this is indeed sometimes the case. Nonetheless, good science, even when it leads to superficially far-fetched or counterintuitive conclusions, is rooted in measurements that show these conclusions to be true, or in problems that call for the apparently crazy solutions we conjecture might be real.

Many elements combine to form the foundation of good scientific thinking. In *Knocking on Heaven's Door,* I have attempted to convey the significance of rational scientific thought and its materialist premises, as well as the ways in which scientific thinking tests ideas through experiments and discards them when they don't measure up. Scientific thought recognizes that uncertainty isn't failure. It properly evaluates risks and accounts for both short- and long-term influences. It allows for creative thinking in the search for solutions. These are all modes of thought that can lead to advances—both in and out of the laboratory or office. The scientific method helps us understand the edges of the universe, but it can also guide us in critical decisions for this world that we now live in. Our society needs to absorb these principles and teach them to future generations.

We shouldn't be afraid to ask big questions or to consider grand concepts. One of my physics collaborators, Matthew Johnson, got it right when he exclaimed, "Never before has there been such an arsenal of ideas." But we don't yet know the answers and are waiting for experimental tests. Sometimes answers come more quickly than expected—as when the cosmic microwave background taught us about the early exponential expansion of the universe. And sometimes they take longer—as with the LHC, which still has us waiting.

We should soon know more about the makeup and forces of the universe, as well as why matter has the properties it does. We also hope to learn more about the missing stuff that we call "dark." So, as our "prequel" ends, let's return to the line from the Beatles song that accompanied the introduction to my earlier book, *Warped Passages*: "Got to be good-looking 'cause he's so hard to see." New phenomena and understanding might be challenging to find, but the wait and challenges will be worth it.

ACKNOWLEDGMENTS

This book covers a lot of ground and I was fortunate to have had many wonderfully generous and thoughtful people providing guidance throughout. Knowing I could count on keen minds to reflect on even early incarnations of this work was a powerful incentive when moving forward. I am especially grateful to Andreas Machl, Luboš Motl, and Cormac McCarthy, all of whom read more than one draft of the book and provided valuable feedback during its different stages. Cormac's high standards, patience, and belief in "my project," Luboš's precision as a physicist and care for science communication, and Andreas's wisdom, enthusiasm, and consistent support were invaluable.

Others' edits, input, and enthusiasm also mattered a great deal. Anna Christina Büchmann was delightfully insightful, smart, and kind with her suggestions and contributions; Jen Sacks helped me through moments of indecision with wisdom and care; Polly Shulman provided important direction and encouragement early on; Brad Farkas's interest and sharp editorial pen helped solidify my enterprise; and the keen eye and overwhelming skill of my British editor, Will Sulkin, improved some key chapters at a critical stage. Thanks, too, are owed to Bob Cahn, Kevin Herwig, Dilani Kahawala, David Krohn, and Jim Stone for their proofreading and suggestions after reading a more final draft.

For helping get details of both the LHC machine and the ATLAS and CMS experiments correct, I am very grateful to the physicists Fabiola Gianotti and Tiziano Camporesi, who know their detectors as well as is humanly possible. And who could be better than Lyn Evans for reading over my writing about the LHC and its history? Thanks also to Doug Finkbeiner, Howie Haber, John Huth, Tom Imbo, Ami Katz, Matthew Kleban, Albion Lawrence, Joe Lykken, John Mason, Rene Ong, Brian Shuve, Robert Wilson, and Fabio Zwirner who also generously commented on some of the physics sections. Thanks as well to my 2010 and 2011 Harvard freshman seminar classes for their input about their understanding of the LHC.

Religion and science was somewhat new territory for me, which I could tread much more confidently equipped with the advice and wisdom of Owen Gingerich, Linda Gregerson, Sam Haselby, and Dave Thom. I am also grateful to others who helped with science history—Ann Blair, Sofia Talas, and Tom Levenson—all of whom made my story more accurate.

Topics like risk and uncertainty can be risky (and uncertain). Thanks to Noah Feldman, Joe Fragola, Victoria Gray, Joe Kroll, Curt McMullen, Jamie Robins, Jeannie Suk, attendees at the Harvard Law School Colloquium, and particularly Jonathan Wiener for sharing their expertise, and also to earlier conversations with Cass Sunstein. Creativity can be another slippery topic and I'm grateful to Karen Barbarossa, Paul Graham, Lia Halloran, Gary Lauder, Liz Lerman, Peter Mays, and Elizabeth Streb for sharing their insights. Special thanks also to Scott Derrickson for his conversations that were key to the first chapter, and for correcting me when his memory was better than mine. Thanks to the organizers of 2010 Techonomy for inviting me to join the opening panel—preparing for it contributed to the book's conclusions. Thanks, too, to the others whose conversations were mentioned in the text. Thanks also to Alfred Assin, Rodney Brooks, David Fenton, Kevin McGarvey, Sesha Pratap, Dana Randall, Andy Singleton, and Kevin Slavin for their generous feedback and thoughts, and to A.M. Homes and Rick Kot for advice and encouragement.

I am grateful to several others for encouraging me early on in the somewhat challenging enterprise I'd set out to pursue. Thanks to John Brockman and Ecco's Dan Halpern for getting this book off the ground, and to Matt Weiland, and his assistant, Shanna Milkey, for helping connect the pieces. Thanks, too, to the others at Ecco who helped make this book a reality, and to Andrew Wylie for shepherding the final stages. I am also pleased to have worked with the great illustrating team of Tommy McCall, Ana Becker, and Richert Schnorr, who conveyed complicated ideas with clear and precise pictures.

Finally, thanks to my research collaborators and fellow physicists for all they've taught me. Thanks to my family for encouraging my love of rationality. Thanks to my friends for their patience and support. And thanks to those—mentioned or not—who have helped shape my ideas along the way.

ENDNOTES

1. I will often approximate this as 27 kilometers.
2. The Large Hadron Collider is quite big, but it is used to study infinitesimal distances. The reasons for its large size are described later on when we discuss the LHC in detail.
3. Unlike in the movie, Herman Hupfield's famous song "As Time Goes By" written in 1931 began with an unmistakable reference to people's familiarity with the latest physics developments:

    This day and age we're living in
    Gives cause for apprehension,
    With speed and new invention,
    And things like fourth dimension,
    Yet we get a little weary
    From Mr. Einstein's theory

4. Fielding, Henry. *Tom Jones*. (Oxford: Oxford World Classics, 1986).
5. Quantum mechanics can have macroscopic effects in carefully prepared systems or when measurements apply to high statistics situations, or very precise devices so that small effects can emerge. However, that does not invalidate using an approximate classical theory for most ordinary phenomena. It depends on precision as Chapter 12 will further address. The effective theory approach allows for the approximation and makes precise when it is inadequate.
6. I will sometimes employ exponential notation, which I will use here to explain what I mean in the middle in terms of powers of ten. The size of the universe is 10^{27} meters. This number is a one followed by 27 zeroes, or one thousand trillion trillion. The smallest imaginable scale is 10^{-35} meters. This number is a decimal point followed by thirty-four zeroes followed by a one, or one hundredth of one billionth of one trillionth of one trillionth. (You can see why exponential notation is easier.) Our size is about 10^1. The exponent here is 1, which is reasonably close to the middle between 27 and -35.
7. Levenson, Tom. *Measure for Measure: A Musical History of Science* (Simon & Schuster, 1994).
8. During the Inquisition, the Romans didn't include Tycho's books in their Index,

as would have been expected based on his Lutheran faith, because they wanted his framework to keep the Earth stationary yet consistent with Galileo's observations.

9. Hooke, Robert. *An Attempt to Prove the Motion of the Earth from Observations* (1674), quoted in Owen Gingerich, *Truth in Science: Proof, Persuasion, and the Galileo Affair, Perspectives on Science and Christian Faith*, vol. 55.

10. Rilke, Rainer Maria. *Duino Elegies* (1922).

11. Doyle, Arthur Conan. *The Sign of the Four* (originally published in 1890 in Lippincott's Monthly Magazine, chapter 1), in which Sherlock Holmes comments on Watson's pamphlet, "A Study in Scarlet."

12. Browne, Sir Thomas. *Religio Medici* (1643, pt. 1, section 9).

13. Augustine. *The Literal Meaning of Genesis*, vol. 1, books 1–6, trans. and ed. by John Hammond Taylor, S. J. (New York: Newman Press, 1982). Book 1, chapter 19, 38, pp. 42–43.

14. Augustine. *On Christian Doctrine*, trans. by D. W. Robertson (Basingstoke: Macmillan, 1958).

15. Augustine. *Confessions*, trans. by R. S. Pine-Coffin (Harmondsworth: Penguin, 1961).

16. Stillman, Drake. *Discoveries and Opinions of Galileo* (Doubleday Anchor Books, 1957) p. 181.

17. Ibid., pp. 179–180.

18. Ibid., p. 186.

19. Galileo, 1632. *Science & Religion: Opposing Viewpoints*, ed. Janelle Rohr (Greenhaven Press, 1988), p. 21.

20. See, for exmple, Gopnik, Alison. *The Philosophical Baby* (Picador, 2010).

21. Matthew 7:7–8.

22. Blackwell, Richard J. *Galileo, Bellarmine, and the Bible* (University of Notre Dame Press, 1991).

23. Quoted in Gerald Holton, "Johannes Kepler's Universe: Its Physics and Metaphysics," *American Journal of Physics* 24 (May 1956): 340–351.

24. Calvin, John. *Institutes of Christian Religion*, trans. by F. L. Battles in *A Reformation Reader*, Denis R. Janz, ed. (Minneapolis: Fortress Press, 1999).

25. For example, in ancient Greece, stadia didn't have a fixed length since they were based on different body part lengths in different regions and in different times.

26. There is, of course, an electromagnetic field, but there is virtually no actual matter.

27. Momentum is a quantity that is approximated by the product of mass and veloc-

ity at small speeds but is equal to the energy divided by the speed of light for objects moving at relativistic velocities.

28. Gamow, George. *One, Two, Three . . . Infinity: Facts and Speculations of Science* (Viking Adult, September 1947).

29. Note that this figure corresponds to a more precise version of unification than was true for the original Georgi-Glashow theory, in which the lines almost converged, but didn't quite meet. This imperfect unification was demonstrated only later on, with better measurements of the forces' interaction strengths.

30. Although it comes close, we now know that unification won't occur within the Standard Model. However, unification can happen in modifications of the Standard Model, such as the supersymmetric models considered in Chapter 17.

31. Feynman, Richard. The QED Lecture at University of Auckland (New Zealand, 1979). See also: *Richard Feynman Lectures, Proving the Obviously Untrue.*

32. Quoted, for example, in Richard Rhodes, *The Making of the Atomic Bomb* (Simon & Schuster, 1986).

33. Particle physicists measure energy in units of electronvolts and those are the units I will use throughout. An electronvolt (eV) is the energy acquired by a free electron when accelerated through an electric potential difference of one volt. More commonly, I will refer to the units GeV, which is a billion electronvolts, and a TeV, which is a trillion electronvolts.

34. Ironically, the plot of Dan Brown's *Angels and Demons* centers on antimatter, whereas the LHC is the first CERN collider for which the initial states are purely matter.

35. Overbye, Dennis. "Collider Sets Record and Europe Takes U. S. Lead." *New York Times*, December 9, 2009.

36. In 1997, the European Physical Society recognized Robert Brout, François Englert, and Peter Higgs for their achievement, and the three were once again awarded in 2004 with the Wolf Prize in Physics. François Englert, Robert Brout, Peter Higgs, Gerald Guralnik, C. R. Hagen, and Tom Kibble all received the J. J. Sakurai Prize for Theoretical Particle Physics from the American Physical Society in 2010. I will refer only to Higgs and Peter Higgs throughout the text, as my focus is the physical mechanism and not the personalities. Of course if the Higgs is discovered, only three at most will receive a Nobel Prize and priority issues will be important. For an overview of the situation, see, for example, Luis Álvarez-Gaumé and John Ellis, "Eyes on a Prize Particle," *Nature Physics* 7 (January 2011).

37. It is ambiguous whether the Standard Model should also include the very

heavy right-handed neutrinos that are likely to exist and play a role in neutrino masses.

38. Its original purpose was to accelerate protons and antiprotons, but currently only protons, in its current use as the SPS accelerator at the LHC.

39. *Physical Review D,* 035009 (2008).

40. http://lsag.web.cern.ch/lsag/LSAG-Report.pdf.

41. See, for example, Taibbi, Matt. "The Big Takeover: How Wall Street Insiders are Using the Bailout to Stage a Revolution," *Rolling Stone,* March 2009.

42. This point is also addressed, for example, in J. D. Graham and J. B. Wiener, *Risk vs. Risk: Tradeoffs in Protecting Health and Environment* (Harvard University Press, 1995), especially Chapter 11.

43. See also, for example, Slovic, Paul. "Perception of Risk," *Science* 236, 280–285, no. 4799 (1987). Tversky, Amos, and Daniel Kahneman, "Availability: A heuristic for judging frequency and probability," *Cognitive Psychology* 5 (1973): 207–232. Sunstein, Cass R., and Timur Kuran. "Availability Cascades and Risk Regulation," *Stanford Law Review* 51 (1999):683–768. Slovic, Paul "If I Look at the Mass I Will Never Act: Psychic Numbing and Genocide," *Judgment and Decision Making* 2, no. 2 (2007): 79–95.

44. See also, for example, Kousky, Carolyn, and Roger Cooke. *The Unholy Trinity: Fat Tails, Tail Dependence, and Micro-Correlations,* RFF Discussion Paper 09-36-REV (November 2009). Kunreuther, Howard, and M. Useem. *Learning from Catastrophes: Strategies for Reaction and Response* (Upper Saddle River, NJ: Wharton School Publishing). Kunreuther, Howard. *Reflections and Guiding Principles for Dealing with Societal Risks,* in *The Irrational Economist: Overcoming Irrational Decisions in a Dangerous World,* E. Michel-Kerjan and P. Slovic, eds., New York Public Affairs Books 2010. Weitzman, Martin L., *On Modeling and Interpreting the Economics of Catastrophic Climate Change,* Review of Economics and Statistics, 2009.

45. See, for example, Joe Nocera's cover story on "Risk Mismanagement" in the *New York Times Sunday Magazine,* January 4, 2009.

46. The problem of irreversibility has been addressed by some economists, including Arrow, Kenneth J., and Anthony C. Fisher, "Environmental Preservation, Uncertainty, and Irreversibility," *Quarterly Journal of Economics,* 88 (1974): 312–319. Gollier, Christian, and Nicolas Treich, "Decision Making under Uncertainty: The Economics of the Precautionary Principle," *Journal of Risk and Uncertainty* 27, no. 7 (2003). Wiener, Jonathan B. "Global Environmental Regulation," *Yale Law Journal* 108 (1999): 677–800.

47. E.g., Richard Posner, *Catastrophe: Risk and Response* (Oxford University Press, 2004).

48. Leonhardt, David. "The Fed Missed This Bubble: Will It See a New One?" *New York Times,* January 5, 2010.

49. In this book, I use the term "systematic uncertainty," rather than the more commonly used term "systematic error." Errors are often associated with mistakes, whereas uncertainty refers to the inevitable level of imprecision, given your apparatus.

50. Again, people commonly use the term statistical error to refer to an uncertain measurement due to finite statistics.

51. Kristof, Nicholas. "New Alarm Bells About Chemicals and Cancer," *New York Times,* May 6, 2010.

52. This quote has also been attributed to Robert Storm Peterson and Niels Bohr.

53. This table includes separate entries for left- and right-handed particles. These particles are distinguished by their chirality, which for massless particles tells the spin along the direction of motion. Masses mix the two—such as a left- and right-handed electron. The precise distinguishing feature is less important for this table than the difference in their interactions. If all particles were massless, the weak force that changed up-type into down-type quarks and charged into neutral leptons would act only on left-handed particles. The strong and electromagnetic forces, on the other hand, act on both, with only the quarks charged under the strong force.

54. The three types of neutrinos get paired via the weak force with the three charged leptons. However, once they are produced, neutrinos can oscillate into each other, no longer remaining identified solely by the charged lepton with which they are paired. The neutrinos will sometimes be labeled simply with numbers to refer to their relative mass and sometimes with labels referring to the charged lepton according to the context.

55. If the initial *b* meson is neutral, you instead see a track that originates from the decay point, with no precursor track from the neutral initial state.

56. The interaction among the *W*, the top quark, and the bottom quark is however the reason the top can decay into a bottom and a *W*.

57. One can also define a relativistic mass that depends on momentum and energy, but the implication is the same.

58. Notice that this spread distinguishes bosons and fermions, classes of particles distinguished by quantum mechanics. Force carriers and the hypothetical Higgs particles are bosons. All other Standard Model particles are fermions.

59. Quoted in Stewart, Ian. *Why Beauty Is Truth* (Basic Books, 2007).

60. On WNYC's *The Takeaway,* March 31, 2007.

61. Sometimes people also debate whether right-handed neutrinos belong in the Standard Model. Even if present, they are likely to be extremely heavy and not very important for lower-energy processes.

62. http://xxx.lanl.gov/PS_cache/arxiv/pdf/1101/1101.1628v1.pdf.

63. This is discussed in much greater detail in *Warped Passages*.

64. Again, this is discussed at length in *Warped Passages*. The original paper is Lisa Randall and Raman Sundrum, *Physical Review Letters* 83 (1999):4690–4693.

65. Arkani-Hamed, Nima, Savas Dimopoulos, Gia Dvali, *Physics Letters* B429 (1998): 263–272; Arkani-Hamed, Nima Savas Dimopoulos, Gia Dvali, *Physical Review* D59:086004, 1999.

66. Randall, Lisa, and Raman Sundrum, *Physical Review Letters* 83 (1999):3370–3373.

67. Original short film *Powers of Ten* by Ray Eames and Charles Eames, 1968; *Powers of Ten: A Flip Book* by Charles and Ray Eames (W. H. Freeman Publishers, 1998); also Philip Morrison and Phylis Morrison and the office of Charles and Ray Eames, *Powers of Ten: About the Relative Sizes of Things in the Universe* (W. H. Freeman Publishers, 1982).

68. See e.g., Alan Guth's *The Inflationary Universe* (Perseus Books, 1997) for a more extensive discussion of this point.

69. Some dark matter particles are their own antiparticles, in which case they need to find other similar particles.

70. Dr. Mihaly Csikszentmihalyi pioneered the concept of flow to describe this phenomenon in his book *Flow: The Psychology of Optimal Experience* (Random House, 2002).

71. Brooks, David. "Genius: The Modern View," *New York Times*, April 30, 2009.

72. Gladwell, Malcolm. *Outliers: The Story of Success* (Little Brown & Co., 2008).

73. Gell-Mann, Murray. *The Quark and the Jaguar: Adventures in the Simple and the Complex* (W.H. Freeman & Company, 1994).

74. *Teacher's Edition of Current Science 49*, no. 14 (January 6–10, 1964).

75. *Verborgene Universen* in German.

76. In German, "rand" means "edge" and "all" means "universe."

77. See, too, for example, Susan Jacoby, *The Age of American Unreason* (Pantheon, 2008).

INDEX

Page references in *italic* indicate illustrations